Lectures on Fractal Geometry and Dynamical Systems

STUDENT MATHEMATICAL LIBRARY
Volume 52

Lectures on Fractal Geometry and Dynamical Systems

Yakov Pesin

Vaughn Climenhaga

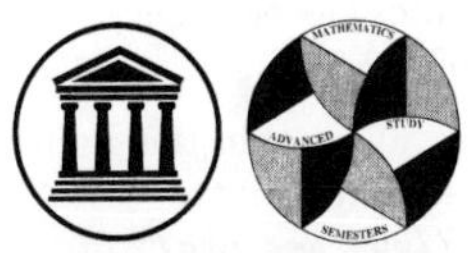

American Mathematical Society

Mathematics Advanced Study Semesters

2000 *Mathematics Subject Classification.* Primary 37–01, 37C45;
Secondary 37B10, 37D20, 37E05.

This work is partially supported by NSF grant 0754911.
The photograph of the Australian coastline in Figure 1.2 is courtesy of
NASA Earth Observatory: `http://earthobservatory.nasa.gov/`
`NaturalHazards/view.php?id=11595`
The map of France in Figure 1.15 is taken from the CIA World Factbook.

For additional information and updates on this book, visit
www.ams.org/bookpages/stml-52

Library of Congress Cataloging-in-Publication Data

Pesin, Ya. B.
    Lectures on fractal geometry and dynamical systems / Yakov Pesin, Vaughn
Climenhaga.
       p. cm. — (Student mathematical library ; v. 52)
    Includes bibliographical references and index.
    ISBN 978-0-8218-4889-0 (alk. paper)
    1. Fractals. 2. Dynamics. I. Climenhaga, Vaughn, 1982– II. Title.

QA614.86 .P47   2009

             2009028324

Contents

Foreword: MASS and REU at Penn State University

This book is part of a collection published jointly by the American Mathematical Society and the MASS (Mathematics Advanced Study Semesters) program as a part of the Student Mathematical Library series. The books in the collection are based on lecture notes for advanced undergraduate topics courses taught at the MASS and/or Penn State summer REU (Research Experiences for Undergraduates). Each book presents a self-contained exposition of a non-standard mathematical topic, often related to current research areas, which is accessible to undergraduate students familiar with an equivalent of two years of standard college mathematics, and is suitable as a text for an upper division undergraduate course.

Started in 1996, MASS is a semester-long program for advanced undergraduate students from across the USA. The program's curriculum amounts to sixteen credit hours. It includes three core courses from the general areas of algebra/number theory, geometry/topology, and analysis/dynamical systems, custom designed every year; an interdisciplinary seminar; and a special colloquium. In addition, every participant completes three research projects, one for each core course. The participants are fully immersed into mathematics, and this, as well as intensive interaction among the students, usually leads to a dramatic increase in their mathematical enthusiasm and achievement. The program is unique for its kind in the United States.

The summer mathematical REU program is formally independent of MASS, but there is a significant interaction between the two: about half of the REU participants stay for the MASS semester in the fall. This makes it possible to offer research projects that require more than seven weeks (the length of the REU program) for completion. The summer program includes the MASS Fest, a two- to three-day conference at the end of the REU at which the participants present their research and that also serves as a MASS alumni reunion. A nonstandard feature of the Penn State REU is that, along with research projects, the participants are taught one or two intense topics courses.

Detailed information about the MASS and REU programs at Penn State can be found on the website `www.math.psu.edu/mass`.

Preface

This book emerged from the course in fractal geometry and dynamical systems, with emphasis on chaotic dynamics, that I taught in the fall semester of 2008 as part of the MASS program at Penn State University.

Both fractal geometry and dynamical systems have a long history of development that is associated with many great names: Poincaré, Kolmogorov, Smale (in dynamical systems), and Cantor, Hausdorff, Besicovitch (in fractal geometry), to name a few. These two areas interact with each other since many dynamical systems (even some very simple ones) often produce fractal sets that are a source of irregular "chaotic" motions in the system.

A unifying factor for merging dynamical systems with fractal geometry is self-similarity. On the one hand, self-similarity, along with complicated geometric structure, is a crucial feature of fractal sets. On the other hand, it is related to various symmetries in dynamical systems (e.g., rescaling of time or space). This is extremely important in applications, as symmetry is an attribute of many physical laws which govern the processes described by dynamical systems.

Numerous examples of scaling and self-similarity resulting in appearance of fractals and chaotic motions are explored in the fascinating book by Schroeder [**Sch91**]. Motivated in part by this book, I designed and taught a course—for a group of undergraduate and graduate students majoring in various areas of science—whose goal was to describe the necessary mathematical tools to study many of the examples in Schroeder's book. An expanded and modified version of this previous course has become the course for MASS students that I mentioned above.

This book is aimed at undergraduate students, and requires only standard knowledge in analysis and differential equations, but the topics covered do not fall into the traditional undergraduate curriculum and may be demanding. To help the reader cope with this, we give formal definitions of notions that are not part of the standard undergraduate curriculum (e.g., of topology, metric space and measure) and we briefly discuss them. Furthermore, many crucial new concepts are introduced through examples so that the reader can get some motivation for their necessity as well as some intuition of their meaning and role.

The focus of the book is on ideas rather then on complicated techniques. Consequently, the proofs of some statements, which require rather technical arguments, are restricted to some particular cases that, while allowing for simpler methods, still capture all the essential elements of the general case. Moreover, to help the reader get a broader view of the subject, we included some results whose proofs go far beyond the scope of the book. Naturally, these proofs are omitted.

Currently, there are some textbooks for undergraduate students that introduce the reader to the dynamical system theory (see for example, [**Dev92**] and [**HK03**]) and to fractal geometry (see for example, [**Fal03**]), but none of them presents a systematic study of their interplay and connections to the theory of chaos. This book is meant to cover this gap.

Chapter 1 of this book starts with a discussion of the principal threefold cord of dynamics, fractals, and chaos. Here our core example is introduced—a one-dimensional linear Markov map whose biggest invariant set is a fractal and whose "typical" trajectories are chaotic. Although this map is governed by a very simple rule, it exhibits all the principal features of dynamics that are important for our purposes.

After being immersed into the interplay between dynamics and fractal geometry, the reader is invited to a more systematic study of dimension theory and its connections to dynamical systems, which are presented in Chapters 2, 3, and 4. Here the reader finds, among other things, rigorous definitions of various dimensions and descriptions of their basic properties; various methods for computing dimensions of sets, most importantly of Cantor sets; and relations between dimension and some other characteristics of dynamics.

Chapters 5 through 9 are dedicated to two "real-life" examples of dynamical systems—the FitzHugh–Nagumo model and the Lorenz model, where the former describes the propagation of a signal through the axon of a neuron cell and the latter models the behavior of fluid between two plates heated to different temperatures. While the underlying mechanism in the FitzHugh–Nagumo model is a map of the plane, the Lorenz model is a system of differential equations in three-dimensional space. This allows the reader to observe various phenomena naturally arising in dynamical systems with discrete time (maps) as well as with continuous time (flows).

An important feature of these two examples is that each system depends on some parameters (of which one is naturally selected to be the leading parameter) so that the behavior of the system varies (bifurcates) when the leading parameter changes. Thus, the reader becomes familiar in a somewhat natural way with various types of behavior emerging when the parameter changes, including homoclinic orbits, Smale's horseshoes, and "strange" (or "chaotic") attractors.

Let me say a few words on how the book was writen. My coauthor, Vaughn Climenhaga (who at the time of writing the book was a fourth-year graduate student) was the TA for the MASS course that I taught. He was responsible for taking and writing up notes. He did this amazingly fast (usually within one or two days after the lecture) so that the students could have them in "real time". The notes, embellished with many interesting details, examples, and some stories that he added on his own, were so professionally written that, with few exceptions, I had to do only some minor editing before they were posted on the web. These notes have become the ground material for the book. Turning them into the book required adding some new material, restructuring, and editing. Vaughn's participation in this process was at least an equal share, but he also produced all the pictures, the TeX source of the book, etc. I do not think that without him this book would have ever been written.

Yakov Pesin

Chapter 1

Basic Concepts and Examples

Lecture 1

a. A threefold cord: fractals, dynamics, and chaos. The word
"fractal" is one which has wriggled its way into the popular conscious-
ness over the past few decades, to the point where a Google search for
"fractal" yields over 12 million results (at the time of this writing),
more than six times as many as a search for the rather more fundamental mathematical notion of "isomorphism". With a few clicks of
a mouse and without any need to enter the jargon-ridden world of
academic publications, one may find websites devoted to fractals for
kids, a blog featuring the fractal of the day, photo galleries of fractals
occurring in nature, online stores selling posters brightly emblazoned
with computer-generated images of fractals, ... the list goes on.

Faced with this jungle of information, we may rightly ask, echo-
ing Paul Gauguin, "What are fractals? Where do they come from?
Where do we go with them?"

The answers to the second and third questions, at least as far
as we are concerned, will have to do with the other two strands of
the threefold cord holding this book together—namely, dynamical

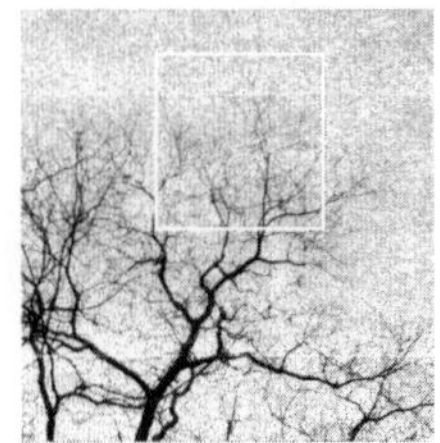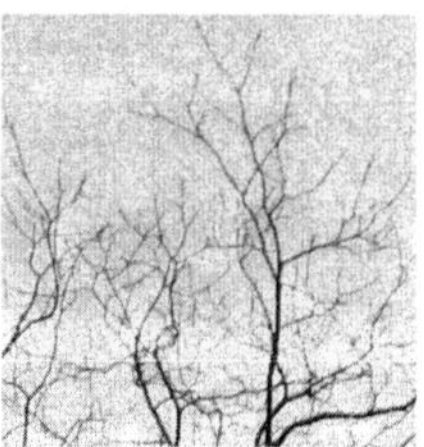

Figure 1.1. Self-similarity as observed in a tree in winter.

systems and chaos.[1] As an initial, naïve formulation, we may say that in most cases where a dynamical system exhibits chaotic behaviour, this behaviour is associated with the presence of a fractal. For our purposes, fractals will come from particular dynamical systems, and will lead us to an understanding of certain aspects of chaos.

But all in good time. We must begin by addressing the first question, "What are fractals?"

b. Fractals: intricate geometry and self-similarity. Consider an oak tree in the dead of winter,[2] viewed from a good distance away, as in the first panel of Figure 1.1. Its trunk rises from the ground to the point where it bifurcates into two large boughs; each of these boughs leads away from the centre of the tree and eventually sends off smaller branches of its own. Walking closer to the tree, one sees that these branches in turn send off still smaller branches, which were not visible from further away, and more careful inspection reveals a similar branching structure all the way down to the level of tiny twigs only an inch or two long. The various scales are shown in the second and third panels of Figure 1.1.

The key points to observe are as follows. First, the tree has a complicated and intricate shape, which is not well captured by more familiar geometric objects, such as lines, circles, polygons, and so on. Second, we see the same sort of shape on all scales: whether we view the tree from fifty yards away or from fifty inches, we will see a

---

[1]Chaos theory has also entered the popular imagination in its own right recently, thanks in part to its mention in movies such as *Jurassic Park*.

[2]One may also consider the tree in summer, of course, but the leaves get in the way of easy observation.

Figure 1.2. A fractal coastline (photograph courtesy of NASA Earth Observatory).

branching structure in which the largest branch (or trunk) in our field of view splits into smaller branches, which then divide themselves, and so on. This *self-similarity* is one of the key characteristics of fractals.

The previous paragraph described a qualitative self-similarity in the structure of the tree; in fact, the self-similarity is quantitative as well. Let us denote the diameter of a particular branch by d, and consider the point where it splits into two branches, whose diameters we denote d_1 and d_2. Taking into account that the amount of water passing through the branch at any given time is proportional to the area of the cross-section of the branch, we find, to a very good approximation, that

$$(1.1) \qquad d^2 = d_1^2 + d_2^2,$$

whatever size branch we started with.

We might imagine a tree where the sizes of the branches are related not by (1.1) but by the more general equation

$$d^\alpha = d_1^\alpha + d_2^\alpha,$$

where the exponent α may be different from 2. Indeed, if we consider the bronchial tree, the network of passageways leading into the lungs, then we find a similar qualitative picture[3]—the tree branches recursively across a wide range of scales—but this time the exponent is closer to $\alpha = 3$. The idea that an exponent such as α can help us to distinguish between qualitatively similar fractals is an important one, which we will take up in more detail later.

Another striking example of a natural fractal may be seen by looking at a high-resolution satellite image (or detailed map) of a

---

[3]This and other examples are described in [**Sch91**].

coastline, as in Figure 1.2, which shows a portion of the northwest coast of Australia. The boundary between land and sea does not follow a nice, simple path, but rather twists and turns back and forth; each bay and peninsula is adorned with still smaller bays and peninsulas, and given a map of an unfamiliar coast, we would be hard pressed to identify the scale at which the map was printed if we were not told what it was.

The two threads connecting the objects in these examples are their complicated geometry and some sort of self-similarity. Recall that two geometric figures (for example, two triangles) are *similar* if one can be obtained from the other by a combination of rigid motions and rescaling. A fractal exhibits a sort of similarity with itself; if we rescale a part of the image to the size of the whole, we obtain something that looks nearly the same as the original.

We now make these notions more precise. Simple geometric shapes, such as circles, triangles, squares, etc., have boundaries which are smooth curves, or are at least piecewise smooth. That is to say, if we write the boundary parametrically as

$$\mathbf{r}(t) = (x(t), y(t)),$$

then for the shapes we are familiar with, x and y are piecewise differentiable functions from $\mathbb{R}$ to $\mathbb{R}$, so that the tangent vector $\mathbf{r}'(t)$ exists for all but a few isolated values of t. By contrast, we will see that a fractal "curve", such as a coastline, is continuous everywhere but differentiable *nowhere*.

As an example of this initially rather unsightly behaviour, we consider the *von Koch curve*, defined as follows. Taking the interval $[0, 1]$, remove the middle third $(1/3, 2/3)$, and replace it with the other two sides of the equilateral triangle for which it is the base. One obtains the piecewise linear curve at the top of Figure 1.3; this is the basic pattern from which we will build our fractal.

Observe that the second curve in Figure 1.3 consists of four copies of the first, each of which has been scaled to $1/3$ its original size and then used to replace one of the four line segments in the original pattern. The new curve contains 16 line segments, each of length $1/9$. Replacing each of these segments with an appropriately scaled

Figure 1.3. The first few steps in the construction of the von Koch curve.

copy of the basic pattern, we obtain the third curve in the figure, and so on.

Each step in this construction—each curve in Figure 1.3—is piecewise linear. We may consider their parametrisations $f_1, f_2, f_3, \ldots$, each of which is a piecewise linear map from $[0,1]$ to $\mathbb{R}^2$. It is not too difficult to show that the sequence $\{f_n\}_{n=1}^{\infty}$ converges uniformly, and hence the limit $f\colon [0,1] \to \mathbb{R}^2$ exists and is continuous. The von Koch curve is the image of this function f, the end of the limiting process whose first few steps are shown.

Although each of the functions f_n is piecewise linear, their limit f is not differentiable anywhere, and hence the von Koch curve, despite being continuous, does not admit a tangent vector at any point. This is a manifestation of the complicated and intricate geometry we referred to earlier; since the self-similarity of the curve is evident from the construction, we may justifiably call this object a fractal.

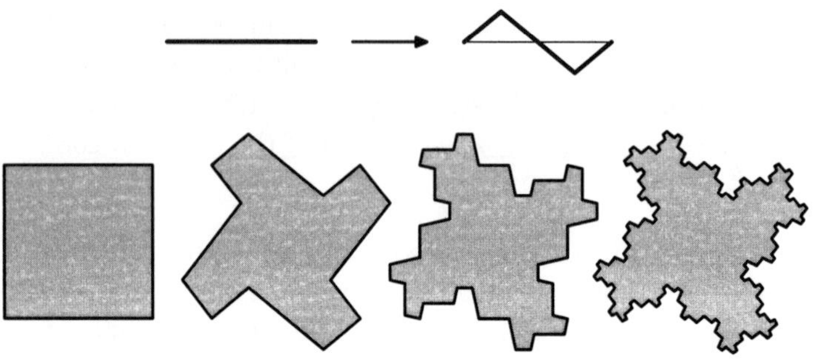

Figure 1.4. A fractal island.

It is natural to characterise a curve by its length, and so we may ask how long the von Koch curve is. One may easily verify that the nth step in the iterative procedure leading to the von Koch curve is a piecewise linear curve containing 4^n line segments, each of length $(1/3)^n$. At this stage of the iteration, then, the entire curve has length $(4/3)^n$; but this quantity grows without bound as n goes to infinity! The only conclusion we can reach is that the von Koch curve has infinite length, despite being contained in a bounded region of the plane. This sort of behaviour is in fact quite common for fractal curves.

Indeed, consider the iterative procedure illustrated in Figure 1.4, wherein each side of the square is replaced with the zig-zag shown, which comprises four line segments of length slightly greater than 1/4. Each of these segments is then replaced with an appropriately scaled version of the zig-zag pattern, and so on; the first few steps of the iteration are shown. Note that at each step, we add exactly as much area as we remove, and so the area of each "island" is equal to 1. However, a similar calculation to the one above shows that the limiting fractal island has a coastline of infinite length, despite having unit area.

These last two examples show that the usual ways of characterising and measuring geometric objects—length, area, volume, etc.—are insufficient to deal with fractals. Both the von Koch curve and the coastline of the fractal island have infinite length, but zero area, and similar results may be obtained for fractal curves constructed beginning from different generators; thus, we will need new tools in order to study them properly. First, though, we briefly turn our attention to dynamical systems, the second strand of the threefold cord.

c. Dynamics: things that move (or don't). In some sense, anything that moves is a dynamical system (and for that matter, so is everything that doesn't move). Somewhat more helpfully, we may consider any set X with a map f taking X to itself; that is, f assigns to each $x \in X$ an element $f(x) \in X$. If we think of each point in X as specifying a particular configuration of some system, then f is merely an encoding of the rule by which the system evolves from one state to the next. Some states evolve to other states under the action of f, while others may be fixed; if every point x is fixed, then f is the identity map, and nothing moves. But this is, of course, a rather trivial case.

We refer to the point $f(x)$ as the *image* of x under the action of f. The essential feature of a dynamical system is that each image $f(x)$ is also an element of X, and thus lies in the domain of f; that is, the map f takes X *into itself*, and so we can iterate it. Having found the image of a point x, we can then take the image of $f(x)$ in turn, which will be denoted $f^2(x) = f(f(x))$. Continuing the iteration, we obtain $f^3(x) = f(f^2(x))$, and in general, $f^{n+1}(x) = f(f^n(x))$.

In light of the key role iterative processes played in our earlier examples of fractals, the reader may feel justified in suspecting that the presence of an iterative process in this description of a dynamical system has something to do with the promised connection between the two; we will see later that this is often the case.

The sequence of points $x, f(x), f^2(x), \ldots$ is referred to as the *trajectory* (or *orbit*) of x. If we think of each iteration of the map f as specifying how the system evolves from one time step to the next, then it makes sense to think of the number of iterations n as the amount of time which has elapsed, and the trajectory is simply a list of the states through which the system passes as time goes on. This describes what is known as a *discrete-time dynamical system*. One may also consider continuous-time dynamical systems, in which the time variable may take any real value, but we will defer discussion of these systems until Chapter 9.

In and of themselves, sets are rather bland objects (with apologies to the reader specialising in set theory), and so we usually consider dynamical systems defined on sets X which possess some additional structure. In particular, if we hope to have anything to do with fractals, which are geometric objects, the set X should possess some geometric structure, and so Euclidean space is a natural place to begin.

Three familiar classes of maps on the plane $\mathbb{R}^2$ are rotation around a point $\mathbf{x}$ by an angle θ, translation by a vector $\mathbf{v}$, and reflection in a line ℓ. Together with the set of glide reflections, these are all the rigid motions of the plane, and may all be iterated and interpreted as dynamical systems.

Each of these maps can be written in the form $\mathbf{x} \mapsto A\mathbf{x} + \mathbf{v}$, where A is a 2×2 matrix with the property that $\|A\mathbf{x}\| = \|\mathbf{x}\|$ for every $\mathbf{x} \in \mathbb{R}^2$. Thus each of these maps preserves distances, and so $d(f^n(x), f^n(y)) = d(x, y)$ for all n. However, we do not need to restrict ourselves to isometries; *any* matrix $A = \left(\begin{smallmatrix} a & b \\ c & d \end{smallmatrix}\right)$ defines a dynamical system on $\mathbb{R}^2$ by

$$f: \quad \mathbb{R}^2 \to \mathbb{R}^2,$$

$$\begin{pmatrix} x_1 \\ x_2 \end{pmatrix} \mapsto \begin{pmatrix} a & b \\ c & d \end{pmatrix} \begin{pmatrix} x_1 \\ x_2 \end{pmatrix}.$$

The properties of this dynamical system are largely governed by the eigenvalues of A. If both eigenvalues lie inside the unit circle ($|\lambda| < 1$), then every trajectory converges to $\mathbf{0}$ (see Appendix); if both eigenvalues lie outside the unit circle ($|\lambda| > 1$), then every trajectory diverges to ∞ (except the one starting at the origin). If A has one eigenvalue outside the unit circle and one inside, the situation is more subtle; we will return to this case in Lecture 22(b).

The determinant of A also has ramifications for the properties of the dynamical system defined by A. For example, if the determinant is equal to 1, then f is *area-preserving*; that is, the image $f(A) = \{f(x) \mid x \in A\}$ of a domain $A \subset \mathbb{R}^2$ has the same area as A itself. Consequently, *every* image $f^n(A)$ has the same area as A.

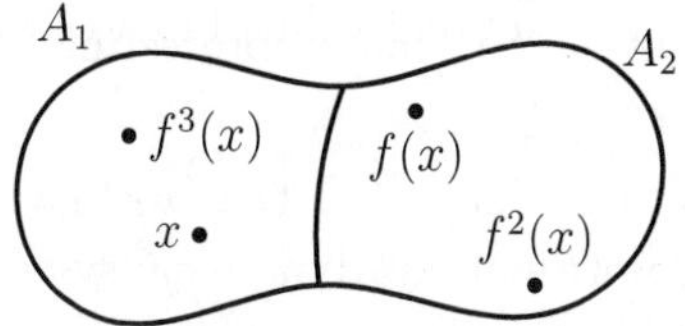

Figure 1.5. Coding a trajectory of f.

We may also consider non-linear maps from the plane to itself; in fact, most of the interesting examples are of this sort. So for now, let $f\colon \mathbb{R}^2 \to \mathbb{R}^2$ be any continuous map of the plane into itself. Let $A \subset \mathbb{R}^2$ be some domain—perhaps a disc, perhaps something rather more complicated—and suppose that the image of each point in A is itself in A. Thus A is mapped into itself by the action of f, and so we can write $f\colon A \to A$. It follows that for any point $x \in A$, the entire trajectory of x lies within A.

In principle, if we have precise knowledge of the map f and the initial point x, then we can precisely compute each point $f^n(x)$ in the trajectory of x. There is no randomness in the action of f; it is entirely deterministic, and given sufficient patience and computing power, we can predict the future. Or so it seems. . . .

Suppose we divide A into two subdomains A_1 and A_2 in such a way that every point $x \in A$ lies in exactly one of the two. Now instead of describing the trajectory of a point x by giving the precise location of each iterate $f^n(x)$, we may instead "blur our vision" and only record whether it lies in A_1 or A_2. In this way we assign to a point x a sequence of 1's and 2's, known as a *coding* of x (really, of the trajectory of x). For example, the trajectory shown in Figure 1.5 has a coding which begins with the symbols 1221. . . .

It is natural to ask if we can go in the other direction: Given a sequence of 1's and 2's, can we find a point x whose trajectory is coded by that sequence? If we can, is it unique, or might there be several such points?

The answer to these questions is somewhat involved, and depends heavily on the particular system f and on the choice of partition

$\{A_1, A_2\}$. We will see that in many interesting cases, the answer to both questions is yes.

Suppose for the moment, then, that we have such a correspondence between trajectories of our dynamical system and sequences of 1's and 2's. Imagine taking a coin and flipping it repeatedly. After each flip, write down the number 1 if the coin comes up heads and the number 2 if it comes up tails. In this manner we obtain a sequence of 1's and 2's which is entirely random, and which codes the trajectory of some point x.

This brings us to a rather jarring conclusion: the trajectory of this point x will appear to hop at random between A_1 and A_2, just as the outcome of the coin toss hops at random between heads and tails. In other words, even if we know which partition element (A_1 or A_2) the points $x, f(x), \ldots, f^n(x)$ lie in, we cannot say for sure whether $f^{n+1}(x)$ will lie in A_1 or A_2; the best we can do is to give the probabilities of these two events. But we said earlier that f is wholly deterministic, with no randomness whatsoever; where, then, does this random-looking behaviour come from?

We will eventually resolve this paradox, but first we need to make the concepts involved more precise. For the time being, we merely observe that this initially unpalatable coexistence of deterministic and random behaviour is at the heart of the theory of chaos; indeed, it was to describe such situations that James Yorke first coined the somewhat controversial term "deterministic chaos". We will see in due course how such behaviour arises from dynamical systems associated with fractal sets.

Lecture 2

a. Dynamical systems: terminology and notation. Let us slow down now and take a more leisurely look at some of the concepts that will be foundational to our discussion of dynamical systems before moving on to consider some apparently simple, but ultimately extremely challenging and enlightening, examples.

We begin with the d-dimensional Euclidean space $\mathbb{R}^d$, that is, the collection of d-tuples of real numbers. As our dynamical system, we may consider any map f which takes one element of $\mathbb{R}^d$ and gives us back another.

It may happen that f is not defined on all of $\mathbb{R}^d$ but only on some domain $D \subset \mathbb{R}^d$. For instance, the rule that lets us go from $\mathbf{x}$ to $f(\mathbf{x})$ may only make sense when $\mathbf{x}$ is a vector of length no greater than R; in this case, the domain of definition is the ball of radius R centred at the origin.

We say that $f(x)$ is *the* image of x, and we also say that x is *a* preimage of $f(x)$. The choice of article is important; while the map f must send x to a unique point $f(x)$, it is quite possible that there is some point $y \neq x$ with $f(y) = f(x)$, in which case y is also a preimage of $f(x)$.

We will also speak of the image of a set: if $A \subset D$ lies within the domain of definition, then the image of A is

$$f(A) = \{f(x) \mid x \in A\}.$$

Of particular importance is the image of the domain D; this image $f(D)$ is also known as the *range* of f.

If the range of f lies inside the domain of definition, then D is mapped into itself by f, and we can apply f again, and again, and again, *ad infinitum*, without ever leaving D. Thus, we may consider not only f, but the map obtained by applying f twice. We denote this by f^2 and write $f^2(x) = (f \circ f)(x) = f(f(x))$. Similarly, we may consider f^3, f^4, and in general f^n, defined as

$$f^n = \underbrace{f \circ f \circ \cdots \circ f}_{n \text{ times}}.$$

The ability to iterate f means that we are in fact considering a whole family of maps $f^n \colon D \to D$. It is immediate from the definition that

$$(1.2) \qquad\qquad f^{n+m} = f^n \circ f^m = f^m \circ f^n$$

whenever m and n are non-negative integers. This is the so-called *semi-group property*, which allows us to translate questions about the behaviour of iterates of f to questions about the action of a particular semi-group[4] (in this case, the natural numbers $\mathbb{N}$) on the domain D, leading to a more algebraic approach which is sometimes useful.

The sequence of points $x, f(x), f^2(x), \dots$ is called the *trajectory* (or *orbit*) of x. The number n of iterations plays the role of time: the near future corresponds to small values of n, the far future to large values. It may at first seem somewhat unnatural to think of time as moving in discrete increments, instead of a steady stream as we are accustomed to. However, in practical terms, any measurement we may wish to make can only be carried out at discrete times; any two observations will be separated by a small interval, which may be vanishingly short or mind-numbingly long, but which nevertheless has the effect that our data is always collected with respect to a set of discrete time steps.

Exercise 1.1. Describe the behaviour of the trajectories of the following maps on $\mathbb{R}$:

(a) $f(x) = |2x - 1|$.

(b) $g(x) = |x - 2|$.

A set A for which $f(A) = A$ is called *f-invariant* (or simply *invariant*). If the domain D is invariant, then every point $x \in D$ has at least one preimage. If in addition the map f is *one-to-one*—if there are no two points which have the same image—then this preimage is *unique*, and we denote it by $f^{-1}(x)$. We say that f is *invertible*, and its inverse is the map $f^{-1} \colon D \to D$. Iterating this inverse gives us the points $f^{-2}(x)$, $f^{-3}(x)$, and so on, which are the images of x under the maps

$$f^{-n} = \underbrace{f^{-1} \circ f^{-1} \circ \cdots \circ f^{-1}}_{n \text{ times}}.$$

Thus for invertible maps, the trajectory is defined not just for non-negative values of n, but over the entire set of integers, and is a doubly infinite sequence of points in X. Furthermore, the group property (1.2) holds for *any* integers m and n, whether positive or negative.[5]

---

[4] We must speak of semi-groups rather than groups because we have so far defined the iterates f^n only for non-negative values of n.

[5] And we have a true group action, rather than the action of a semi-group.

If the map f is not one-to-one (which is the case for many important systems), different initial conditions may eventually wind up following the same trajectory, and we cannot run time backwards; the past determines the future, but the future does not necessarily determine the past, since different initial conditions may lead to the same outcome.

Even in this time-irreversible case, we still may (and often do) consider preimages of points—and also of sets. Given a map f (which may or may not be invertible) on a domain D and a set $A \subset D$, the preimage of A is defined as

$$(1.3) \qquad f^{-1}(A) = \{x \in D \mid f(x) \in A\},$$

which consists of all the points whose images lie in A. Note that although there may be several preimages of a point, the preimage of a set is uniquely defined.[6]

b. Population models and the logistic map.

b.1. *A rather unrealistic population model.* Consider a population of duck-billed platypi (or bacteria, or whatever species you fancy), whose size will be represented by a variable x. Given the size of the population at the present time, we want to predict the size of next year's population (or perhaps the next hour's, in the case of bacteria). So if there are x platypi this year, there will be $f(x)$ next year, where f is a suitable function which models the change in the platypus population from year to year. Of course, since we cannot have a negative number of platypi, we must restrict x to lie in the interval $[0, \infty)$, which will be the domain of definition for f.

What form should f take? As a first (simplistic) approximation, we may suppose that the platypi reproduce at a constant rate, and so if there are x of them this year, there will be rx next year, where $r > 1$ is a real number, and $r - 1$ represents the proportion of newborns each year.

We would like to understand what the trajectories of the system look like for various possible starting populations. To this end, we

---

[6]Of course, if we think of a point x as a set $\{x\}$ with only one element, then its preimage $f^{-1}(\{x\})$ as defined in (1.3) is unique as a set.

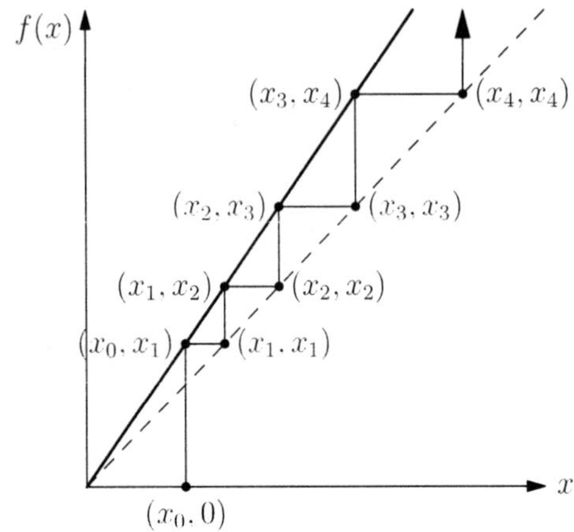

Figure 1.6. Cobweb diagram for a simple population model.

use a *cobweb diagram*,[7] as in Figure 1.6, which shows the graph of f. If x_0 is the initial value of x, then the next point in the trajectory is $f(x_0)$, which we denote by x_1. We may find this value by following the vertical line through $(x_0, 0)$, which intersects the graph of f at the point $(x_0, f(x_0)) = (x_0, x_1)$. Following the horizontal line through this point until it intersects the bisectrix $y = x$, we reach the point (x_1, x_1), and now our x-coordinate is $x_1 = f(x_0)$, the next point in the trajectory after x_0.

In order to find the next point in the trajectory after x_1, we repeat this process: first follow the vertical line through (x_1, x_1) to its intersection with the graph of f, the point $(x_1, f(x_1)) = (x_1, x_2)$, and then move horizontally to (x_2, x_2).

In general, we write $x_n = f^n(x_0)$ for the points of the trajectory, and we see that one obtains x_n from x_{n-1} by moving vertically to the graph and then horizontally to the bisectrix. This gives a simple graphical procedure that allows us to investigate the qualitative properties of the trajectory of x_0.

In this case, we see that for any initial population size $x_0 \neq 0$, the population size grows without bound; we say that the trajectory

---

[7]Also known as a *Verhulst diagram*, after the Belgian mathematician Pierre Verhulst.

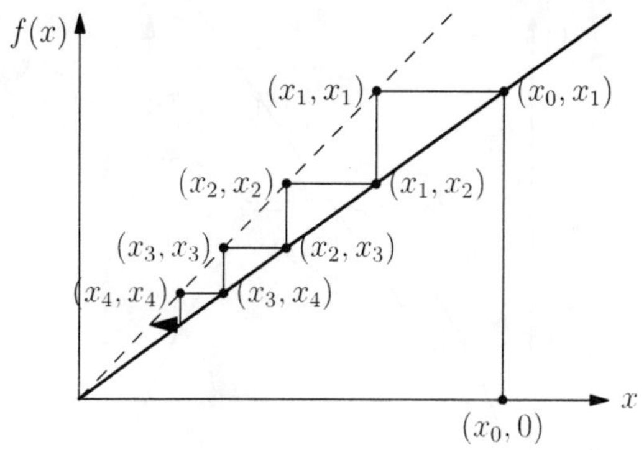

Figure 1.7. A dying population.

goes to infinity. The case $x_0 = 0$ is different, reflecting the fact that if there are no platypi to begin with, then no new ones will be born: nothing begets nothing. We say that 0 is a *fixed point* for the map f.

Algebraically, a fixed point is a point x such that $f(x) = x$, and for any such point we see that the trajectory never moves. Graphically, fixed points may be described as the points where the graph of f intersects the bisectrix $y = x$.

An important feature of this particular fixed point is that it is *unstable*; even a very small population x_0 will eventually grow to be arbitrarily large. Fixed points with this property, for which the trajectories of nearby points are driven away, are also called *repelling*.

b.2. *A model which could be realistic.* Of course, as everyone knows, platypi are not immortal. *Alles Fleisch es ist wir Gras*, and our model needs to take into account the population reduction caused by death by disease, predation, out-of-season platypus hunting, etc. This will have the effect of changing the value of the parameter r, reducing it by counteracting the increase in population provided by the year's births. If it reduces it to the point where $r < 1$, then the graph of f is as shown in Figure 1.7, and the cobweb diagram clearly illustrates the fate of the platypus colony.

In this case, 0 is still a fixed point, but it is now *stable*; a value of x_0 near 0 will lead to a trajectory which converges to 0. Fixed

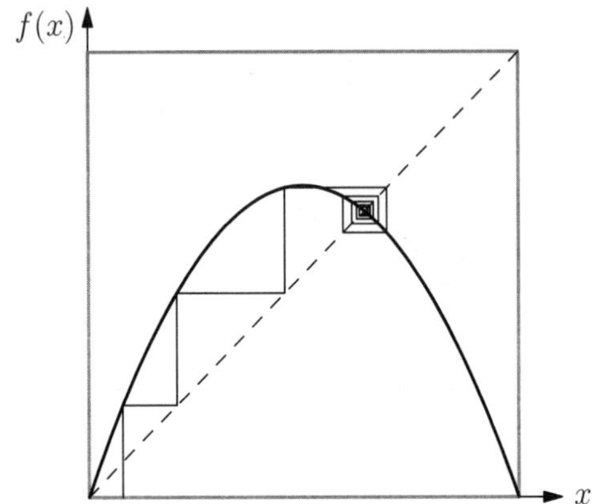

Figure 1.8. The logistic map with $r = 2.8$.

points with this property, for which the trajectories of nearby points are drawn to the fixed point, are also called *attracting*.

Note that from the mathematical point of view, there is a simple relationship between the case $r > 1$ and the case $r < 1$—they are inverses. Indeed, if we write $f_r \colon x \mapsto rx$, then it is easy to see that $f_r^{-1} = f_{1/r}$. One also sees that the slope of the graph at the fixed point 0 determines whether that fixed point is attracting or repelling, and that an attracting fixed point for f is a repelling fixed point for f^{-1}, and vice versa.

b.3. *An innocent-looking model.* As any biologist or ecologist will no doubt protest quite vigorously, the preceding models are so simplistic as to be entirely unrealistic. Among other weaknesses, they fail to take into account the fact that resources are limited, and whatever river our platypi find themselves in can only support a finite population size before starvation or overcrowding leads to disaster.

To address this shortcoming, we introduce a new term into our equation. Suppose environmental factors determine some maximum population P, which corresponds, for instance, to the amount of food (or some other resource) available. Then the population cannot grow beyond P; furthermore, if the population reaches P, all the food will be eaten and the platypi will starve, sending the next year's population to 0. We model this situation with the formula $x \mapsto rx(P - x)$. In order to keep the equations as simple as possible, though, we consider the quantity $\tilde{x} = x/P$, which stands for the *proportion* of the maximum population P, and so lies between 0 and 1. Then we have

$$\tilde{f}(\tilde{x}) = f(\tilde{x}P) = r\tilde{x}P(P - \tilde{x}P) = \tilde{r}\tilde{x}(1 - \tilde{x}),$$

where we write $\tilde{r} = rP^2$. For simplicity of notation we will from now on drop the tilde and simply write the dynamical system in question as

(1.4)
$$f\colon [0, 1] \to [0, 1],$$
$$x \mapsto rx(1 - x),$$

where r is a parameter encoding information about the system, such as reproduction rate, mortality rate, etc. The map in (1.4) is known as the *logistic map*, and its graph for the value $r = 2.8$ is shown in Figure 1.8, along with a typical trajectory.[8]

Unlike the example $x \mapsto rx$ examined earlier, the logistic map displays a startling intricacy when we begin to track the behaviour of typical trajectories for various values of the parameter r. Indeed, the amount of literature on the logistic map is such that one could easily devote an entire year's course to the subject without exhausting the corpus of present knowledge, and the logistic map (along with its relatives) is still an area of active research.

Leaving behind for the time being any biological interpretations of the model, let us focus on its mathematical structure. It turns out that the map f is equivalent to the map $g\colon y \mapsto y^2 + c$ in the following sense: given a fixed value of the parameter r, we can find a value of $c \in \mathbb{R}$, an interval $I \subset \mathbb{R}$, and a change of coordinates $h\colon [0, 1] \to I$ such that $g \circ h = h \circ f$. That is, the following diagram commutes:

(1.5)
$$\begin{array}{ccc} [0, 1] & \xrightarrow{\ f\ } & [0, 1] \\ \downarrow{\scriptstyle h} & & \downarrow{\scriptstyle h} \\ I & \xrightarrow{\ g\ } & I \end{array}$$

---

[8]This map was proposed in a seminal paper by the biologist Robert May as a discrete-time demographic model similar to the logistic differential equation created by Pierre Verhulst.

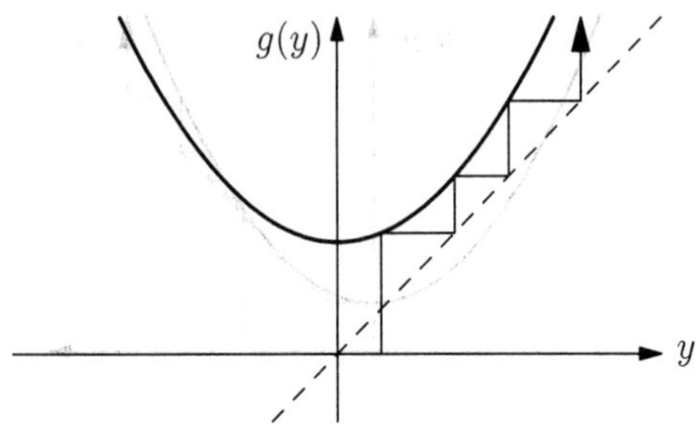

Figure 1.9. Trajectories escaping to infinity.

By "commutes", we mean that we may follow a point in $[0, 1]$ by first applying f and then using h to pass to a point in I, or by first passing to a point in I via h, and then applying g, and we will reach the same point whichever way we go.

An invertible map such as h which intertwines the dynamics of two systems in this manner is called a *conjugacy*, and the two maps f and g are said to be *conjugate*.[9] Using the conjugacy h, any question we have about the dynamics of the map f can be translated into a question about the dynamics of g.

Exercise 1.2. Find an explicit change of coordinates which demonstrates the conjugacy in (1.5). Which values of c correspond to values of r that could occur in the model?

For large enough values of c, the graph of g lies entirely above the bisectrix, and every trajectory escapes to infinity, as shown in Figure 1.9. The parabola moves down as c decreases, and eventually, for some critical value of c, becomes tangent to the bisectrix, as shown in Figure 1.10.

The point of tangency p is a fixed point for g. As is evident from the cobweb diagram, trajectories which start a little bit to the left of p are attracted to it, while trajectories which start just to the right are repelled and go to infinity. Thus in this case we have a fixed point which is *neutral*: it is neither an attractor nor a repeller.

---

[9]In fact, since the change of coordinates h can be chosen to be continuous, we will eventually speak of a *topological conjugacy*.

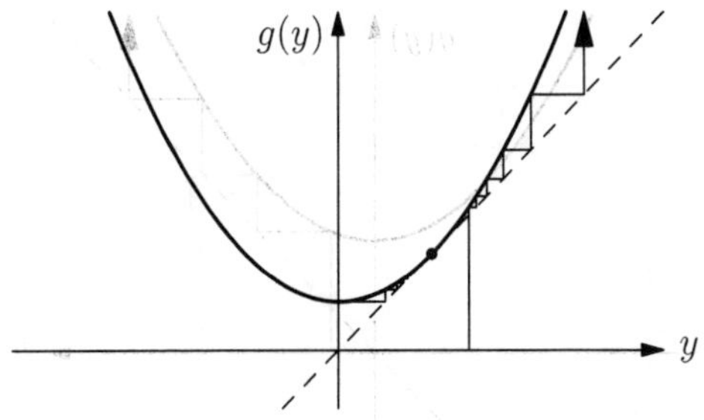

Figure 1.10. A fixed point which is neither attracting nor repelling.

Moving the initial point further to the left, one sees that for large enough negative values of x_0, the next point in the trajectory leaps to the right of p and then the trajectory goes to infinity. The point of transition between the two sorts of behaviour is $x_0 = -p$, which leads to $x_1 = g(x_0) = p$, and so the trajectory becomes trapped on the fixed point p.

Thus we have completely classified the asymptotic behaviour of trajectories for this particular map; points in $[-p, p]$ are attracted to the fixed point p, while all other points go to $+\infty$ under repeated iterations.

As we will see, the picture becomes vastly more complicated than this if we continue to decrease the parameter c.

Exercise 1.3. Describe all trajectories of the map g generated by the function $g(y) = y^2 + c$ for all the values of $c \geq 0$.

Lecture 3

a. A linear map with chaotic behaviour and the middle-third Cantor set. Aside from being quite unrealistic, the linear population model in the previous lecture did not display any chaotic behaviour in the sense of Lecture 1; we cannot find a reasonable partition of the phase space for which most trajectories have a coding which appears random. This is actually a feature of any linear map; the theory of *Jordan normal form*, which is one of the most important results in basic linear algebra, offers a complete classification of linear maps in

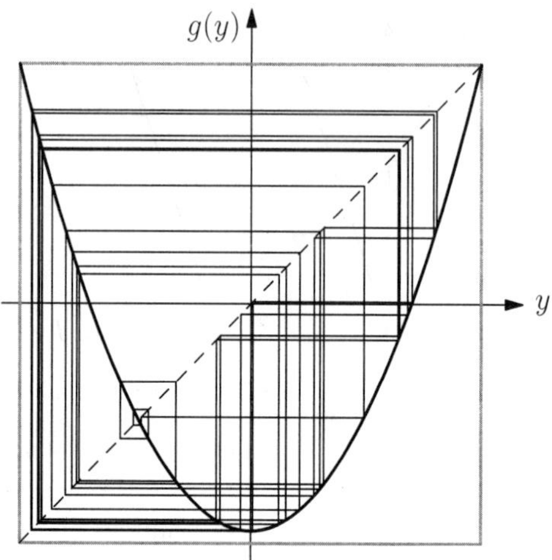

Figure 1.11. Chaotic behaviour in the logistic map for $c = -1.85$.

$\mathbb{R}^d$ and describes all the possible behaviours, none of which display any real complexity.

By contrast, the logistic map $g\colon y \mapsto y^2 + c$ displays a variety of complex behaviours as we consider different parameter values, a quality which makes it eminently worthy of further study. In the previous lecture, we described its behaviour for large values of c and took a very brief look at how that behaviour becomes more intricate as c decreases. In fact, for most values of c near -2, the logistic map exhibits fully chaotic behaviour, in the sense discussed in Lecture 1; a typical trajectory for the value $c = -1.85$ is shown in Figure 1.11.

We will have more to say about the logistic map later on, in Chapter 6. For the time being, we remark that a good deal of its complex behaviour can be attributed to its non-linearity. That same non-linearity, though, makes the map far more difficult to study; linear models are simply more tractable than non-linear ones. For this reason, we will first spend some time studying a piecewise linear map which displays chaotic behaviour. This map will be easier to study because of its linearity, but the fact that it is only piecewise linear, rather than fully linear, will still permit the existence of the chaotic behaviour we wish to understand.

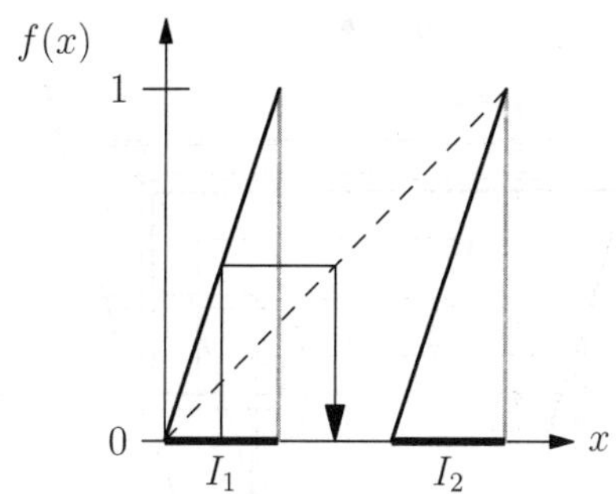

Figure 1.12. A piecewise linear map with chaotic behaviour.

Consider the map f shown in Figure 1.12, which is defined linearly on each of the intervals $I_1 = [0, 1/3]$ and $I_2 = [2/3, 1]$ so that the image of both intervals is $f(I_1) = f(I_2) = [0, 1]$. Thus the domain of definition of f is $D = I_1 \cup I_2$, and the range is $[0, 1]$. Notice that the range does not lie inside the domain of definition; $I_1 \cup I_2$ is not mapped into itself by f, and so f is not defined at every point in the range. The cobweb diagram in the figure shows one iteration in the trajectory of the point $1/6$, whose image lies outside the domain of definition, and hence cannot be iterated further.

If we cannot iterate the map f, then we cannot study the dynamics, and so we must determine which points admit a second iteration. That is, what is the domain on which the map $f^2 = f \circ f$ is defined?

In order for $f^2(x_0)$ to be defined, both x_0 and $f(x_0)$ must lie in the domain of f; that is, we must have

$$x_0 \in D \cap f^{-1}(D) = \{x \mid x \in D \text{ and } f(x) \in D\}.$$

This is shown graphically in Figure 1.13, a sort of inverse cobweb diagram. Placing the domain D along the vertical axis, we find its preimage $f^{-1}(D)$ by first following each horizontal line through D to *all* of the points where it intersects the graph of f and then moving vertically from these intersection points to the x-axis. We see that the domain on which f^2 is defined consists of four closed intervals,

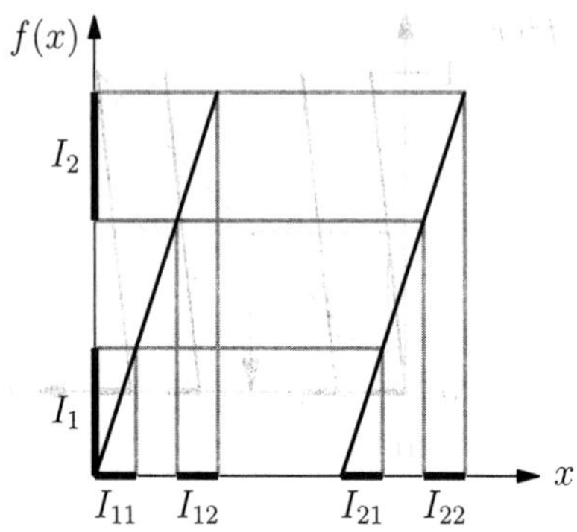

Figure 1.13. Finding the preimage $f^{-1}(D)$.

each of length $1/9$. Writing these as

$$I_{11} = \left[0, \frac{1}{9}\right], \quad I_{21} = \left[\frac{2}{3}, \frac{7}{9}\right], \quad I_{12} = \left[\frac{2}{9}, \frac{1}{3}\right], \quad I_{22} = \left[\frac{8}{9}, 1\right],$$

we see that $f(I_{11}) = f(I_{21}) = I_1$, that $f(I_{12}) = f(I_{22}) = I_2$, and that $f^2(I_{11}) = f^2(I_{12}) = f^2(I_{21}) = f^2(I_{22}) = [0, 1]$. Observe that the intervals $I_{w_1 w_2}$ (where the indices w_1 and w_2 are either 1 or 2) may be defined in terms of the action of f as follows:

$$(1.6) \qquad\qquad I_{w_1 w_2} = I_{w_1} \cap f^{-1}(I_{w_2}).$$

We obtained the domain of f by removing the (open) middle third from the interval $[0, 1]$, leaving two closed intervals of length $1/3$. From these, we obtained the domain of f^2 by removing the (open) middle third of each, leaving four closed intervals of length $1/9$. The graph of f^2 is shown in Figure 1.14. In this picture we already see the beginnings of self-similarity, and the same reasoning shows that the domain of f^3 will likewise consist of eight closed intervals, each of length $1/27$.

In general, an inductive argument shows that the domain of f^n consists of 2^n closed intervals, each of length 3^{-n}. Following (1.6), we may denote these by

$$(1.7) \qquad I_{w_1 w_2 \ldots w_n} = I_{w_1} \cap f^{-1}(I_{w_2}) \cap \cdots \cap f^{-(n-1)}(I_{w_n}),$$

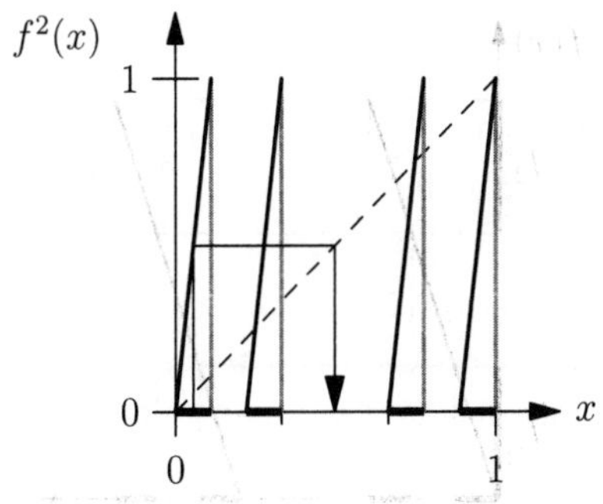

Figure 1.14. The second iterate of f.

where each w_k is either 1 or 2. Observe that for a fixed value of n, any two such intervals are disjoint; that is,

$$I_{v_1\ldots v_n} \cap I_{w_1\ldots w_n} = \emptyset$$

whenever $(v_1,\ldots,v_n) \neq (w_1,\ldots,w_n)$. Increasing n by one, we see from the construction that

$$f^{-n}(D) \cap I_{w_1\ldots w_n} = I_{w_1\ldots w_n 1} \cup I_{w_1\ldots w_n 2}.$$

Thus the domain of definition of the nth iterate f^n may be written

$$D_n = f^{-(n-1)}(D) = \bigcup_{w_1\ldots w_n} I_{w_1\ldots w_n},$$

where the union is taken over all n-tuples with values in $\{1,2\}$. Letting n run to infinity, we see that the domain on which *every* iterate f^n is defined is

$$(1.8) \qquad C = \bigcap_{n\geq 1}\left(\bigcup_{w_1\ldots w_n} I_{w_1\ldots w_n}\right),$$

which is the *standard middle-third Cantor set*. As far as the dynamics of the map f are concerned, the key property of C is that it is a *repeller*; that is, there exists a neighbourhood V of C (in this case the interval $[0,1]$) such that C consists of precisely those points whose orbits remain in V:

$$C = \bigcup_{n\geq 1} f^{-n}(V).$$

Equivalently, we see that C is the largest invariant set in $[0, 1]$. In the first place, it is genuinely invariant, unlike the approximations D_n, for which we have $f(D_n) = D_{n+1} \not\subset D_n$; for the Cantor set itself, we have $f(C) = C$. Furthermore, if $A \subset D$ is any invariant set, $f(A) = A$, then we must have $A \subset C$.

Thus if we wish to study the dynamics of f, the "proper" domain to consider is the Cantor set C. To get a first idea of how the dynamics of $f\colon C \to C$ behave, consider two distinct points $x, y \in C$ which are very close together, so that $d(x, y)$ is very small. How far apart are their images? It is not too hard to see that we have

$$d(f(x), f(y)) = 3d(x, y),$$

so that the distance between x and y is increased by a factor of 3. For higher iterates f^n, we have

$$d(f^n(x), f^n(y)) = 3^n d(x, y),$$

provided that the trajectories have stayed close up until that point; in particular, this will be true if $f^k(x)$ and $f^k(y)$ lie in the same interval I_1 or I_2 for each $1 \leq k < n$.

The significance of this result is that small errors are magnified: if x represents the true state of the system but we instead measure it as y due to a very small experimental error, then the trajectory we predict will diverge exponentially quickly from the true trajectory. In this case, the notion of instability, which we introduced earlier for fixed points, applies to every trajectory of the system; whatever trajectory we look at, nearby trajectories will be repelled at an exponential rate. This phenomenon represents a property known as *sensitive dependence on initial conditions*, which is a weak form of chaotic behaviour.[10]

b. The Cantor set and symbolic dynamics. When Georg Cantor first conceived his eponymous set, he was hoping to settle the continuum hypothesis by constructing a subset of the interval whose cardinality lay strictly between that of the integers and that of the

---

[10]Although trajectories of the linear map $x \mapsto rx$ (for $r > 1$) also diverge at an exponential rate, they do so by escaping to infinity. One of the key features which makes the present behaviour chaotic is that this divergence happens while the trajectories remain bounded.

real line. As we shall see, this turned out not to be the case; nevertheless, the Cantor set has become an object of fundamental importance to a number of different areas in mathematics, both in dynamics and elsewhere, although we shall be most concerned with the dynamical applications.

We start with an innocent-looking question. How big is the Cantor set? Of course, we need a notion of "bigness", and so we may first try to compute the "length" of the Cantor set. To this end, observe that at the first iteration, D_1 comprises two intervals of length $1/3$, and so has total length $2/3$. At the next iteration, four intervals of length $1/9$ give D_2 a total length of $4/9$. In general, D_n is a union of intervals with total length $(2/3)^n$. Since this goes to 0 as $n \to \infty$, we must consider the "length" of C to be 0. Alternately, we may look at the lengths of the intervals which are removed at each step and see that they sum to 1.

From a probabilistic point of view, this means that if we choose a point in the interval $[0,1]$ at random, the probability of picking a point on the Cantor set is precisely zero. It would seem that length is not the proper way to measure how big C is.

Since C is not "big enough" to have positive length, we may try measuring it in a different way, by counting the number of points it contains. We immediately see that it has infinitely many points, and so we next ask whether it is countable or uncountable. To answer this question, we observe, as Cantor did, that each point $x \in C$ uniquely determines a sequence $w_1, w_2, \ldots$, where each w_k is either 1 or 2, by the rule

$$x \in I_{w_1} \cap I_{w_1 w_2} \cap \cdots \cap I_{w_1 \ldots w_n} \cap \cdots \,,$$

where all we are doing is asking which interval $I_{w_1 \ldots w_n}$ contains x at each step n of the iteration. This defines a map from C to the space of symbolic sequences

$$\Sigma_2^+ = \{1, 2\}^{\mathbb{N}} = \{(w_k)_{k=1}^{\infty} \mid w_k = 1 \text{ or } 2 \text{ for every } k \geq 1\}.$$

Furthermore, the correspondence is bijective; given any sequence $w \in \Sigma_2^+$, the intersection

$$\bigcap_{n=1}^{N} I_{w_1 \ldots w_n} = I_{w_1 \ldots w_N}$$

is an interval whose length goes to 0 as $N \to \infty$, and so the intersection $\bigcap_{n=1}^{\infty} I_{w_1 \dots w_n}$ consists of precisely one point. It follows that the sequence $w_1, w_2, \dots$ comes from exactly one point x, and we have demonstrated that the following *coding map* is a bijection:

$$h \colon \Sigma_2^+ \to C,$$

$$w = (w_1, w_2, \dots) \mapsto \bigcap_{n \geq 1} I_{w_1 \dots w_n}.$$

Exercise 1.4. Using binary expansions of real numbers, show that Σ_2^+ has the same cardinality as $[0, 1]$, and hence that the Cantor set C does as well.[11]

In fact, the coding map h does more than just establish a bijection between Σ_2^+ and C, which only shows that the two are the same from a set-theoretic point of view. The correspondence runs deeper than that, to an equivalence between the *dynamics* of the two sets as well.

Of course, at this point we have not put any dynamics on the set Σ_2^+, and so we must define a map $\sigma \colon \Sigma_2^+ \to \Sigma_2^+$ in order for the previous claim to make any sense. Recalling the definition of the sets $I_{w_1 \dots w_n}$ in (1.7), we see that the coding of a point $w = (w_n)_{n \in \mathbb{N}} \in \Sigma_2^+$ can be written

$$h(w) = \bigcap_{n=1}^{\infty} f^{-(n-1)}(I_{w_n}) = I_{w_1} \cap f^{-1}(I_{w_2}) \cap f^{-2}(I_{w_3}) \cap \cdots,$$

and so

$$
\begin{aligned}
f(h(w)) &= I_{w_2} \cap f^{-1}(I_{w_3}) \cap f^{-2}(I_{w_4}) \cap \cdots \\
&= h(w')
\end{aligned}
$$
(1.9)

where we write $w' = (w_2, w_3, \dots)$, and use the fact that $f(I_{w_1}) = [0, 1]$, and also that $f(f^{-1}(X)) = X$ for any set X in the range of f.[12] The map which takes w to w' is particularly simple: all we have to

---

[11]We will see later that cardinality is a zero-dimensional measure, and so this result says that C is somehow "bigger than zero-dimensional". Similarly, length is a one-dimensional measure, and the earlier result regarding the "length" of C means that C is "smaller than one-dimensional". We will formulate these statements more precisely when we discuss Hausdorff dimension in Lecture 7.

[12]Note that the similar-looking statement $f^{-1}(f(X)) = X$ is *not* true in general.

do is drop the first symbol, w_1, and shift all the others one position to the left. This is the *shift map*

$$\sigma \colon \Sigma_2^+ \to \Sigma_2^+,$$

$$(w_1, w_2, w_3, \dots) \mapsto (w_2, w_3, w_4, \dots),$$

with which (1.9) can be written in the form

$$(1.10) \qquad\qquad f \circ h = h \circ \sigma.$$

Thus we have shown that the following diagram commutes (as we did earlier in (1.5)):

$$(1.11)$$

$$
\begin{array}{ccc}
\Sigma_2^+ & \xrightarrow{\ \sigma\ } & \Sigma_2^+ \\
\downarrow{\scriptstyle h} & & \downarrow{\scriptstyle h} \\
C & \xrightarrow{\ f\ } & C
\end{array}
$$

Hence the coding map h is a conjugacy between the shift σ and the map f, which allows us to draw conclusions about the dynamics of f based on analogous results for the dynamics of σ.

For example, we may ask how many periodic points of a given order f has; that is, how many solutions there are to the equation $f^m(x) = x$ for a fixed integer m. Two obvious periodic points are 0 and 1, which are fixed by f and are thus immediately periodic. It is not so obvious what happens for larger values of m, but we may obtain the answer relatively easily by passing to the symbolic setting. Here we see that any fixed point must have $w_2 = w_1$, and similarly $w_{n+1} = w_n$ for every n. Thus the only fixed points are $(1, 1, 1, \dots)$ and $(2, 2, 2, \dots)$, which correspond to 0 and 1, respectively. For $m = 2$, the equation $\sigma^2(w) = w$ tells us that we may choose w_1 and w_2 to be either 1 or 2, but that after that we must have

$$
w_n = \begin{cases} w_1 & n \text{ odd}, \\ w_2 & n \text{ even}. \end{cases}
$$

Thus there are four points with $\sigma^2(\omega) = \omega$; in addition to the two mentioned above, we have $(1, 2, 1, 2, \dots)$ and $(2, 1, 2, 1, \dots)$.

In general, any sequence w which repeats after m digits will satisfy $f^m(w) = w$, and since there are 2^m such sequences, we have 2^m

periodic points of period m.[13] Passing to C via the conjugacy given by h, we see that f also has 2^m periodic points of period m, and so the set of periodic points is countably infinite.

Exercise 1.5. Argue directly from the definition of f that the set of periodic points is countably infinite. Can you obtain a result on the number of periodic points with a given period without using symbolic dynamics?

Lecture 4

a. Some point-set topology. In this lecture, we pause to recall some of the basic definitions of point-set topology, metric spaces, and Lebesgue measure. These will be needed for our subsequent discussions of Cantor sets, symbolic space, and dimension theory.

We begin with a (rather hasty) sketch of the basic topological definitions without reference to a metric.

Definition 1.1. Let X be a set and $\mathcal{T}$ a collection of subsets of X such that:

(1) $\emptyset, X \in \mathcal{T}$.

(2) If $U, V \in \mathcal{T}$, then $U \cap V \in \mathcal{T}$.

(3) If $\{U_\alpha\} \subset \mathcal{T}$, then $\bigcup_\alpha U_\alpha \in \mathcal{T}$, where the union may be taken over any collection of sets in $\mathcal{T}$, countable or not.

Then the pair $(X, \mathcal{T})$ is a *topological space*, and $\mathcal{T}$ is the *topology* on X. The sets in $\mathcal{T}$ are referred to as *open sets*; given a point $x \in X$, an open set containing x is a *neighbourhood* of x. Similarly, an open set containing $E \subset X$ is a *neighbourhood* of E.

One of the most familiar topological spaces is $\mathbb{R}$, where a set U is open if and only if it can be written as a countable union of open intervals,

$$(1.12) \qquad\qquad U = \bigcup_{n=1}^{\infty} (a_n, b_n).$$

---

[13]Of course, some of these will also be periodic points of lower order. We must do a little more work if we are to count points with *primitive* period m.

It can easily be checked that the collection $\mathcal{T}$ of such sets satisfies the three properties given above. In fact, this definition codifies these three properties as the *only* axioms which need to be satisfied in order for topological concepts such as convergence, compactness, and continuity to be used. Once we know which sets are open, all these concepts are in play, wherever our underlying space came from.

A sequence $x_n \in X$ *converges* to $x \in X$ if for every neighbourhood U of x, there exists N such that $x \in U$ for all $n \geq N$.

A set $E \subset X$ is *compact* if every open cover of E has a finite subcover; that is, given an arbitrary collection of open sets whose union contains E, there exists a finite subcollection whose union already contains E. In many cases (but not all), this is equivalent to the requirement that any sequence $x_n \in E$ contains a subsequence which converges to a point in E. For sets in $\mathbb{R}^d$, it is also equivalent to the requirement that E be closed (see below) and bounded.

If X and Y are two topological spaces, a map $f \colon X \to Y$ is *continuous* if given any open set $U \subset Y$, the preimage $f^{-1}(U) = \{x \in x \mid f(x) \in U\}$ is open in X. A bijection f (that is, a one-to-one map whose range is all of Y) such that both f and f^{-1} are continuous is called a *homeomorphism*; this is the natural equivalence relation in the category of topological spaces.

In order to give all these topological definitions, it also suffices to know which sets are *closed*; we say that $A \subset X$ is closed if its complement $X \setminus A$ is open. There are various equivalent definitions of closed sets. For example, $x \in X$ is an *accumulation point* for $A \subset X$ if every neighbourhood of x contains a point in A other than x itself (which may or may not be in A). Then the *closure* of A, denoted $\overline{A}$, is the set A together with all its accumulation points, and A is closed if and only if $A = \overline{A}$; the closure $\overline{A}$ may also be characterised as the smallest closed set which contains A. The dual notion to the closure is the *interior* of a set A, characterised as the largest open set contained in A; equivalently,

$$\operatorname{int} A = X \setminus \left(\overline{X \setminus A} \right) = \{x \in A \mid U \subset A \text{ for some open } U \ni x\}.$$

In what follows, one of our fundamental examples of a topological space will be the middle-third Cantor set $C \subset \mathbb{R}$. The topology $\mathcal{T}$ on

$\mathbb{R}$ given in (1.12) determines a topology $\mathcal{T}'$ on the Cantor set $C \subset \mathbb{R}$ (indeed, on any subset of $\mathbb{R}$) as follows: a set $U \subset C$ is open (lies in $\mathcal{T}'$) if and only if there exists $V \in \mathcal{T}$ such that $U = V \cap C$.

With this induced topology, the Cantor set has several important topological properties, which we now describe.

Given a topological space X, a set $A \subset X$ is *disconnected* if there exist disjoint open sets $U, V \subset X$ such that $A \subset U \cup V$, and is *connected* if no such U, V exist. If for any $x, y \in A$, there exist disjoint open sets U, V such that $x \in U$, $y \in V$, and $A \subset U \cup V$, we say that A is *totally disconnected*.

Finally, a set $A \subset X$ is *perfect* if every point $x \in A$ is an accumulation point for A.

Exercise 1.6. Show that the middle-third Cantor set is compact, perfect, and totally disconnected.

It is in many cases important to have a notion of when a set is somehow "large" in a topological sense. When can a set $A \subset X$ be said to contain "most" of X, from a topological point of view? One possibility is as follows: we say that A is *dense* if $\overline{A} = X$, or *dense in* $E \subset X$ if $\overline{A} = E$.

Exercise 1.7. Show that the set of rational points in $[0,1]$ is dense.

Exercise 1.8. Show that a set $A \subset X$ is dense in X if and only if it intersects every open set $U \subset X$. Show that the set of endpoints of removed intervals in the middle-third Cantor set is a dense subset of the Cantor set.

Since open sets are guaranteed to contain certain parts of X (a neighbourhood of x, for any x in the set), being open is in some sense a "large" property. Thus a set which is both open and dense is very large, topologically speaking. This motivates the following definition: a set $A \subset X$ is *residual* if it is the intersection of a countable number of open and dense sets. If a property holds on a residual set of points, then it may be thought of as holding on very nearly all of X; such a property is said to be true *generically*.

These notions of topological size are qualitative rather than quantitative; thus, they are not terribly useful for much of the more detailed work we eventually wish to carry out. We have no way to say that a particular neighbourhood of x is "small", or to compare neighbourhoods of different points. Consequently, many concepts from calculus and real analysis cannot be stated in the context of a general topological space; we can go a long way and obtain many analytic results using only topological methods, but we cannot do everything. The definition of a topological space is general enough to permit some behaviour which is rather pathological from the point of view of the standard topology on $\mathbb{R}$.

b. Metric spaces. A particular class of topological spaces, which are in some sense better behaved and do not exhibit as much pathological behaviour, consists of those spaces whose topology is induced by a *metric*; that is, a distance function $d\colon X \times X \to [0, \infty)$ such that the following hold for all $x, y, z \in X$:

(1) $d(x, y) \geq 0$, with equality if and only if $x = y$.

(2) $d(x, y) = d(y, x)$.

(3) $d(x, y) \leq d(x, z) + d(z, y)$ (triangle inequality).

For example, the usual distance function

$$(1.13) \qquad\qquad d(x, y) = |x - y|$$

on $\mathbb{R}$ makes the real line a metric space.

Given a space X with a metric d, we may consider the set of all points which lie within a fixed distance $r > 0$ of a point $x \in X$:

$$B(x, r) = \{y \in X \mid d(x, y) < r\}.$$

This is the *open ball of radius r centred at x*. We say that a set $U \subset X$ is *open* if for every $x \in U$ there exists $r > 0$ such that $B(x, r) \subset U$; that is, if a sufficiently small ball around x is contained in U for every $x \in U$.

If U_1 and U_2 are open sets, it is easy to verify that their union $U_1 \cup U_2$ is open as well; indeed, this holds for the union of *any* collection of open sets, no matter how large. One may also check that the intersection $U_1 \cap U_2$ is open as well, and that this property carries over to *finite* intersections $U_1 \cap \cdots \cap U_n$, but not to infinite intersections, as the example $U_n = (-1/n, 1/n) \subset \mathbb{R}$ illustrates. Upon the observation that the empty set and X are both open, we see that the collection of open sets as just defined satisfies the properties of a topology as given at the beginning of the lecture.

Exercise 1.9. Show that the metric (1.13) on $\mathbb{R}$ induces the topology in which the open sets are given by (1.12).

Given any set $E \subset X$, we may consider the *r-neighbourhood of* E, defined as

$$\bigcup_{x \in E} B(x, r).$$

This is just the set of all points in X which lie within a distance r of some point in E, and is in some sense a "fattening" of the set E. For example, if E is the ball $B(0, a) \subset \mathbb{R}^d$, then the r-neighbourhood of E is just the larger ball $B(0, a+r)$. If E is a set with a more complicated geometry, such as the Cantor set, then its r-neighbourhoods will in some sense have a simpler geometric structure than E itself.

For metric spaces, we may formulate an equivalent definition of closed sets: a set E is closed if for every sequence $(x_n)_{n \in \mathbb{N}} \subset E$ which converges to some point $x \in X$, we have in fact $x \in E$. This is often expressed as the statement that E contains its limit points.

From this last statement, or from the definition, it follows that arbitrary intersections of closed sets are closed, as are finite unions of closed sets. Infinite unions of closed sets may not be closed—consider $E_n = [1/n, 1]$.

In the context of metric spaces, the definition of a continuous map (and hence of a homeomorphism) can be phrased in a slightly more familiar way.

Exercise 1.10. Given two metric spaces (X, d) and (Y, ρ), show that $f: X \to Y$ is continuous if and only if for all $x \in X$ and $\varepsilon > 0$ there exists $\delta > 0$ such that $\rho(f(x), f(y)) < \varepsilon$ whenever $y \in X$ is such that $d(x, y) < \delta$.

The natural equivalence relation on the class of metric spaces is *isometric equivalence*; two metric spaces (X, d) and (Y, ρ) are isometrically equivalent if there exists a bijection $f: X \to Y$ such that

Figure 1.15. Vacationing in France.

$\rho(f(x), f(y)) = d(x, y)$ for all $x, y \in X$. However, since metric spaces also carry a natural topological structure, it is often of interest to understand when two metric spaces are equivalent *as topological spaces*, rather than as metric spaces.

Exercise 1.11. Show that if X is any metric space and $E \subset X$ is non-empty, compact, perfect, and totally disconnected, then E is homeomorphic to the middle-third Cantor set.

It is also of interest to understand how different metrics can induce entirely different topologies. To illustrate the various exotic topologies that can be induced by unconventional metrics, let X be the unit disc in $\mathbb{R}^2$, and introduce a metric d by the formula

$$(1.14) \qquad d(\mathbf{x}, \mathbf{y}) = \begin{cases} \|\mathbf{x} - \mathbf{y}\| & \text{if } \mathbf{x} \text{ is a scalar multiple of } \mathbf{y}, \\ \|\mathbf{x}\| + \|\mathbf{y}\| & \text{otherwise}, \end{cases}$$

where $\|\cdot\|$ denotes the usual norm on $\mathbb{R}^2$, $\|\mathbf{x}\| = \sqrt{x_1^2 + x_2^2}$. This is not quite so unnatural an example to consider as it may appear. Suppose you are on holiday in France (see Figure 1.15), and wish to take the high-speed train (the TGV) from Marseille (on the Mediterranean) to Nantes (near the Atlantic). Because there is no direct TGV line between the two cities, you must go via Paris (which is, of course, the origin in this example), and so as far as your travels are concerned, the distance from Marseille to Nantes is found by adding together the distance from Marseille to Paris ($\|\mathbf{x}\|$) and the distance from Paris to Nantes ($\|\mathbf{y}\|$).

On the other hand, if you wish to go from Marseille to Lyon, then there is no need to go all the way to Paris first, since the two cities lie on the same branch of the rail system, and so the distance is given by the usual formula ($\|\mathbf{x} - \mathbf{y}\|$).

The new metric induces a rather different topology on the disc than the one we are used to. For example, given any $\mathbf{x} \neq \mathbf{0}$ in the disc, the interval

$$((1 - \varepsilon)\mathbf{x}, (1 + \varepsilon)\mathbf{x}) = \{r\mathbf{x} \mid 1 - \varepsilon < r < 1 + \varepsilon\}$$

is open for every $\varepsilon > 0$, although these sets were neither open nor closed in the usual topology.

A striking distinction between the two spaces is given by the notion of *separability*: a topological space is *separable* if it has a countable dense subset. With the usual metric, the disc is a separable metric space (consider points with rational coordinates), but in this new metric, there is no countable dense subset, and so the space is not separable.

c. Lebesgue measure. Each of the Cantor sets we have studied so far inherits a metric, and thus a topology, from the real line, as indeed does every subset of $\mathbb{R}$. Topological notions provide one, very coarse, way of classifying subsets of $\mathbb{R}$, while metric notions provide another, somewhat more discriminating, tool. Another tool, which in some ways is coarser and in other ways more precise, is *Lebesgue measure*, which generalises the notion of "length" to sets which are not intervals.

We will write $\mathrm{Leb}(Z)$ for the Lebesgue measure of a set $Z \subset \mathbb{R}$; if Z is an interval $[a, b]$, then the Lebesgue measure of Z is just the length of the interval,

$$(1.15) \qquad\qquad \mathrm{Leb}([a, b]) = b - a.$$

Endpoints do not contribute to the Lebesgue measure of an interval; (1.15) also applies to intervals of the form (a, b), $(a, b]$, and $[a, b)$.

If I_1 and I_2 are two intervals, they may either overlap or be disjoint. If they overlap, then their union is again an interval, and so $\mathrm{Leb}(I_1 \cup I_2)$ is given by (1.15). If they are disjoint, we define

$$\mathrm{Leb}(I_1 \cup I_2) = \mathrm{Leb}(I_1) + \mathrm{Leb}(I_2).$$

In fact, we define Lebesgue measure this way for any set Z which is a union of countably many disjoint intervals:

$$(1.16) \qquad \mathrm{Leb}\left(\bigcup_i I_i\right) = \sum_i \mathrm{Leb}(I_i).$$

Each partial sum over finitely many disjoint intervals is less than 1, and so the infinite sum converges.

In order to generalise (1.16) to sets which are not countable unions of intervals, we cover them with such unions. Since any particular union may cover more than just the set we are interested in, we define the Lebesgue measure of a set $Z \subset \mathbb{R}$ as the greatest lower bound over all such covers:

$$(1.17)$$

$$\mathrm{Leb}(Z) = \inf\left\{\sum_i \mathrm{Leb}(I_i) \,\middle|\, \{I_i\} \text{ is a cover of } Z \text{ by open intervals}\right\}.$$

In fact, when we give a less superficial treatment of measure theory in Chapter 3, we will see that what we have just defined is actually an *outer measure*, unless we require Z to be a *measurable set*. For the time being, though, this rough idea will serve us well as motivation for other definitions and discussions, and we will return to the technical details in due course.

Before moving on, we note that we may now make more precise our statement in Lecture 3 that the "length" of the Cantor set is zero. As Lebesgue measure is the generalisation of length, we ought to say that the Cantor set has Lebesgue measure zero, or $\mathrm{Leb}(C) = 0$.

Exercise 1.12. Let $Z \subset \mathbb{R}$ be countable, and show that $\mathrm{Leb}(Z) = 0$. In particular, the set of rational numbers has Lebesgue measure zero.

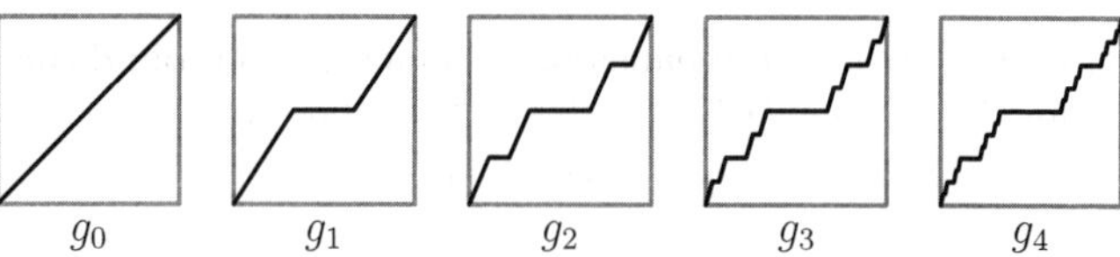

Figure 1.16. Building the devil's staircase.

Exercise 1.13. Define a sequence of functions $g_n \colon [0,1] \to [0,1]$ as follows (see Figure 1.16):

(1) $g_n(0) = 0$, $g_n(1) = 1$.

(2) g_n has constant slope 3^n on the intervals $I_{w_1 \ldots w_n}$ (the component intervals of the nth step in the iterative construction of the middle-third Cantor set).

(3) g_n has slope 0 on $[0,1] \setminus \left(\bigcup_{w_1 \ldots w_n} I_{w_1 \ldots w_n} \right)$ (the intervals which are removed from the middle-third Cantor set before step n).

By taking the limit as $n \to \infty$, obtain a non-decreasing continuous function $g \colon [0,1] \to [0,1]$ such that $g(0) = 0$ and $g(1) = 1$. Give an explicit description of a set $Z \subset [0,1]$ such that g is differentiable at every $x \in Z$ with $g'(x) = 0$, and $[0,1] \setminus Z$ has Lebesgue measure zero.

The function g constructed in Exercise 1.13 is known as the *devil's staircase* (or sometimes as the *Cantor function*) and is rather pathological when compared with the sorts of functions one usually encounters in introductory calculus and analysis. It is non-increasing everywhere except on the middle-third Cantor set, which has Lebesgue measure zero, and it is continuous everywhere on the unit interval; nevertheless, it increases in value as we go from one end of the interval to the other.

One might be forgiven for thinking that a function with such strange behaviour could not arise in any natural setting and could only be the result of a deliberate construction, as in Exercise 1.13. One would, however, be mistaken, as the following exercise shows.

Exercise 1.14. Define two affine maps f_1 and f_2 from $\mathbb{R}$ to itself by

$$f_1(x) = 3x, \qquad f_2(x) = 3(x-1) + 1.$$

Thus f_1 and f_2 are linear expansions by a factor of 3 around the points $x = 0$ and $x = 1$, respectively. Given an initial condition $x_0 \in [0, 1]$, construct a sequence $x_1, x_2, \ldots$ as follows:

(1) Select one of the two maps f_1 or f_2 with equal probability (by flipping a coin, for example).

(2) If f_i was selected, let $x_1 = f_i(x_0)$.

(3) Return to step (1) and select a map f_j at random (independently of which map was chosen at first); let $x_2 = f_j(x_1)$.

(4) By successive random choices and iterations, construct $x_3, x_4, \ldots$.

Let $P(x)$ be the probability that a sequence constructed in the above fashion starting at $x_0 = x$ escapes to infinity; $x_n \to +\infty$. Show that $P \colon [0, 1] \to [0, 1]$ is the Cantor function constructed in Exercise 1.13.

Lecture 5

a. The topological structure of symbolic space and the Cantor set. Having seen that h respects the dynamics of f and σ, it is natural to ask what other aspects of the sets C and Σ_2^+ are preserved by the conjugacy. We saw in the previous lecture that the Cantor set C inherits a metric, and hence a topology, from the real line. If we define a metric on Σ_2^+, then we will have a topology there too, and we may ask whether h takes convergent sequences in Σ_2^+ to convergent sequences in C, and vice versa. This will expand the range of questions about the dynamics of f which can be answered by looking at the symbolic case to include questions of a topological nature—that is, questions involving convergence.

To this end, fix a real number $a > 1$, and given two sequences $v, w \in \Sigma_2^+$, define the distance between them by

$$(1.18) \qquad d_a(v, w) = \sum_{k \geq 1} \frac{|v_k - w_k|}{a^j}.$$

Note that since each numerator $|v_k - w_k|$ is either 0 or 1, this series converges absolutely. We may easily verify that $d = d_a$ satisfies the axioms of a metric from the previous lecture, each of which follows immediately from its counterpart for the usual distance on $\mathbb{R}$.

Exercise 1.15. Define two different distance functions d' and d'' on Σ_2^+ by

$$d'(s,t) = \frac{1}{t(v,w)+1},$$

$$d''(s,t) = e^{-t(v,w)},$$

where $t(v,w)$ is the minimum value of the index k for which $v_k \neq w_k$, and $d'(w,w) = d''(w,w) = 0$. Show that d' and d'' are metrics, and that they define the same topology as d.

Exercise 1.16. Consider the symbolic dynamical system (Σ_2^+, σ). Fix a number $\varepsilon > 0$ and a point $w \in \Sigma_2^+$. Show that there exists $N > 0$ such that for any $n \geq N$ one can find a point $v \in \Sigma_2^+$ for which $d(v,w) \leq \varepsilon$ and $d_a(\sigma^n(v), \sigma^n(w)) = 1$.

What are the open and closed sets in the symbolic space Σ_2^+? We defined a metric (1.18) on Σ_2^+, and so all the usual topological notions make sense, but what do the open and closed sets actually look like?

For the sake of this discussion, assume that $a > 2$, and consider the ball $B(w,r)$ centred at a point $w \in \Sigma_2^+$ with radius $r = 1/a > 0$. How do we tell if another point $v \in \Sigma_2^+$ is in $B(v,r)$? The distance between the two points is given by (1.18), and we see immediately that if $v_1 \neq w_1$, the first term alone means that the sum is $\geq 1/a$. Conversely, if $v_1 = w_1$, then the first term in the sum vanishes, and the distance is at most

$$\sum_{k=2}^{\infty} \left(\frac{1}{a}\right)^k = \frac{1}{a}\frac{1}{a-1} < \frac{1}{a},$$

where the last inequality uses the fact that $a > 2$. Thus we see that

$$B(w,r) = \{v \in \Sigma_2^+ \mid v_1 = w_1\}.$$

There are exactly two possibilities for w_1, and so there are exactly two possible sets of this form:

$$C_1 = \{v = (1, v_2, v_3, \dots)\},$$
$$C_2 = \{v = (2, v_2, v_3, \dots)\}.$$

We refer to C_1 and C_2 as *cylinders of length* 1; each contains all sequences in Σ_2^+ for which the first term matches a particular specification. If we demand that the first n terms follow a particular itinerary, we obtain a *cylinder of length n*:

$$(1.19) \qquad C_{w_1 \dots w_n} = \{ v \in \Sigma_2^+ \mid v_j = w_j \text{ for every } 1 \le j \le n \}.$$

Following the above argument, we see that these are exactly the balls of radius $1/a^n$, provided $a > 2$. There is a one-to-one correspondence between cylinders of length n and n-tuples with entries in $\{1, 2\}$.

If we take $1 < a \le 2$, then in order to contain an entire n-cylinder, a ball must have radius greater than

$$\frac{1}{a^{n+1}} + \frac{1}{a^{n+2}} + \cdots = \frac{1}{a^n(a-1)} \ge \frac{1}{a^n},$$

and so it will also contain sequences from other n-cylinders. Thus cylinders are no longer balls; however, the topology turns out to be the same, thanks to the following result.

Proposition 1.2. *Cylinders are open.*

Proof. Given a cylinder $C_{w_1 \dots w_n}$ and a point $w \in C_{w_1 \dots w_n}$, we may choose any $r < 1/a^n$, and then we see, as above, that $d(v, w) \ge r$ unless all the terms with $k \le n$ vanish; that is, unless $v_k = w_k$ for all $1 \le k \le n$. Thus $d(v, w) < r$ implies $v \in C_{w_1 \dots w_n}$, and so $B(w, r) \subset C_{w_1 \dots w_n}$. $\qquad \square$

That's not the end of the story, though. . .

Proposition 1.3. *Cylinders are closed.*

Proof. Let $C_{w_1 \dots w_n}$ be a cylinder in Σ_2^+, and suppose $(w^m)_{m \in \mathbb{N}} \subset C_{w_1 \dots w_n}$ is a sequence which converges to $v \in \Sigma_2^+$ as $m \to \infty$. Then $d(w^m, v) \to 0$, and in particular, each term in the sum (1.18) must go to 0. Thus $\lim_{m \to \infty} w_k^m = v_k$ for every $k \ge 1$, and since $w_k^m = w_k$ for every $1 \le k \le n$ and all m, we have $v_k = w_k$ for $1 \le k \le n$, and so $v \in C_{w_1 \dots w_n}$. It follows that $C_{w_1 \dots w_n}$ is closed. $\qquad \square$

Thus cylinders are both open *and* closed, a somewhat unfamiliar phenomenon if our only experience is with the topology of $\mathbb{R}$. The feature of the topology of Σ_2^+ which permits this behaviour is the fact that the cylinders of a given length are all disjoint, and their union is the whole space: we say that they *partition* Σ_2^+. This gives an alternate proof of Proposition 1.3, once Proposition 1.2 is known; the

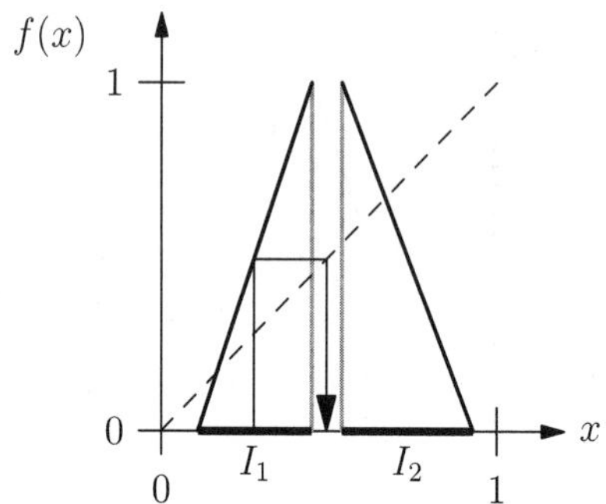

Figure 1.17. Another piecewise linear map with chaotic behaviour.

complement of an n-cylinder $C_{w_1...w_n}$ is a union of $2^n - 1$ n-cylinders, each of which is open, and hence $C_{w_1...w_n}$ is closed.

Furthermore, for any two points $v, w \in \Sigma_2^+$, we can find a cylinder which contains v but not w; since this cylinder is both open and closed, this implies that Σ_2^+ is totally disconnected. It is not hard to show that it is also compact and perfect, and hence by the result of Exercise 1.11, it is homeomorphic to the middle-third Cantor set C. Indeed, we have already encountered the map which exhibits this equivalence.

Proposition 1.4. *The coding map $h: \Sigma_2^+ \to C$ is a homeomorphism.*

Proof. Recall that h is a bijection, and is defined by the inclusion

$$x \in I_{w_1} \cap I_{w_1 w_2} \cap \cdots,$$

and so we see that $h(C_{w_1...w_n}) = I_{w_1...w_n}$ for every cylinder in Σ_2^+. Since the sets $I_{w_1...w_n}$ are all closed in C (being closed intervals), we have shown that h and h^{-1} both take closed sets to closed sets, which suffices to show that h is a homeomorphism. $\square$

We saw earlier that the coding map respects the dynamics of the two systems $f: C \to C$ and $\sigma: \Sigma_2^+ \to \Sigma_2^+$, and Proposition 1.4 shows that it respects topology as well. A map such as h, which is both a homeomorphism and a conjugacy, is called a *topological conjugacy*, and the two maps σ and f are called *topologically conjugate*. Thus from either a dynamical or topological point of view, we may as well

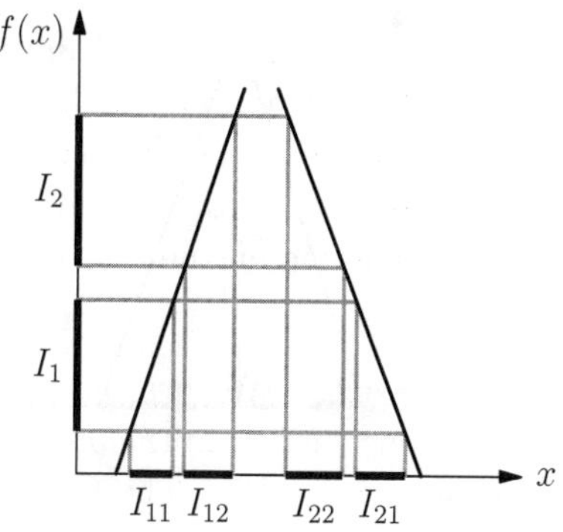

Figure 1.18. The domain of definition of f^2.

study whichever is better suited to the problem at hand, knowing that our results will be valid for the other as well.

b. What the coding map doesn't do. Despite the fact that the map f and the shift σ are topologically conjugate, the two systems are not equivalent in every aspect; Σ_2^+ does not capture quite everything there is to know about the Cantor set C. To convince ourselves of this, let us consider a more general class of dynamical systems defined in the interval. Fix two disjoint closed intervals $I_1, I_2 \subset [0, 1]$, and define a piecewise linear map $f \colon I_1 \cup I_2 \to [0, 1]$ as shown in Figure 1.17, so that $f(I_1) = f(I_2) = [0, 1]$ (note that for our purposes, each branch of f may be either increasing or decreasing).

If we try to iterate f more than once, we run into the same problem as before; some points in I_1 or I_2 have images which do not lie in either interval, and so cannot be iterated again. This leads us down exactly the same path as in Lecture 3; the domain of definition of f^2 is a union of four intervals, as shown in Figure 1.18, and so on for $f^3, f^4, \ldots$. The only difference in this case is that the intervals may be of varying lengths, but the combinatorial and topological structure is identical to that in the previous analysis, and we again get *a* Cantor set (rather than *the* middle-third Cantor set), for which we have a coding map and symbolic dynamics just as before.

Thus we see that Σ_2^+ models not just the dynamics of our original map on the middle-third Cantor set, but the dynamics of *any* map defined in this fashion (note that the original definition is just a special case of this one, with $I_1 = [0, 1/3]$ and $I_2 = [2/3, 1]$). Indeed, all these maps (and Cantor sets) have the same dynamical and topological structure; however, Σ_2^+ does not capture certain metric properties of the system, which vary depending on the choice of I_1 and I_2.

c. Geometry of Cantor sets. We turn our attention now to the geometric aspects of the construction in the previous section; that is, those which are not captured by Σ_2^+. The Cantor set C is defined by the same formula (1.8) as the middle-third Cantor set; the only difference is that while in that case the basic intervals $I_{w_1 \ldots w_n}$ had length 3^{-n} and were of the form $[a/3^n, (a+1)/3^n]$, now they may have variable lengths and locations.

How long are the basic intervals $I_{w_1 \ldots w_n}$? Beginning with $n = 1$, let λ_1 and λ_2 denote the lengths of I_1 and I_2, respectively. Then writing $|I|$ for the length of the interval I, we have $|I_i| = \lambda_i$ for $i = 1, 2$.

For $n = 2$, we examine the four intervals shown in Figure 1.14 and recall that the ratio $|f(I_{i_1 i_2})|/|I_{i_1 i_2}|$ is given by the slope of f in $I_{i_1 i_2}$, which the comments above show to be $1/\lambda_{i_1}$. Thus the intervals $I_{i_1 i_2}$ have lengths given by

$$|I_{11}| = \lambda_1^2, \quad |I_{21}| = \lambda_2 \lambda_1, \quad |I_{12}| = \lambda_1 \lambda_2, \quad |I_{22}| = \lambda_2^2,$$

where we use the fact that $f(I_{i_1 i_2}) = I_{i_2}$.

This generalises immediately to a formula for all values of n:

$$(1.20) \qquad\qquad |I_{w_1 \ldots w_n}| = \lambda_{i_1} \cdots \lambda_{i_n}.$$

Now that we know how long the basic intervals $I_{w_1 \ldots w_n}$ ought to be, we can try to carry out the construction of the set C (or one like it) without reference to the dynamics of f. To this end, consider the following geometric construction (which is a particular case of the Moran constructions we will introduce and study in Lecture 10):

(1) Begin by choosing two disjoint closed intervals $I_1, I_2 \subset [0, 1]$ and two *ratio coefficients* $\lambda_1, \lambda_2 > 0$ with $\lambda_1 + \lambda_2 < 1$.

(2) Put two disjoint closed intervals at arbitrary locations inside I_1, whose lengths are $\lambda_1|I_1|$ and $\lambda_2|I_1|$; denote these by I_{11} and I_{12}, respectively.

(3) Construct $I_{21}, I_{22} \subset I_2$ in a similar manner.

(4) Repeat steps (2) and (3) within each of the intervals $I_{w_1 w_2}$ to construct eight disjoint closed intervals $I_{w_1 w_2 w_3}$; iterate this procedure to produce intervals $I_{w_1 \ldots w_n}$ with length given by

$$(1.21) \qquad |I_{w_1 \ldots w_n}| = |I_{w_1}| \left(\prod_{k=2}^{n} \lambda_{w_k} \right).$$

(5) Construct a Cantor set C as the limit of this iterative procedure, just as in (1.8).

We will call C a *geometrically constructed Cantor set* on the line. The primary difference between C and the dynamically defined Cantor sets we saw before is that in the initial dynamical procedure, the position of the intervals $I_{w_1 \ldots w_n}$ was determined by the dynamics, whereas here they are free to be placed anywhere within $I_{w_1 \ldots w_{n-1}}$, provided they are disjoint. We can construct a coding map $h \colon \Sigma_2^+ \to C$ exactly as we did before, by considering the intersection $\bigcap_{n \geq 1} I_{w_1 \ldots w_n}$, and we get the following diagram:

$$
\begin{array}{ccc}
\Sigma_2^+ & \xrightarrow{\ \sigma\ } & \Sigma_2^+ \\
\Big\downarrow{\scriptstyle h} & & \Big\downarrow{\scriptstyle h} \\
C & & C
\end{array}
$$

The lack of any dynamics on $C \subset [0,1]$ means that the diagram does not close as it did in the case discussed in Lecture 3. We can, however, define a map $f \colon C \to C$ so that the diagram closes and commutes by simply taking $f = h \circ \sigma \circ h^{-1}$.

We have much less information about this artifically constructed map f than we had before, when we began with f and used it to construct C. All we can say in this case is that f is continuous, since each of σ, h, and h^{-1} is, and that it has all the dynamical properties of the shift map σ which we discussed before, such as a dense set of periodic orbits.

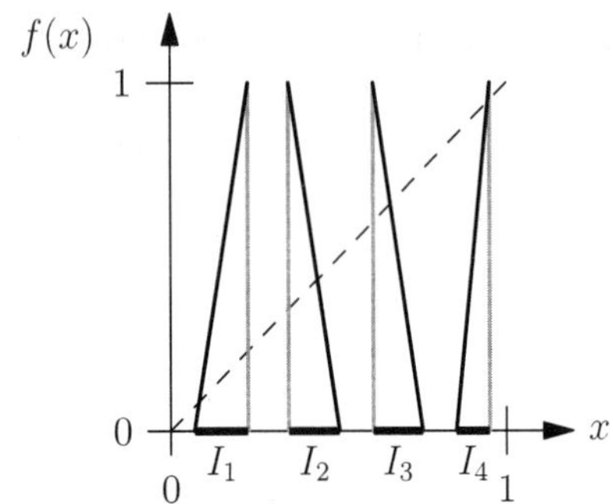

Figure 1.19. A piecewise linear map with four branches.

Exercise 1.17. Let C be a geometrically constructed Cantor set on the line. Show that the coding map $h\colon \Sigma_2^+ \to C$ is a Hölder function (see Appendix). Can this function be Lipschitz? Is the map $f = h \circ \sigma \circ h^{-1}$ Hölder continuous?

The geometric construction outlined above may just as well be carried out with more than two intervals. If we begin with disjoint intervals $I_1, \ldots, I_k \subset [0, 1]$ and ratio coefficients $\lambda_1, \ldots, \lambda_k > 0$ with $\Sigma_i \lambda_i < 1$, then we may build $I_{w_1 \ldots w_n}$ as before, with length given by (1.20), and define a Cantor set C by (1.8).

As in the case $k = 2$, a particular case of this construction is given by a dynamically defined Cantor set obtained as the maximal invariant set (the repeller) of a piecewise linear map with k branches, as shown in Figure 1.19. The purely geometric procedure just described is more general, since it "forgets" about the map f and allows the intervals $I_{w_1 \ldots w_n}$ to be placed arbitrarily within $I_{w_1 \ldots w_{n-1}}$ (as long as they are disjoint).

The geometric construction also generalises the dynamical definition by allowing the lengths of the initial basic intervals I_j to be arbitrary, provided subsequent basic intervals have lengths given by (1.21). For a dynamically defined Cantor set, we must have $|I_j| = \lambda_j$; the lengths of the initial basic intervals are equal to the ratio coefficients.

We can find a coding map for a geometrically constructed (or dynamically defined) Cantor set with $k > 2$, but the presence of more intervals at each step means that the *alphabet* for the symbolic space is larger—$\{1, \ldots, k\}$ instead of $\{1, 2\}$, where k is the number of branches of the map or the number of basic intervals at the first step. Writing

$$\Sigma_k^+ = \{1, \ldots, k\}^{\mathbb{N}} = \{w = (w_1, w_2, \ldots) \mid w_j \in \{1, \ldots, k\}\},$$

the coding map $h \colon \Sigma_k^+ \to C$ is defined as before and is again a homeomorphism. If we obtained C as the repeller for a map f, then h respects the dynamics of f; if we obtained C via a purely geometric construction, we may again place some dynamics on C by the formula $f = h \circ \sigma \circ h^{-1}$.

These examples illustrate the use of dynamical systems and geometric constructions as tools to study each other, which will be a prominent theme of this book. Many dynamical systems can be better understood by examining the appropriate geometric construction, and similarly, many geometric constructions are best viewed as arising from a particular dynamical system.

At this point, however, we have not yet developed the proper tools to study the geometric properties of the various Cantor sets we have encountered. Each has the power of the continuum (that is, it can be put into a bijective correspondence with the set of real numbers) and yet has zero length, in a sense which is made precise by the notion of Lebesgue measure. If we wish to use these sets as a tool to study the associated maps, we must somehow characterise the sets themselves, but we do not yet have the means to do so.

As before, different Cantor sets have the same coding, and so some structure is certainly lost in passing to the symbolic point of view. In the end, the fact that we cannot completely restore the set C from knowledge of Σ_k^+ will not cause us to lose much sleep because the crucial dynamical information is preserved. We will see rather more complicated examples which are modeled by these same symbolic dynamics, which turn out to contain the essence of the chaotic behaviour.

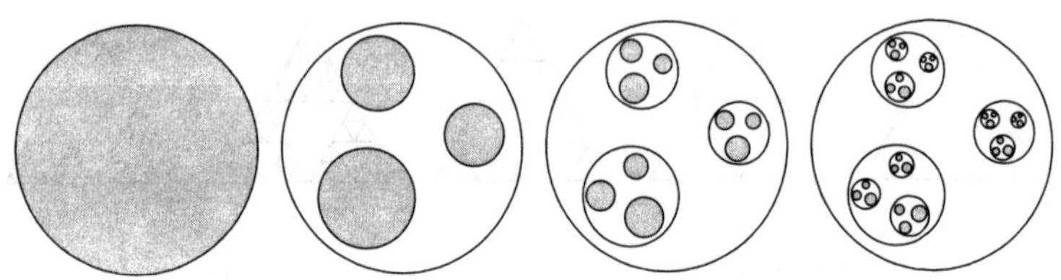

Figure 1.20. A general Cantor-like construction in $\mathbb{R}^2$.

Lecture 6

a. More general constructions.

a.1. *Higher dimensions.* Before examining possible ways of characterising Cantor sets, let us stretch our legs a bit and examine some of the other creatures in the zoo. So far we have been planted firmly in front of the cage labeled "one-dimensional constructions", but there is no reason why we could not consider examples in higher dimensions as well.

To this end, let $D_1, \ldots, D_k$ be disjoint closed discs contained in the unit disc in $\mathbb{R}^2$; the case $k = 3$ is illustrated in Figure 1.20. Choose ratio coefficients $\lambda_1, \ldots, \lambda_k$, and carry out an iterative procedure as before; within the disc D_{w_1}, place disjoint discs $D_{w_1 w_2}$ whose diameters are $\lambda_{w_2} \operatorname{diam}(D_{w_1})$, and so on. Taking the union over all discs corresponding to words of length n, and then taking the intersection over all $n \geq 1$, we obtain a Cantor set as before, in (1.8).

Of course, there is nothing special about discs, or about two dimensions, in this construction. The same procedure goes through for any domain in $\mathbb{R}^2$, or indeed in any $\mathbb{R}^d$, and the end result will always be homeomorphic to Σ_k^+. Thus we see that all these various Cantor sets have the same topology, despite our feeling that they must be somehow different geometrically. This reinforces our earlier point that Σ_k^+ carries no information about the geometry of the Cantor sets it models.

Exercise 1.18. Show that Σ_k^+ and Σ_m^+ are homeomorphic for any $k, m \geq 2$, and construct an explicit homeomorphism between them. (Note that this homeomorphism does not respect the dynamics of the shift.)

Figure 1.21. Constructing the Sierpiński gasket.

a.2. *Connected examples.* A similar construction, which has a little more built-in regularity, may be carried out by dividing an equilateral triangle into four smaller triangles, each similar to the first and congruent to each other, removing the middle triangle, and then iterating the procedure on the remaining three. The first few steps of the process are shown in Figure 1.21.

The fractal set C obtained as the limit of this procedure is known as the *Sierpiński gasket*.[14] It may also be constructed via the following algorithm, which Manfred Schroeder playfully dubs "Sir Pinski's game" [**Sch91**]. Given a point $\mathbf{x}$ inside the equilateral triangle, we define $f(\mathbf{x})$ by first finding the nearest vertex of the triangle to $\mathbf{x}$, and then doubling the distance from $\mathbf{x}$ to that vertex, as shown in Figure 1.22. Repeating the process takes us to the point $f^2(\mathbf{x})$, and so on, until we leave the triangle, at which point the trajectory will go off to infinity. The game, then, is to choose an initial point $\mathbf{x}$ whose trajectory remains in the triangle for as long as possible.

The reader may verify that the winning points, whose trajectory never leaves the triangle, are precisely the points in the Sierpiński gasket. The map f is of the same sort as we encountered earlier, when we looked at piecewise linear maps on the interval; in this case, f is a piecewise affine map on the plane, and the Sierpiński gasket is the maximal invariant set, the repeller, for f.

We can produce a coding map $h\colon \Sigma_3^+ \to C$ by labeling the triangles at each step of the iteration with the appropriate sequence of 1's, 2's, and 3's, and associating to each infinite sequence in Σ_3^+ the corresponding infinite intersection of nested triangles, which is just a single point. As in the one-dimensional case, the coding map completes the commutative diagram (1.11), replacing Σ_2^+ with Σ_3^+.

---

[14]Or as the *Sierpiński triangle* or *Sierpiński sieve*; there is also a *Sierpiński carpet*, whose construction is similar but non-equivalent.

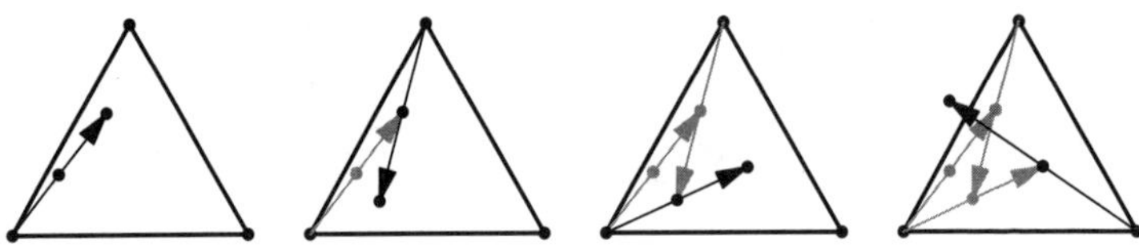

Figure 1.22. Sir Pinski's game.

The careful reader will by this point be howling in protest that the map f is not well defined everywhere. Indeed it is not; what are we to do with the points in C which are equidistant from two vertices of the triangle? f is supposed to double the distance from the nearest vertex, but what if that vertex is not unique? There are only three points in C which encounter this problem, but there are six more which are mapped into one of those three, and in general, there are 3^n points in C for which f^n is not uniquely defined.

This technicality arises since the domains on which f is defined are not disjoint, which had been a requirement for all our constructions up to this point. Consequently, the Sierpiński gasket is quite different topologically from the Cantor sets we have studied thus far; it is connected, while they were totally disconnected. Furthermore, the dynamics are different, since the coding map is not one-to-one on the whole space.

One encounters a similar difficulty when dealing with decimal representations of the real numbers. Such representations are not unique for certain numbers, namely those whose decimal expansion terminates. Both here and in the Sierpiński gasket, the trouble occurs on a countable set of points; we will later see that from the point of view of dimension theory (with which we will be primarily concerned), countable sets can be treated as negligible.

We will come back to this detail later when we discuss Moran constructions in Lecture 10. For the time being, we choose to ignore this seemingly troublesome quirk of the Sierpiński gasket, and divert prying eyes elsewhere by bedazzling the reader with a higher-dimensional version of the same thing. Instead of a triangle, begin with a tetrahedron in $\mathbb{R}^3$, decompose it into five congruent tetrahedra,

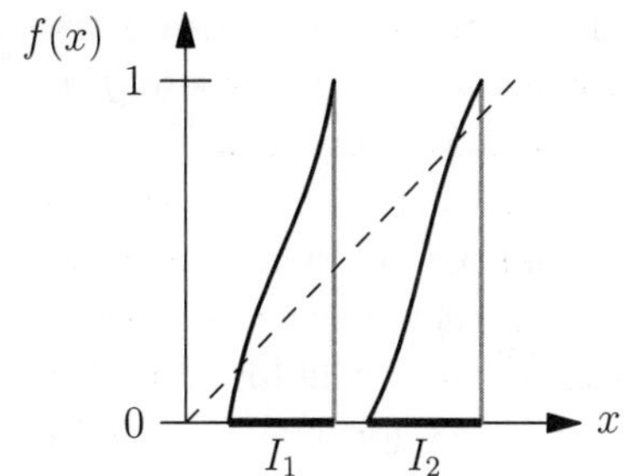

Figure 1.23. A non-linear interval map.

and remove the middle one. Iterating this procedure, one obtains a fractal sometimes known as the *Sierpiński sponge.*

a.3. *Non-conformal constructions.* We obtain another related construction by modifying the two-dimensional construction shown in Figure 1.20. Rather than shrinking the discs equally in all directions, we could contract by a factor of λ in one direction and μ in another, so that the building blocks at successive iterations are increasingly eccentric ellipses. The topological characterisation in terms of Σ_k^+ still goes through, but the geometry is patently different from what came before. The construction in Figure 1.20 was *conformal*—all directions were treated equally—by taking $\lambda \neq \mu$, we obtain a non-conformal construction, whose geometry is quite different (see Figure 7.5).

a.4. *Non-linear constructions.* All our examples up to this point have been linear; the building blocks at any given step of the construction are just scaled-down copies of those at the previous step. This will not always be the case in the examples of interest; most of the truly interesting phenomena in the real world, after all, are not particularly linear. Thus we may return to the one-dimensional setting and consider the sort of map $f \colon I_1 \cup I_2 \to [0,1]$ shown in Figure 1.23, which maps both I_1 and I_2 homeomorphically but not necessarily linearly onto $[0,1]$. Such a map is called a *one-dimensional full-branched Markov map.*[15] If f is piecewise continuously differentiable, with $|f'(x)| \geq a > 1$, where a is fixed, then it is called *expanding.*

---

[15]We say *full-branched* to distinguish it from the case examined later in Lecture 15 and shown in Figure 3.3, where the branches of the map may not extend the entire length of $[0,1]$.

Nearly everything from our previous discussion of piecewise linear maps goes through in the case where f is an expanding one-dimensional full-branched Markov map. We can follow a Cantor construction to obtain a repeller C for f, which is a maximal invariant set and is homeomorphic to symbolic space via the coding map $h \colon \Sigma_2^+ \to C$. This coding map also conjugates the dynamics of σ and f. The only thing that fails is the formula (1.20) for the lengths of the intervals $I_{w_1 \ldots w_n}$; a new formula can be found, but it is rather more complicated.

Exercise 1.19. Let Λ be the middle-third Cantor set and let C be the repeller for a one-dimensional (not necessarily linear) full-branched Markov map $f \colon I_1 \cup I_2 \to [0,1]$. Show that there exists a one-to-one continuous map h from Λ onto C, and that h is a homeomorphism if and only if I_1 and I_2 are disjoint.

Exercise 1.20. Let Λ be the Sierpinski gasket and let C be the repeller for a one-dimensional (not necessarily linear) full-branched Markov map $f \colon I_1 \cup I_2 \cup I_3 \to [0,1]$ where the subintervals I_1, I_2 and I_3 are disjoint. Show that there exists a continuous map h from C onto Λ which is one-to-one at all but countably many points.

b. Making sense of it all. Up to this point, we have been behaving like Adam, merely wandering around the Garden and naming all the animals; now we must become Linnaeus, and make some attempt at *classifying* the fractal fauna we find around us. For the various fractal sets we have described are in fact different from each other geometrically, but our usual measuring sticks are not properly equipped to distinguish them. Every example we have encountered is uncountable, and so cardinality alone is insufficient.

Topological tools also fail to properly analyse the exhibits in the zoo. While some sets are connected (such as the Sierpiński gasket) and others are totally disconnected (such as the various Cantor sets), this classification is too crude to distinguish between the repellers associated with different full-branched Markov maps, or between the gaskets generated from different triangles, or between fractal curves built from different generators.

We might try to analyse various fractals with the notion of length, or more properly, Lebesgue measure; however, we found that the Cantor sets all have length zero, and it turns out that the fractal curves (the von Koch curve, the fractal coastlines) have infinite length. Trying a two-dimensional notion of size, one finds that these curves have zero area, and so no help is forthcoming from length, area, volume, and so on. We seem to be at an impasse.

The way out of our predicament is provided by the idea of *fractal dimension*, a notion which is at once tremendously important and frustratingly elusive. Its importance will become apparent when we see how readily it lets us make sense of the thicket of examples we have presented thus far; its elusiveness is due to the fact that, put starkly, it is not defined!

Let us explain this last statement. The concept of fractal dimension was discussed from a "practical" point of view by Benoit Mandelbrot in his landmark 1982 book *The Fractal Geometry of Nature* [**Man82**]. In that work, he examined a wide variety of examples and described certain numbers which are associated to the scaling properties of the systems in these examples and which may reasonably be referred to as fractal dimension; moreover, he demonstrated that unlike our usual idea of dimension, these numbers need not be integers.

Mandelbrot's book, though, is primarily concerned with exhibiting the utility of fractals as tools in various scientific contexts, rather than with mathematical minutiae. Thus while this book was instrumental in making fractals an important part of science, where they have proved their worth in a dazzlingly wide variety of scientific models (to say nothing of the fractal artwork which has sprung up in the decades following Mandelbrot's work), it does not contain a single unifying definition of just what exactly the fractal dimension of a set actually *is*.

In fact, the notion of dimension that we will use to characterise fractal sets predates Mandelbrot by over half a century; one important definition is due to Felix Hausdorff in 1919.[16] Now referred to as

---

[16]Although we should point out that Constantin Carathéodory introduced a somewhat more general notion even earlier, in 1914.

the *Hausdorff dimension*, it is one of the fundamental geometric characteristics of a set, and we will see that it lets us distinguish between the various Cantor sets we have met.

We will also see that the Hausdorff dimension is rather difficult to compute; partly because of this, there are dozens of alternative definitions of fractal dimension, some of which are often more tractable. In particular, we will be concerned with the notions of upper box dimension and lower box dimension, which are also candidates for the title of "fractal dimension".[17] The situation is made quite messy by the fact that these various dimensional quantities may not coincide; indeed, we will see concrete examples for which they are different.

One of our primary goals, then, will be to clarify the relations between the various notions of fractal dimension. These quantities are fundamental characteristics of many dynamical systems, giving a geometrical characterisation of chaotic behaviour, and so it is important to understand how they fit together. Mercifully, we will see that in many important cases, all the reasonable definitions of fractal dimension lead to the same value, and the mess cleans itself up; however, the variety of dimensional notions goes some way towards explaining Mandelbrot's omission of a mathematically rigorous definition.

---

[17]There are many other dimensional quantities, such as correlation dimension and information dimension, which are considered in more advanced studies of the subject.

Chapter 2

Fundamentals of Dimension Theory

Lecture 7

a. Definition of Hausdorff dimension. In this lecture, we define the notion of Hausdorff dimension for a set $Z \subset \mathbb{R}^d$. This definition requires some work to set up, and so we first take some time to go through the necessary preliminaries.

Given a set $Z \subset \mathbb{R}^d$, we consider a collection $\mathcal{U} = \{U_i\}$ of open sets in $\mathbb{R}^d$ which cover Z; that is, for which $\bigcup_i U_i \supset Z$.

Such a collection is known as an *open cover*; we will usually simply refer to a *cover*, with the implicit assumption that every element of the cover is an open set. The picture to keep in mind is a collection of open balls (whose radii may vary), although more general open sets are allowed. We denote the diameter of a set U_i by

$$\operatorname{diam} U_i = \sup\{d(x,y) \mid x,y \in U_i\}$$

and the diameter of a cover by

$$\operatorname{diam} \mathcal{U} = \sup_{U_i \in \mathcal{U}} \operatorname{diam} U_i.$$

Fix $\varepsilon > 0$. If $\operatorname{diam} \mathcal{U} \leq \varepsilon$, that is, if every $U_i \in \mathcal{U}$ has $\operatorname{diam} U_i \leq \varepsilon$, then we say that $\mathcal{U}$ is an ε-*cover*.

We will only consider covers with at most countably many elements. The reason for this is that we will use the cover $\mathcal{U}$ to "measure" the set Z, by assigning each element of the cover a certain (positive) weight, and then summing these weights over all elements of the cover. Since an uncountable collection of positive numbers cannot have a finite sum, we must take $\mathcal{U}$ to be countable.

Denote by $\mathcal{D}(Z, \varepsilon)$ the collection of all countable ε-covers of Z. In what follows, we consider a fixed $\alpha \geq 0$, $\varepsilon > 0$, and $\mathcal{U} \in \mathcal{D}(Z, \varepsilon)$. We define (somewhat arbitrarily, it may seem) a "potential" for the entire cover $\mathcal{U}$ by the formula

$$(2.1) \qquad \sum_i (\operatorname{diam} U_i)^\alpha.$$

The reader may justifiably feel that we have just pulled a rabbit from a hat without any explanation of where it came from;[1] that will come in due course. For the time being, observe merely that (2.1) gives us a way of assigning a number, sometimes called the "potential", to any ε-cover $\mathcal{U}$.

Now we want to use this number to characterise the set Z, to measure how "big" it is, in some sense. The difficulty is that Z admits many open covers, each of which has a different potential, and it is not immediately clear which one to use. Which of the many possible numbers obtained from (2.1) properly measures Z?

By adding unnecessary extra sets to our cover, we can make the quantity in (2.1) arbitrarily large; thus it seems that large values of the potential are somehow to be disregarded, and we should look for the cover which minimises (2.1). Since such an optimal cover may not exist (the minimum may not be achieved!), we consider the greatest lower bound of such quantities, and write

$$(2.2) \qquad m(Z, \alpha, \varepsilon) = \inf_{\mathcal{D}(Z,\varepsilon)} \sum_i (\operatorname{diam} U_i)^\alpha.$$

Note the similarity in form between this equation and the definition of Lebesgue measure in (1.17).

---

[1] Indeed, the reader who does not feel this has either seen these definitions before, or is not paying enough attention. Wake up!

Now we have a function m which depends on the set Z and the parameters α and ε. Observe that given $\varepsilon_1 > \varepsilon_2 > 0$, any ε_2-cover is also an ε_1-cover. Thus the set of covers over which the infimum in (2.2) is taken for ε_2 is a subset of the set of covers for ε_1, and it is then immediate that

$$m(Z, \alpha, \varepsilon_2) \geq m(Z, \alpha, \varepsilon_1).$$

This shows that $m(Z, \alpha, \varepsilon)$ is monotonic as a function of ε, and hence the limit

$$(2.3) \qquad m(Z, \alpha) = \lim_{\varepsilon \to 0} m(Z, \alpha, \varepsilon)$$

exists, although it may be ∞, and indeed often is, as we shall see.

Fixing a particular value of $\alpha \geq 0$, we have a *set function* $m(\cdot, \alpha)$. This is a real-valued function defined on the space of *all* subsets of $\mathbb{R}^d$, which assigns to a subset $Z \subset \mathbb{R}^d$ the value $m(Z, \alpha)$ defined as above. The next proposition summarises its basic properties.

Proposition 2.1. *The set function $m(\cdot, \alpha)\colon Z \mapsto m(Z, \alpha)$ satisfies the following properties.*

(1) Normalisation: *$m(\emptyset, \alpha) = 0$ for all $\alpha > 0$, where $\emptyset$ is the empty set.*

(2) Monotonicity: *$m(Z_1, \alpha) \leq m(Z_2, \alpha)$ whenever $Z_1 \subset Z_2$.*

(3) Countable subadditivity: *Given any finite or countable collection of subsets Z_j, we have*

$$(2.4) \qquad m\left(\bigcup_j Z_j, \alpha\right) \leq \sum_j m(Z_j, \alpha).$$

Proof. (1) follows immediately upon observing that any open set, of any diameter, covers the empty set.

(2) uses the same idea as in the proof of monotonicity of $m(Z, \alpha, \cdot)$; an ε-cover of Z_2 is an ε-cover of Z_1, and hence the infimum in $m(Z_2, \alpha, \varepsilon)$ is being taken over a smaller set.

(3) is slightly more involved, and requires the following lemma.

Lemma 2.2. *Fix $Z \subset \mathbb{R}^2$ and $\alpha \geq 0$ such that $m(Z,\alpha) < \infty$. For every $\delta > 0$ and $\varepsilon > 0$, there exists an open ε-cover $\mathcal{U} = \{U_i\}$ of Z such that $|m(Z,\alpha) - \sum_i (\operatorname{diam} U_i)^\alpha| \leq \delta$.*

Proof. Follows immediately from the definitions of limit and infimum. $\qquad\qquad\square$

To prove (2.4) using this lemma, we first observe that if any of the values $m(Z_j, \alpha)$ is infinite, then their sum is infinite, and the inequality is trivial. Thus we assume they are all finite; fixing $\delta > 0$ and writing $Z = \bigcup_j Z_j$, we may apply the lemma to each Z_j to obtain ε-covers $\mathcal{U}_j = \{U_{ji}\}$ (for arbitrarily small $\varepsilon > 0$) such that

$$\left| m(Z_j, \alpha) - \sum_i (\operatorname{diam} U_{ji})^\alpha \right| \leq \frac{\delta}{2^j}.$$

We see that $\mathcal{U} = \bigcup_j \mathcal{U}_j$ is an open cover of Z, and since each of the U_{ji} has diameter $\leq \varepsilon$, it is actually an ε-cover. Thus

$$m(Z, \alpha, \varepsilon) \leq \sum_{i,j} (\operatorname{diam} U_{ji})^\alpha = \sum_j \left(\sum_i (\operatorname{diam} U_{ji})^\alpha \right)$$

$$\leq \sum_j \left(m(Z_j, \alpha) + \frac{\delta}{2^j} \right) = \sum_j m(Z_j, \alpha) + \delta.$$

This holds for all $\delta > 0$ and for all $\varepsilon > 0$; hence (2.4) holds. $\qquad\square$

So far the parameter α has been listed among the *dramatis personae*, but has done little more than linger at the edge of the stage, constant and unchanging. Its appearance on centre stage will finally bring us to the definition of Hausdorff dimension.

To that end, let us consider $m(Z, \cdot) \colon [0, +\infty) \to [0, +\infty]$ as a function of α. As they say, a picture is worth a thousand words, and so we try to draw its graph.

There are three possibilities for the value of $m(Z, \alpha)$ at any given α: it may be 0, it may be ∞, or it may be finite. The former two are not particularly interesting; after all, our main grievance with the ideas of cardinality, length, area, etc. as tools for classifying fractals was that they always returned answers which were either 0 or ∞.

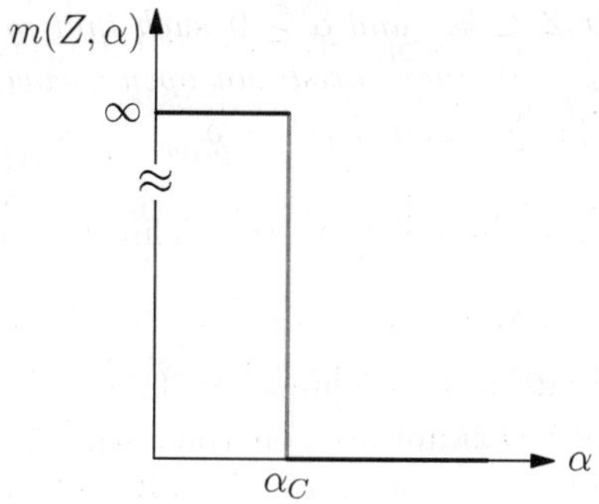

Figure 2.1. The graph of $m(Z, \cdot)$.

The third possibility, that $0 < m(Z, \alpha) < \infty$ for a particular value of α, turns out to have rather drastic consequences, as we see in the following two propositions.

Proposition 2.3. *If $\alpha \geq 0$ is such that $m(Z, \alpha) < \infty$, then $m(Z, \beta) = 0$ for every $\beta > \alpha$.*

Proof. A straightforward computation shows that

$$m(Z, \beta, \varepsilon) = \inf_{\mathcal{U}} \sum_i (\mathrm{diam}\, U_i)^\beta$$

$$= \inf_{\mathcal{U}} \sum_i (\mathrm{diam}\, U_i)^{\beta - \alpha} (\mathrm{diam}\, U_i)^\alpha$$

$$\leq \inf_{\mathcal{U}} \sum_i \varepsilon^{\beta - \alpha} (\mathrm{diam}\, U_i)^\alpha$$

$$= \varepsilon^{\beta - \alpha} m(Z, \alpha, \varepsilon),$$

and since $\beta - \alpha > 0$, we have $\varepsilon^{\beta - \alpha} \to 0$. Since $m(Z, \alpha, \varepsilon) \leq m(Z, \alpha) < \infty$, this implies that

$$m(Z, \beta) = \lim_{\varepsilon \to 0} m(Z, \beta, \varepsilon) = 0. \qquad \square$$

As an immediate consequence of this proposition, we have the following dual statement:

Proposition 2.4. *If $\alpha \geq 0$ is such that $m(Z, \alpha) > 0$, then $m(Z, \beta) = \infty$ for every $\beta < \alpha$.*

It follows from these propositions that the graph of $m(Z, \cdot)$ is as shown in Figure 2.1; below some critical value α_C, the function takes infinite values, and for $\alpha > \alpha_C$, we have $m(Z, \alpha) = 0$. Thus the function $m(Z, \cdot)$ is entirely determined by the location of α_C and the value of $m(Z, \alpha_C)$; the latter may lie anywhere in $[0, \infty]$, while the former may take any value in $[0, \infty)$. Just which values they take, of course, depends on the set Z; this is the whole point.

We are now in a position to complete our definition. The Hausdorff dimension of a set Z, denoted $\dim_H Z$, is the critical value α_C at which the function $m(Z, \cdot)$ passes from ∞ to 0. Thus, we have

$$\dim_H Z = \sup\{\alpha \in [0, \infty) \mid m(Z, \alpha) = \infty\}$$
$$= \inf\{\alpha \in [0, \infty) \mid m(Z, \alpha) = 0\}.$$

So we have the definition! But in what sense is this the "dimension" of the set Z? Does it agree with our usual intuitive understanding of dimension? What properties does it have? How do we actually compute it for specific examples? What does it have to do with fractals? Where in the world does the function $m(Z, \alpha, \varepsilon)$ *come* from? We will address these questions in the next lecture and see that they do in fact have satisfactory answers.

b. Hausdorff dimension of the middle-third Cantor set. Before we get too far ahead of ourselves, though, we take some time to illustrate the definition of Hausdorff dimension by computing it for the middle-third Cantor set C. We start with the observation that the set function $m(Z, \alpha)$ has nice scaling properties under certain geometric transformations.

Definition 2.5. A map $f \colon \mathbb{R}^d \to \mathbb{R}^d$ is a *similarity transformation* if there exists a real number $\lambda > 0$, called the *scaling factor*, such that

$$d(f(\mathbf{x}), f(\mathbf{y})) = \lambda d(\mathbf{x}, \mathbf{y})$$

for every $\mathbf{x}, \mathbf{y} \in \mathbb{R}^d$. We say that two sets are *similar* if one is the image of the other under a similarity transformation.

All similarity transformations are linear maps; a general similarity transformation of $\mathbb{R}^d$ can be written as the composition of a suitable isometry with the map $\mathbf{x} \mapsto \lambda \mathbf{x}$.

Exercise 2.1. Given a set $Z \subset \mathbb{R}^d$ and a real number $\lambda > 0$, let Z' be a set that is similar to Z, with scaling factor λ. Show that

$$(2.5) \qquad m(Z', \alpha) = \lambda^\alpha m(Z, \alpha).$$

The middle-third Cantor set is the disjoint union of two scaled-down copies of itself: $C = C_1 \cup C_2$, where both $C_1 = I_1 \cap C$ and $C_2 = I_2 \cap C$ are similar to C with scaling factor $1/3$. It follows from (2.5) that

$$m(I_1, \alpha) = m(I_2, \alpha) = \left(\frac{1}{3}\right)^\alpha m(C, \alpha).$$

Using the subadditivity property (2.4), this gives

$$(2.6) \qquad m(C, \alpha) \le m(I_1, \alpha) + m(I_2, \alpha) = \frac{2}{3^\alpha} m(C, \alpha).$$

Exercise 2.2. Using the fact that I_1 and I_2 are disjoint and separated by a positive distance, show that equality holds in (2.6).

As a consequence of Exercise 2.2, we see that

$$(2.7) \qquad m(C, \alpha) = 2(3^{-\alpha})m(C, \alpha).$$

Assuming that $0 < m(C, \alpha) < \infty$ for some value of α, we conclude that $3^\alpha = 2$, and hence $\alpha = \log 2/\log 3$.

If we knew that the set function $m(C, \cdot)$ always passed through a finite value on its voyage from ∞ to 0—that is, if we knew that $0 < m(C, \dim_H C) < \infty$—then the above argument would imply that $\dim_H C = \log 2/\log 3$. However, for some sets Z it may happen that $m(Z, \alpha)$ is equal to either 0 or ∞ for *every* $\alpha \ge 0$, and never takes a finite value. Fortunately, this turns out not to be the case for the middle-third Cantor set.

Proposition 2.6. *Let C be the middle-third Cantor set, and $\alpha = \log 2/\log 3$. Then $m(C, \alpha) = 1$.*

Sketch of proof. By covering C with the basic intervals $I_{w_1 \ldots w_n}$, it is not hard to show that $m(C, \alpha, \varepsilon) \le 1$ for every $\varepsilon > 0$. To prove the reverse inequality, one must show that $\sum_i (\operatorname{diam} U_i)^\alpha \ge 1$ for every open cover $\mathcal{U}$. In the case where every element of $\mathcal{U}$ is a basic interval, this turns out not to be that hard. However, one must deal

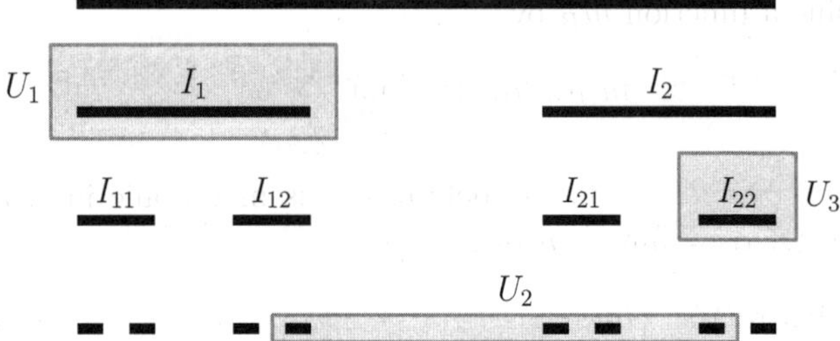

Figure 2.2. An open cover of the middle-third Cantor set.

with covers such as the one shown in Figure 2.2, where U_1 and U_3 each reduce to a single basic interval, while U_2 is more problematic. The full proof of this is somewhat subtle (although not prohibitively difficult), and so we defer a more rigorous treatment of the matter until the proof of Moran's theorem in Lecture 10. $\qquad\qquad\square$

Proposition 2.6 shows that in this case, our heuristic arguments do in fact give the correct value for the Hausdorff dimension, so that the Hausdorff dimension of the middle-third Cantor set is $\log 2/\log 3$.

Exercise 2.3. Use the heuristic arguments given in this section to guess the Hausdorff dimension of the Sierpiński gasket and the von Koch curve. Note that without computing the value of $m(Z, \alpha)$ or giving some further argument, this does not yet constitute a proof that the Hausdorff dimension is what you think it is.

c. Alternative definitions of Hausdorff dimension. The definition of Hausdorff dimension in the previous lecture involves arbitrary open covers; however, we might consider using a different class of sets as the elements of our covers. Thus we now show that it suffices to consider covers by open balls $B(x, r)$ with $r \leq \varepsilon$.

To this end, given $Z \subset \mathbb{R}^d$, let $\mathcal{B}(Z, \varepsilon)$ denote the collection of all countable sets

$$\{(x_i, r_i)\} \subset Z \times (0, \varepsilon)$$

with the property that

$$Z \subset \bigcup_i B(x_i, r_i),$$

and define a function m_B by

$$(2.8) \qquad m_B(Z, \alpha, \varepsilon) = \inf_{\mathcal{B}(Z,\varepsilon)} \sum_i r_i^\alpha,$$

where $\alpha \geq 0$ and $\varepsilon > 0$. As before, m_B is monotonic in ε, and we write $m_B(Z, \alpha) = \lim_{\varepsilon \to 0} m_B(Z, \alpha, \varepsilon)$.

Proposition 2.7. *Given a set $Z \subset \mathbb{R}^d$, we have*

$$\dim_H Z = \inf\{\alpha \geq 0 \mid m_B(Z, \alpha) = 0\}$$
$$= \sup\{\alpha \geq 0 \mid m_B(Z, \alpha) = \infty\}.$$

Proof. Observe that any cover of Z by balls of radius less than ε is a 2ε-cover by open sets, and hence $\mathcal{B}(Z, \varepsilon) \subset \mathcal{D}(Z, 2\varepsilon)$. Thus

$$m_B(Z, \alpha, \varepsilon) = \inf_{\mathcal{B}(Z,\varepsilon)} \sum_i r_i^\alpha = 2^{-\alpha} \inf_{\mathcal{B}(Z,\varepsilon)} \sum_i (\operatorname{diam} B(x, r_i))^\alpha$$
$$\geq 2^{-\alpha} \inf_{\mathcal{D}(Z,\varepsilon)} \sum_i (\operatorname{diam} U_i)^\alpha = 2^{-\alpha} m(Z, \alpha, 2\varepsilon),$$

since the infimum can only decrease if we consider a larger collection of covers.

Furthermore, given an arbitrary open cover $\mathcal{U} \in \mathcal{D}(Z, \varepsilon)$, we may take one point x_i from each U_i, and set $r_i = \operatorname{diam} U_i$, to obtain an open cover $\mathcal{U}' \in \mathcal{B}(Z, \varepsilon)$, since $U_i \subset B(x_i, r_i)$ for each i. It follows that

$$m(Z, \alpha, \varepsilon) = \inf_{\mathcal{D}(Z,\varepsilon)} \sum_i (\operatorname{diam} U_i)^\alpha$$
$$\geq \inf_{\mathcal{B}(Z,\varepsilon)} \sum_i r_i^\alpha = m_B(Z, \alpha, \varepsilon).$$

Combining these inequalities, we obtain

$$2^{-\alpha} m(Z, \alpha, 2\varepsilon) \leq m_B(Z, \alpha, \varepsilon) \leq m(Z, \alpha, \varepsilon),$$

and taking the limit as $\varepsilon \to 0$ yields

$$2^\alpha m_B(Z, \alpha) \leq m(Z, \alpha) \leq m_B(Z, \alpha).$$

Thus the critical value α_C is the same for both m and m_B. $\qquad \square$

Remark. In fact, one can go rather further than this. Let $\mathcal{F}$ be *any* collection of subsets of $\mathbb{R}^d$ such that for every $\varepsilon > 0$, there exists a cover of $\mathbb{R}^d$ by elements of $\mathcal{F}$ with diameter $< \varepsilon$. Then we may define $m_{\mathcal{F}}(Z, \alpha, \varepsilon)$ as in (2.2) by taking the infimum over all ε-covers of Z whose elements are in $\mathcal{F}$, and carry out the entire definition as before, obtaining the same critical value α_C. Thus we are free to use arbitrary sets when computing Hausdorff dimension, if we so desire.

Observe that although our entire discussion has been in the context of Euclidean space $\mathbb{R}^d$, the definition of Hausdorff dimension works just as well when Z lies in an arbitrary separable metric space X. This is true whether we use covers by open sets, covers by open balls, or covers by some other collection $\mathcal{F}$ of subsets of X. To simplify the exposition, we will develop the properties of Hausdorff dimension (and other dimensional quantities) under the assumption that $Z \subset \mathbb{R}^D$; however, all of this can also be done in the more general setting.

Exercise 2.4. Use the heuristic arguments given in this lecture to guess the Hausdorff dimension of symbolic space (Σ_k^+, d_a), where $k \geq 2$ and $a > 2$.

Lecture 8

a. Properties of Hausdorff dimension. Now that we have a definition (or two) of Hausdorff dimension and have seen what's under the hood and how it works for a rather simple example, let's take this new notion out for a test drive and see how it behaves. Some important properties of Hausdorff dimension can be deduced from the corresponding properties of the set function $m(\cdot, \alpha)$ given in Proposition 2.1:

Proposition 2.8. *The Hausdorff dimension has the following basic properties.*

(1) Normalisation: $\dim_H \emptyset = 0$.

(2) Monotonicity: $\dim_H Z_1 \leq \dim_H Z_2$ *whenever* $Z_1 \subset Z_2$.

(3) Countable stability: $\dim_H \left(\bigcup_j Z_j \right) = \sup_j \dim_H Z_j$, *where* $\{Z_j\}$ *is any countable collection of subsets of* $\mathbb{R}^d$.

Proof. (1) and (2) are direct consequences of the corresponding properties in Proposition 2.1. To see countable stability, we write $Z = \bigcup_j Z_j$, and first observe that since $Z_j \subset Z$ for all j, monotonicity implies $\dim_H Z_j \leq \dim_H Z$, and hence $\sup_j \dim_H Z_j \leq \dim_H Z$. Furthermore, if $\alpha > \sup_j \dim_H Z_j$, then $\alpha > \dim_H Z_j$ for all j, and hence $m(Z_j, \alpha) = 0$ for all j. Thus $m(Z) = 0$, which implies that $\dim_H Z \leq \alpha$ for all $\alpha > \sup_j \dim_H Z_j$, and we are done. $\qquad\square$

A singleton set $Z = \{x\}$ has $m(Z, \alpha, \varepsilon) = 0$ for all $\alpha > 0$, $\varepsilon > 0$, and so applying the third property above, we obtain

Corollary 2.9. *If Z is countable, then* $\dim_H Z = 0$.

Thus the set of rational numbers has Hausdorff dimension zero, despite being dense in the interval, and hence fairly "large" in the topological sense.

So points have zero Hausdorff dimension, which we would expect. What about lines and planes? Do they have the "correct" Hausdorff dimension? Before answering this question, we state two lemmas which codify a common technique for giving upper and lower bounds on the Hausdorff dimension.

Lemma 2.10. *To show that* $\dim_H Z \leq \alpha$, *it suffices to exhibit* $C > 0$ *such that for all* $\varepsilon > 0$, ***there exists*** *an ε-cover* $\mathcal{U} = \{U_i\}$ *with* $\sum_i (\operatorname{diam} U_i)^\alpha \leq C$.

Proof. The given condition guarantees that $m(Z, \alpha, \varepsilon) \leq C$ for all $\varepsilon > 0$; hence $m(Z, \alpha) < \infty$, and the result follows. $\qquad\square$

Lemma 2.11. *To show that* $\dim_H Z \geq \alpha$, *it suffices to exhibit* $C > 0$ *and* $\varepsilon > 0$ *such that* $\sum_i (\operatorname{diam} U_i)^\alpha \geq C$ ***for all*** *ε-covers* $\mathcal{U} = \{U_i\}$.

Proof. The given condition guarantees that $m(Z, \alpha, \varepsilon) \geq C$ for some $\varepsilon > 0$, hence $m(Z, \alpha) > 0$, and the result follows. $\qquad\square$

In general, upper bounds on the Hausdorff dimension are usually easier to obtain than lower bounds. The reason for this is that in order to apply Lemma 2.10, we only needed to construct a family of arbitrarily fine "good" covers with uniformly bounded potentials, whereas in order to get a lower bound by applying Lemma 2.11, we need to deal with *every* ε-cover for some sufficiently small ε. We will wrestle with this in the proof of the following proposition, which applies Lemmas 2.10 and 2.11 to the case of the real line.

Proposition 2.12. $\dim_H \mathbb{R} = 1$.

Proof. Let $Z = [0, 1]$ be the unit interval in $\mathbb{R}$. By part (3) of Proposition 2.8, it suffices to show that $\dim_H Z = 1$. We do this by using Lemma 2.10 to show that $\dim_H Z \leq 1$, and then using Lemma 2.11 to show that $\dim_H Z \geq 1$.

To satisfy the condition of Lemma 2.10, we consider $\varepsilon > 0$ and choose an integer n such that $1/n \leq \varepsilon$. Consider the open intervals $\left(\frac{i}{3n}, \frac{i+1}{3n}\right)$ for $i = 0, \ldots, 3n - 1$; these cover every point of $[0, 1]$ except the endpoints $i/3n$. If we extend each interval to include the two beside it, then we get the intervals $U_i = \left(\frac{i-1}{3n}, \frac{i+2}{3n}\right)$, each of which has length $1/n \leq \varepsilon$, and so $\mathcal{U} = \{U_i\}$ is an ε-cover of Z. It has $3n$ elements, and so we see that

$$\sum_i (\operatorname{diam} U_i) = 3n \cdot \frac{1}{n} = 3.$$

Thus Lemma 2.10 applies with $C = 3$ and $\alpha = 1$, and we have $\dim_H Z \leq 1$.

The other inequality is rather harder, for the reason explained above. Indeed, we cannot apply Lemma 2.11 directly to our present problem with $\alpha = 1$, but we can reach the same result by showing that the lemma applies for every $\alpha < 1$. Indeed, then we will have $\dim_H Z \geq \alpha$ for all $\alpha < 1$, which of course implies that $\dim_H Z \geq 1$, as desired.

To that end, fix $\alpha < 1$. We wish to find $\varepsilon > 0$ such that $\sum_i (\operatorname{diam} U_i)^\alpha > 1$ for every ε-cover $\mathcal{U}$ of $[0, 1]$. For any such $\mathcal{U}$, we have

$$\sum_i (\operatorname{diam} U_i)^\alpha = \sum_i (\operatorname{diam} U_i)(\operatorname{diam} U_i)^{\alpha-1} \geq \left(\sum_i \operatorname{diam} U_i\right) \varepsilon^{\alpha-1}.$$

It is not hard to see that since the sets U_i cover $[0, 1]$, we must have $\sum_i \operatorname{diam} U_i \geq 1$. Indeed, if we write $a_i = \inf U_i$ and $b_i = \sup U_i$, then we have $U_i \subset (a_i, b_i)$, with $\operatorname{diam} U_i = b_i - a_i$ (where we take $[0, b_i)$ and $(a_j, 1]$ in the cases $a_i = 0$, $b_j = 1$), and so the argument reduces to the case where each U_i is an interval.[2] By compactness of $[0, 1]$, we may find $i_1, \ldots, i_n$ such that $a_{i_1} = 0$, $b_{i_n} = 1$, and $a_{i_{k+1}} < b_{i_k}$ for $k = 1, 2, \ldots, n - 1$; it follows that

$$
\sum_i \operatorname{diam} U_i = \sum_i (b_i - a_i) \geq \sum_{k=1}^{n} (b_{i_k} - a_{i_k})
$$

(2.9)

$$
> \left(\sum_{k=1}^{n-1} a_{i_{k+1}} - a_{i_k} \right) + b_{i_n} - a_{i_n} = b_{i_n} - a_{i_1} = 1.
$$

Thus $\sum_i (\operatorname{diam} U_i)^\alpha \geq \varepsilon^{\alpha - 1}$. Since $\alpha < 1$, we have $\varepsilon^{\alpha - 1} > 1$ for sufficiently small $\varepsilon > 0$. Then Lemma 2.11 applies, showing that $\dim_H Z \geq \alpha$. Since $\alpha < 1$ is arbitrary, we have $\dim_H Z \geq 1$, which completes the proof. $\qquad\qquad\square$

This result shows that Hausdorff dimension gives the result we would expect for a line. A similar argument shows that this is also true for the plane, and indeed for any $\mathbb{R}^d$; the key step is (2.9), which will need to be replaced by the inequality $\sum_i (\operatorname{diam} U_i)^n \geq 1$.

Proposition 2.13. $\dim_H \mathbb{R}^d = d$.

Sketch of proof in the case $d = 2$. As before, it is enough to consider the unit square $[0, 1]^2$, and it is relatively straightforward to show that $\dim_H [0, 1]^2 \leq 2$ by producing a suitable family of covers. To give the other inequality, we fix an arbitrary ε-cover $\mathcal{U} = \{U_i\}$, and let $\delta > 0$ be the Lebesgue number for $\mathcal{U}$; that is, δ is such that any set with diameter $\leq \delta$ is contained in some U_i (the existence of such a positive δ is a consequence of the compactness of $[0, 1]^2$). Then choosing $m \in \mathbb{N}$ such that $\sqrt{2}/m < \delta$, we consider the sets

$$
R_{j,k} = \left[\frac{j}{m}, \frac{j+1}{m} \right] \times \left[\frac{k}{m}, \frac{k+1}{m} \right]
$$

for $0 \leq j, k \leq m - 1$. Each of these sets has diameter $< \delta$, and hence is contained in some U_i; thus if we let a_i be the number of sets $R_{j,k}$

---

[2]More generally, we may observe that each U_i is a finite or countable union of intervals, by letting $I_i(x)$ denote the largest open interval containing x which is a subset of U_i, and verifying that $I_i(x)$ and $I_i(y)$ either coincide or are disjoint. There are at most countably many different $I_i(x)$ since each contains a distinct rational number.

contained in U_i, we have $\sum_i a_i \geq m^2$. However, the number of sets $R_{j,k}$ covered by a single U_i cannot exceed $(\operatorname{diam} U_i/(1/m))^2$, and so

$$m^2 \leq \sum_i a_i \leq \sum_i m^2 (\operatorname{diam} U_i)^2;$$

it follows that $\sum_i (\operatorname{diam} U_i)^2 \geq 1$ for every ε-cover $\mathcal{U}$, and hence $\dim_H [0,1]^2 \geq 2$. $\qquad\qquad\qquad\qquad\qquad\qquad\qquad\qquad\qquad\qquad\qquad\quad\square$

In fact, this argument actually shows that if $Z \subset \mathbb{R}^d$ is any open set (indeed, any set with non-empty interior), then $\dim_H Z = d$; this boils down to the fact that open sets have positive Lebesgue measure... but we are getting ahead of ourselves. For the time being, we content ourselves with reiterating that by Propositions 2.12 and 2.13, Hausdorff dimension agrees with our usual definition of dimension for lines, planes, and so on.[3]

But what is our usual definition of dimension? Of course we know that the d-dimensional Euclidean space $\mathbb{R}^d$ has, or ought to have, dimension d, but why? What is it about this space that makes it d-dimensional?

b. Topological dimension. Let X be a topological space, and consider an open cover $\mathcal{U}$ of X. Fix a point $x \in X$, and count the number of elements of the cover which contain x; we call this the *multiplicity of $\mathcal{U}$ at x*, and denote it by $M(\mathcal{U}, x)$. The quantity $M(\mathcal{U}) = \sup_x M(\mathcal{U}, x)$ is called the *multiplicity of $\mathcal{U}$*.

As with our earlier definition of $m(Z, \alpha, \varepsilon)$ via covers, we can make $M(\mathcal{U})$ as large as we like (even infinite) by choosing a cover with "too many" elements. To deal with this, we need the following definition.

Definition 2.14. Let $\mathcal{U}$ and $\mathcal{V}$ be open covers of X; $\mathcal{V}$ is a *refinement* of $\mathcal{U}$ if every $V \in \mathcal{V}$ is contained in some $U \in \mathcal{U}$.

Passing to a refinement allows us to discard unnecessary sets from the cover and also to "clean up" areas where more sets than necessary

---

[3]We will also consider an alternate approach based on the fact that $\mathbb{R}^d$ is the direct product of n copies of $\mathbb{R}$, once we establish something about how dimension behaves under taking direct products.

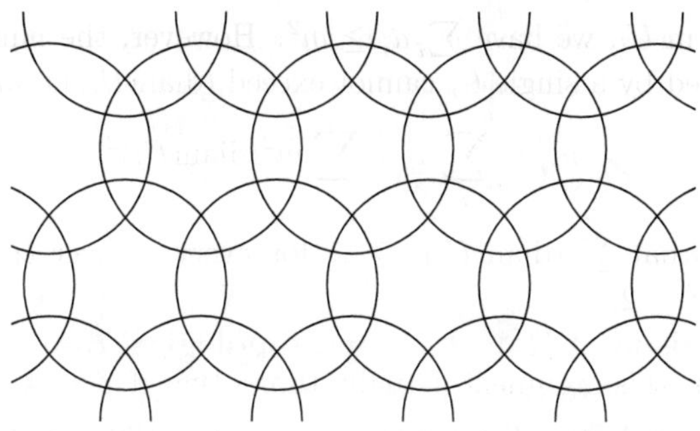

Figure 2.3. A cover of $\mathbb{R}^2$ with multiplicity 3.

cover a single point. We want to choose a refinement $\mathcal{V}$ of $\mathcal{U}$ for which $M(\mathcal{V})$ is minimal. Thus we define

$$M(X) = \sup_{\mathcal{U}} \left(\inf\{M(\mathcal{V}) \mid \mathcal{V} \text{ is a refinement of } \mathcal{U}\} \right)$$

and investigate how this quantity is connected to the dimension of X in the case when $X = \mathbb{R}^d$, where we know what the dimension ought to be. (We must take the supremum over all covers to eliminate certain trivial examples, such as the cover $\mathcal{U} = \{X\}$, for which $M(\mathcal{U}) = 1$.)

In the case $d = 1$, we are just covering the line, and given any open cover $\mathcal{U}$, it is easy to construct a refinement $\mathcal{V}$ with $M(\mathcal{V}) = 2$, as in the proof of Proposition 2.12, by passing to a minimal subcover (note that subcovers are refinements, although the converse is not true in general). Since we must have intersections between the elements of the cover (otherwise $\mathbb{R}$ would be disconnected), we see that $M(\mathbb{R}) = 2$.

In the plane, any open cover admits a refinement which has multiplicity 3, resembling the one shown in Figure 2.3, and it is not too hard to show that this is optimal, so $M(\mathbb{R}^2) = 3$. Similarly, we have $M(\mathbb{R}^d) = d+1$, which connects $M(\mathbb{R}^d)$ to the dimension of $\mathbb{R}^d$. Since all that is required for the definition of M is a topological space, we may make the following definition.

Definition 2.15. The *topological dimension* (or *Lebesgue covering dimension*) of a topological space X is the quantity $M(X) - 1$; that is, the dimension is one less than the maximal multiplicity of an optimal refinement.

The discussion at the beginning of this lecture shows that Hausdorff dimension and topological dimension agree when it comes to the Euclidean spaces $\mathbb{R}^d$; is this always the case? Have we just given two rather different definitions of the same quantity?

Lecture 9

a. Comparison of Hausdorff and topological dimension. One difference in the two dimensions we have defined is immediately apparent; the topological dimension is always an integer, while the Hausdorff dimension has no *a priori* reason to take integer values. Indeed, it can take any non-negative real value (see Exercise 2.20).

Another difference becomes apparent if we look at what notions are used in the definitions; the topological dimension can be defined for any topological space, whether or not it has a metric, while the Hausdorff dimension requires a metric for its definition. If we need to explicitly indicate the metric being used, we will write the Hausdorff dimension of Z with respect to the metric d as $\dim_H^d Z$.

This distinction becomes important when we observe that a single topological space can be equipped with multiple metrics. For example, the usual metric on $\mathbb{R}^d$ is given by Pythagoras' formula

$$(2.10) \qquad d(x, y) = \sqrt{\sum_i (x_i - y_i)^2},$$

but other metrics may be introduced by the formulae

$$(2.11) \qquad \rho(x, y) = \sum_i |x_i - y_i|,$$

$$(2.12) \qquad \sigma(x, y) = \max_i |x_i - y_i|,$$

and it is not hard to check that these metrics all induce the same topology on $\mathbb{R}^d$ (see Exercise 2.5). In particular, they all lead to the same topological dimension; do they all lead to the same Hausdorff dimension? To answer this question, we need some new definitions, giving three senses in which two metrics d_1 and d_2 on $\mathbb{R}^d$ (or more generally, on any metric space X) may be said to be "the same".

Definition 2.16. d_1 and d_2 are *equivalent* (denoted $d_1 \sim d_2$) if the identity map $\mathrm{Id}\colon (X, d_1) \to (X, d_2)$ is a homeomorphism; that is, if for every $x \in X$ and every $\varepsilon > 0$ there exists $\delta > 0$ such that $d_1(x, y) \leq \delta$ implies $d_2(x, y) \leq \varepsilon$, and $d_2(x, y) \leq \delta$ implies $d_1(x, y) \leq \varepsilon$.

d_1 and d_2 are *uniformly equivalent* if the identity map and its inverse are both *uniformly* continuous; that is, if for every $\varepsilon > 0$ there exists $\delta > 0$ (independent of x, y) such that the implications in the previous paragraph hold.

d_1 and d_2 are *strongly equivalent* if there exists $C > 0$ such that for all x, y, we have

$$C^{-1}d_2(x, y) \leq d_1(x, y) \leq Cd_2(x, y).$$

The statement that d_1 and d_2 are equivalent may be rephrased as the statement that every d_1-ball contains a d_2-ball, and vice versa. In particular, since an open set U is one which contains a sufficiently small ball around every point in U, we see that the metrics d_1 and d_2 define precisely the same collection of open sets; that is, they define the same topology.

Exercise 2.5. Show that d, ρ, and σ as defined above on $\mathbb{R}^d$ are all strongly equivalent.

Strong equivalence implies uniform equivalence, which in turn implies equivalence, but neither of the reverse implications holds in general. Thus two metrics may induce the same topology (that is, be equivalent) but fail to be strongly equivalent, and so the metric carries rather more information about the space than the topology does.

Exercise 2.6. Construct a separable metric on $\mathbb{R}^d$, which is

(a) not equivalent to the standard metric on $\mathbb{R}^d$;

(b) equivalent to the standard metric on $\mathbb{R}^d$, but not strongly equivalent.

We now return to the question of where the dimensional information resides: Is it carried by the topology itself? Or does it rely on the extra information which is carried by the metric?

To address this matter, we consider a superficially different question, which is actually quite related. Given two metric spaces (X, d)

and (X', d') and a continuous map $f \colon X \to X'$, what is the relationship between $\dim_H Z$ and $\dim_H f(Z)$?

In general, there may be no relationship. If $f \colon \mathbb{R}^d \to \mathbb{R}^d$ is projection to a subspace, then the Hausdorff dimension can decrease under the action of f. On the other hand, the von Koch curve is a homeomorphic image of the unit interval but has Hausdorff dimension strictly greater than 1, as we will see later, and so the Hausdorff dimension can also increase under the action of f.

If f is Lipschitz, however, the story is different.

Proposition 2.17. *If $f \colon \mathbb{R}^d \to \mathbb{R}^d$ is Lipschitz, then $\dim_H f(Z) \le \dim_H Z$ for every $Z \subset \mathbb{R}^d$.*

Proof. Let $L > 0$ be such that $d(f(x), f(y)) \le L d(x, y)$. Then if $\mathcal{U} = \{U_i\}$ is any ε-cover of Z, we have $\operatorname{diam} f(U_i) \le L \operatorname{diam} U_i$, and so $f(\mathcal{U}) = \{f(U_i)\}$ is an $L\varepsilon$-cover of $f(Z)$, for which

$$\sum_i (\operatorname{diam} f(U_i))^\alpha \le L^\alpha \sum_i (\operatorname{diam} U_i)^\alpha.$$

It follows that $m(f(Z), \alpha, L\varepsilon) \le L^\alpha m(Z, \alpha, \varepsilon)$, whence $m(f(Z), \alpha) \le L^\alpha m(Z, \alpha)$. Thus if $m(Z, \alpha)$ is finite, so is $m(f(Z), \alpha)$, which implies that $\dim_H f(Z) \le \dim_H Z$. $\qquad\square$

Exercise 2.7. Let $f \colon \mathbb{R} \to \mathbb{R}$ be a C^1 function (see Appendix). Show that $\dim_H f(Z) \le \dim_H Z$ for any (not necessarily bounded) set Z.

A bijection f such that both f and f^{-1} are Lipschitz is called *bi-Lipschitz*. It follows from Proposition 2.17 that $\dim_H f(Z) = \dim_H Z$ whenever f is bi-Lipschitz; this fact actually goes some way towards answering our earlier question regarding the dependence of Hausdorff dimension on the metric, as follows.

If $f \colon X \to Y$ is a bijection and ρ is a metric on Y, then the formula $\rho_f(x, y) = \rho(f(x), f(y))$ defines a metric ρ_f on X.

Exercise 2.8. Show that $\dim_H^{\rho_f} Z = \dim_H^\rho f(Z)$.

It follows from Exercise 2.8 and the remarks above that if (X, d) and (Y, ρ) are two metric spaces and $f \colon X \to Y$ is bi-Lipschitz, then

$$\dim_H^{\rho_f} Z = \dim_H^d Z.$$

If $X = Y$ as sets, then d and ρ are two different metrics on the same space; they are strongly equivalent if and only if the identity map f is bi-Lipschitz. Furthermore, in this case $\rho_f = \rho$, and so we see that Hausdorff dimension is preserved by strong equivalence.

The result of the following exercise shows that this is not necessarily the case if the metrics merely induce the same topology.

Exercise 2.9. Compute the Hausdorff dimension of $\mathbb{R}^d$ with respect to the metrics you constructed in Exercise 2.6. If the metric from part (b)—which is equivalent to the standard metric, but not strongly equivalent—does not give $\mathbb{R}^d$ a different Hausdorff dimension, then find another equivalent metric which changes the Hausdorff dimension.

b. Metrics and topologies. So far, all our examples of topological spaces have been metric spaces as well. One may rightly ask, then, if every example arises this way; given a topological space $(X, \mathcal{T})$, can we always find a metric d on X such that the sets in $\mathcal{T}$ are precisely those sets which are unions of d-balls? Such a space is called *metrisable*, and so we may ask, are all topological spaces metrisable?

It turns out that the answer is "no": Some topologies do not come from metrics. But which ones? Given a particular topology, how can we tell whether or not it comes from a metric? To answer this question, we examine properties of metric spaces which do not follow from the axioms of a topological space.

Exercise 2.10. Let (X, d) be a metric space, and fix $x \in X$. Show that the set $\{x\}$ is closed.

Exercise 2.11. Let (X, d) be a metric space, and fix $x, y \in X$. Show that there exist disjoint open sets $U, V \subset X$ such that $x \in U$ and $y \in V$; that is, metric spaces are *Hausdorff*.

Exercise 2.12. Let (X, d) be a metric space, and let $A, B \subset X$ be disjoint closed sets. Show that there exist disjoint open sets $U, V \subset X$ such that $A \subset U$ and $B \subset V$; that is, metric spaces are *normal*.

These three properties are examples of *separation axioms*, which more or less describe what sort of sets (single points, closed sets, etc.)

can be separated by open sets in the manner described above. None of the separation axioms follow from the axioms of a topological space. Indeed, if we consider the *trivial topology* $\mathcal{T} = \{\emptyset, X\}$ on any set X with more than one element, then the first two properties above fail (the third holds vacuously); hence the trivial topology is non-metrisable. Other examples of non-metrisable topologies are not hard to come by. For example, given a set X with at least three elements, take two arbitrary sets $A, B \subset X$, and consider the smallest topology with respect to which both A and B are open:

$$\mathcal{T} = \{\emptyset, A, B, A \cap B, A \cup B, X\}.$$

(Note that $A \cap B$ may coincide with $\emptyset$, and $A \cup B$ may coincide with X.) This topology does not have any of the three properties above, and hence is non-metrisable.

On the other hand, if a topological space *is* both normal and Hausdorff (the latter implies that points are closed), then this is almost enough to make it metrisable.

In order to state what more is needed, we first observe that in any metrisable topology $(X, \mathcal{T})$, an open set can always be written as a union of balls $B(x, r)$. Thus although there are many open sets which are not of the form $B(x, r)$, the collection of balls is sufficient to generate the topology. More generally, any collection of open sets $\mathcal{B} \subset \mathcal{T}$ with the property that any element of $\mathcal{T}$ can be written as a union of elements of $\mathcal{B}$ is known as a *base* (or *basis*) of the topology.[4]

In fact, there are many cases in which the topology can be generated by an even smaller collection of open sets. In the familiar case of $\mathbb{R}^d$, we may consider the collection of all balls of rational radius centred at points with rational coordinates; this forms a *countable* base for the topology. A topological space with a countable base is called *second-countable*, and one sees immediately that every *separable* metric space is second-countable.[5]

---

[4]Equivalently, a base may be characterised by the requirement that every open set contain a member of the base.

[5]Given the choice of terminology, the reader may justifiably suspect that there is a notion of *first-countable space*; indeed there is (though we shall not use it), and it refers to a similar local property of the space.

This flurry of definitions allows us to state (without proof) one of the most important results in basic point set topology, which gives a nearly complete answer to the question of which topological spaces are metrisable.

Theorem 2.18 (Urysohn's metrisation theorem). *If X is a second-countable, normal, Hausdorff topological space, then X is metrisable.*

Actually, slightly stronger versions of this theorem are available, but this will be enough for our purposes. In a nutshell, the moral of the story is that *all topological spaces are metrisable, except for rather weird cases with which we will not concern ourselves.*

So we are interested in topological spaces whose topologies come from some metric. But which metric do we use? As we saw in Example 2.5, the three metrics in (2.10)–(2.12) all lead to the same topology on $\mathbb{R}^d$, and in general, a single topology on a space X may be induced by many different, but equivalent, metrics.

With this in mind, we return to our discussion of the topological and Hausdorff dimensions of a set $Z \subset \mathbb{R}^d$ endowed with a metric ρ. We continue to write d for the standard metric given by (2.10).

The relationship between the two notions of dimension is given by the following deep theorem due to Hausdorff; this pearl of dimension theory was the motivation for his introduction of the notion of Hausdorff dimension.

Theorem 2.19 (Hausdorff's theorem). *Given a set $Z \subset \mathbb{R}^d$, the topological dimension $\dim Z$ and the Hausdorff dimensions $\dim_H^\rho Z$ are related by the following variational principle:*

$$(2.13) \qquad\qquad \dim Z = \inf_{\rho \sim d} \dim_H^\rho Z.$$

That is, the topological dimension of Z is the infimum of the possible Hausdorff dimensions, taken over all metrics ρ which are equivalent to the standard metric d.

Proof. See [**Hau18**]. $\qquad\qquad\qquad\qquad\qquad\qquad\qquad\qquad\qquad\qquad\quad$ $\square$

In fact, we will eventually see an even better result than this in Theorem 2.34.

c. Topology and dimension. Even though the Hausdorff dimension depends on the metric, we can occasionally use it to deduce some purely topological information.

Theorem 2.20. *If $Z \subset \mathbb{R}$ and $\dim_H Z < 1$, then Z is totally disconnected.*

Proof. Comparing (1.17) and (2.2), we see that $\mathrm{Leb}(Z) = m(Z, 1)$, and so the Hausdorff function $m(\cdot, 1)$ is just Lebesgue measure. In particular, we see that if a set Z has Hausdorff dimension strictly less than 1, then Z admits interval covers of arbitrarily small total length, hence $\mathrm{Leb}(Z) = 0$.

To show that Z is totally disconnected, we consider arbitrary points $x, y \in Z$, and produce two disjoint open sets $U, V \subset Z$ such that $x \in U$, $y \in V$, and $U \cup V = Z$. To this end, define a function $f \colon Z \to \mathbb{R}^+$ by $f(z) = d(x, z)$; that is, f measures the distance from x to z.

Clearly, f is a Lipschitz function, and so Proposition 2.17 gives the bound

$$\dim_H f(Z) \leq \dim_H Z < 1,$$

and it follows from our earlier remarks that $\mathrm{Leb}(f(Z)) = 0$. But this implies that $\mathbb{R} \setminus f(Z)$ is dense; indeed, if it were not dense, then $f(Z)$ would contain an interval, and hence have positive Lebesgue measure.

Thus we may find $r \in \mathbb{R} \setminus f(Z)$ such that $0 < r < f(y)$, and we define two open sets by

$$U = f^{-1}([0, r)) = \{ z \in Z \mid d(x, z) < r \},$$
$$V = f^{-1}((r, \infty)) = \{ z \in Z \mid d(x, z) > r \}.$$

Obviously, U and V are disjoint; furthermore, since $r \notin f(Z)$, there is no point $z \in Z$ with $d(x, z) = r$, and hence $U \cup V = Z$. Finally, $x \in U$ and $y \in V$, so x and y do not lie in the same connected component of Z. Since x and y were arbitrary, the desired result follows. $\square$

Exercise 2.13. Show that any topological space X with topological dimension 0 is totally disconnected, provided it satisfies the property in Exercise 2.10. Use this fact to derive Theorem 2.20 as a corollary of Hausdorff's Theorem (which is much deeper).

Exercise 2.14. Show that the converse of Exercise 2.13 fails in general by giving an example of a totally disconnected topological space with topological dimension ≥ 1. Finally, show that the converse *does* hold for a compact metric space X: if X is totally disconnected, then it has topological dimension 0.

Lecture 10

a. Hausdorff dimension of Cantor sets. We return now to the various Cantor sets we constructed in Chapter 1, and address the problem of computing their Hausdorff dimensions. We first give a brief discussion of dynamically defined Cantor sets which have a very regular self-similarity (but we do not prove anything); in the next section, we consider more general geometrically constructed Cantor sets and prove a general result on their Hausdorff dimension.

Let f be a piecewise linear one-dimensional full-branched Markov map, and $\lambda_1, \ldots, \lambda_k$ the lengths of the basic intervals $I_1, \ldots, I_k$. Let C be the maximal invariant set (the repeller) for f. We *assume* that there exists $t \geq 0$ such that $0 < m(C, t) < \infty$.

Define sets $C_1, \ldots, C_k$ by $C_j = C \cap I_j$; each of the sets C_j is similar to C, with scaling factor λ_j, and it follows from Exercise 2.1 that

$$m(C_j, t) = \lambda_j^t \, m(C, t).$$

The result of Exercise 2.2 still holds in this case, and so

$$m(C, t) = \sum_{j=1}^{k} m(C_j, t) = \sum_{j=1}^{k} \lambda_j^t \, m(C, t).$$

By the assumption that $0 < m(C, t) < \infty$, we must have

$$(2.14) \qquad\qquad \sum_{j=1}^{k} \lambda_j^t = 1.$$

We stress that just as with the middle-third Cantor set, this argument does not prove anything in and of itself; all it does is suggest that the solution of (2.14) is a good candidate for $\dim_H C$. In order to prove that t is actually the Hausdorff dimension, we will show that $m(C, t)$ is in fact positive and finite.

b. Moran's Theorem. The argument in the previous section used the fact that C is a union of sets to which it is geometrically similar. For an arbitrary geometrically constructed Cantor set on the line, this is not usually the case; the freedom to place basic intervals at different locations at each step can very easily destroy the strict self-similarity that the above argument requires.

Indeed, the construction described in Lecture 5(c) is quite a general one. We may begin with any number $k \geq 2$ of basic intervals, we may choose as ratio coefficients any positive numbers $\lambda_1, \ldots, \lambda_k$ whose sum is less than 1, and we may place the basic intervals $I_{w_1 \cdots w_{n+1}}$ anywhere we like within the basic interval $I_{w_1 \ldots w_n}$ from the previous step, provided they are disjoint and have lengths given by (1.21). How do these choices affect the Hausdorff dimension $\dim_H C$? Does it matter where we put the intervals? Is the dependence on the ratio coefficients λ_j still given by (2.14)?

These questions were first asked by Abram Besicovitch, one of the founders of the dimension theory of fractals; he posed them to his students in a seminar he organised at Cambridge upon his arrival there from Russia in 1927. One of those students, Patrick Moran, proved the rather remarkable result that (2.14) applies in general, regardless of the spacing of the basic intervals.

Theorem 2.21 (Moran's theorem). *If C is a geometrically constructed Cantor set on the line with ratio coefficients $\lambda_1, \ldots, \lambda_k$, then its Hausdorff dimension $\dim_H C$ is the unique value of t which satisfies* (2.14).

Proof. First we must verify that (2.14) does in fact have a unique solution. The function defined by the left-hand side is continuous, takes the value $k \geq 2$ at $t = 0$, and is equal to $\sum_j \lambda_j < 1$ at $t = 1$; hence by the Intermediate Value Theorem, there exists some $t \in (0, 1)$ such that the function is equal to 1. Furthermore, a simple computation of the derivative shows that it is negative, hence the function is strictly decreasing, and so the solution t is unique.

From now on, t shall denote the unique value for which (2.14) holds. As in the proof of Proposition 2.12, there are two parts to the argument; first we show that $\dim_H C \leq t$, then that $\dim_H C \geq t$.

We begin by obtaining the upper bound, which uses Lemma 2.10. So for every $\varepsilon > 0$, we must find a "good" ε-cover, for which the quantity $\sum_i (\operatorname{diam} U_i)^t$ is bounded independently of ε.

We claim that the cover by basic intervals at an appropriate step n of the iteration is the desired one. Writing $\lambda_{\max} = \max\{\lambda_1, \ldots, \lambda_k\}$, we see that

$$|I_{w_1 \ldots w_n}| = |I_{w_1}| \left(\prod_{j=2}^{n} \lambda_{w_j} \right) \leq |I_{w_1}| \lambda_{\max}{}^{n-1} \leq \lambda_{\max}{}^{n-1},$$

and so if we fix n such that $\lambda_{\max}{}^{n-1} < \varepsilon$, we may consider the ε-cover

$$\mathcal{U} = \{I_{w_1 \ldots w_n} \mid 1 \leq w_j \leq k \text{ for every } 1 \leq j \leq n\}.$$

It follows that

$$
\begin{aligned}
m(C, t, \varepsilon) &\leq \sum_{(w_1, \ldots, w_n)} |I_{w_1 \ldots w_n}|^t \\
&= \sum_{(w_1, \ldots, w_{n-1})} |I_{w_1 \cdots w_{n-1} 1}|^t + \cdots + |I_{w_1 \cdots w_{n-1} k}|^t \\
&= \sum_{(w_1, \ldots, w_{n-1})} |I_{w_1 \cdots w_{n-1}}|^t (\lambda_1^t + \cdots + \lambda_k^t) \\
&= \sum_{(w_1, \ldots, w_{n-1})} |I_{w_1 \cdots w_{n-1}}|^t = \cdots = \sum_{w_1} |I_{w_1}|^t,
\end{aligned}
$$

but this last quantity is a constant, independent of ε, so Lemma 2.10 applies: $m(C, t) \leq \sum_i |I_i|^t < \infty$, therefore $\dim_H C \leq t$.

As usual, the proof that $\dim_H C \geq t$ is harder. We want to apply Lemma 2.11 by showing that for a sufficiently small $\varepsilon > 0$, *every* ε-cover has

$$(2.15) \qquad\qquad \sum_i (\operatorname{diam} U_i)^t \geq K > 0,$$

where K is a constant chosen independently of the cover. The plan of attack is first to establish (2.15) for ε-covers by basic intervals (which may occur at different depths of the construction), and then to use this bound to obtain a bound for *all* ε-covers.

The case of covers by basic intervals is dealt with in the following lemma.

Lemma 2.22. *There exists a constant K, chosen independently of any cover, such that if $\mathcal{U} = \{U_i\}$ is any cover of C such that each U_i is a basic interval, then (2.15) holds.*

Proof. Consider the constant

$$(2.16) \qquad K = \sum_{j=1}^{k} |I_j|^t,$$

and let $\mathcal{U}$ be any cover of C by basic intervals. Because C is compact, $\mathcal{U}$ has a finite subcover; it suffices to establish (2.15) for this finite subcover, and so without loss of generality we may assume that $\mathcal{U}$ is finite. Indeed, we can (and do) take $\mathcal{U}$ to be minimal in the sense that no proper subcollection of $\mathcal{U}$ covers C.

Given a basic interval $I_{w_1 \ldots w_n}$, we refer to n as the *depth* of the interval in the construction. Different elements of $\mathcal{U}$ may lie at different depths of the construction; however, because $\mathcal{U}$ is finite, there exists some n such that the depth of each basic interval in $\mathcal{U}$ is at most n.

Let $I_{w_1 \ldots w_n}$ be a basic interval of maximal depth in $\mathcal{U}$. Since $\mathcal{U}$ is minimal, it does not contain the basic interval $I_{w_1 \ldots w_{n-1}}$ (otherwise we could eliminate $I_{w_1 \ldots w_n}$ and obtain a proper subcover). It follows that each of the basic intervals $I_{w_1 \ldots w_{n-1}j}$ for $j = 1, \ldots, k$ is contained in $\mathcal{U}$.

Thus the sum in (2.15) contains the partial sum

$$\left|I_{w_1 \ldots w_{n-1}1}\right|^t + \cdots + \left|I_{w_1 \ldots w_{n-1}k}\right|^t.$$

By the formula (1.20) for the lengths of the basic intervals and the definition of t, this is equal to

$$\left(|I_{w_1 \ldots w_{n-1}}|\lambda_1\right)^t + \cdots + \left(|I_{w_1 \ldots w_{n-1}}|\lambda_k\right)^t$$
$$= |I_{w_1 \ldots w_{n-1}}|^t(\lambda_1^t + \cdots + \lambda_k^t) = |I_{w_1 \ldots w_{n-1}}|^t,$$

and it follows that the sum in (2.15) is not changed if we replace all the basic intervals of depth n by the corresponding intervals of depth $n - 1$. The result follows by induction. $\qquad \square$

We now show that the case of an arbitrary ε-cover can be reduced to the case of a cover by basic intervals, and find a uniform lower bound K' in terms of $K = \sum_j |I_j|^t$.

To this end, for each $r > 0$, we consider the collection $\mathcal{V}(r)$ of basic intervals $I_{w_1 \ldots w_n}$ whose lengths satisfy

$$(2.17) \qquad r\lambda_{\min} \leq |I_{w_1 \ldots w_n}| \leq \frac{r}{\lambda_{\min}},$$

where $\lambda_{\min} = \min\{\lambda_1, \ldots, \lambda_k\} < 1$. Now given $x \in C$, let $w \in \Sigma_k^+$ be the symbolic sequence corresponding to x, so that

$$\{x\} = \bigcap_{n \geq 1} I_{w_1 \ldots w_n}.$$

The lengths $|I_{w_1 \ldots w_n}|$ of the basic intervals are related by the inequalities

$$\lambda_{\min}|I_{w_1 \ldots w_{n-1}}| \leq |I_{w_1 \ldots w_n}| \leq \lambda_{\max}|I_{w_1 \ldots w_{n-1}}|;$$

consequently, the number of values of n for which $|I_{w_1 \ldots w_n}|$ lies between $r\lambda_{\min}$ and $r/\lambda_{\min}$ is at least 1 and at most

$$M = 2 \log \lambda_{\min} / \log \lambda_{\max}.$$

This is exactly the number of elements of $\mathcal{V}(r)$ which contain x; hence $\mathcal{V}(r)$ is a cover of C with multiplicity bounded by M (which is independent of r).

Lemma 2.23. *Let $U \subset \mathbb{R}$, and write $r = \operatorname{diam} U$. Then U intersects at most $M' = 2M/\lambda_{\min}$ elements of $\mathcal{V}(r)$.*

Proof. Let $a = \inf U$, so $U \subset [a, a+r]$, and choose points $a = x_0 < \cdots < x_m = b \in U$ such that $x_{j+1} - x_j < r\lambda_{\min}$, and $m \leq 1/\lambda_{\min}$. Then if $V \in \mathcal{V}(r)$ intersects U non-trivially, it must intersect one of the x_j, since $\operatorname{diam} V \geq r\lambda_{\min}$. Because the multiplicity of $\mathcal{V}(r)$ is bounded by M, each of the x_j can be contained in at most M such sets V, and it follows that the total number of elements of $\mathcal{V}(r)$ which intersect U is bounded above by $M(m+1) \leq 2Mm \leq 2M/\lambda_{\min}$. $\square$

The exact value of the constant M' in the lemma is unimportant; what matters is that it is independent of r.

Now let $\mathcal{U}$ be any ε-cover of C; for each U_i, write $r_i = \operatorname{diam} U_i$, and let $U_{i,1}, \ldots U_{i,m(i)}$ be the basic intervals in $\mathcal{V}(r_i)$ which intersect

U_i. It follows from the above remarks that $m(i) \leq M'$; furthermore, we see from (2.17) that

$$\operatorname{diam} U_{i,j} \leq \frac{\operatorname{diam} U_i}{\lambda_{\min}},$$

whence we obtain the bound

$$\sum_{j=1}^{m(i)} (\operatorname{diam} U_{i,j})^t \leq \frac{M'}{\lambda_{\min}{}^t} (\operatorname{diam} U_i)^t.$$

Summing over all the elements of $\mathcal{U}$ yields

$$\sum_i (\operatorname{diam} U_i)^t \geq \left(\frac{\lambda_{\min}{}^t}{M'}\right) \sum_i \sum_{j=1}^{m(i)} (\operatorname{diam} U_{i,j})^t,$$

and since $\{U_{i,j}\}$ is a cover of C by basic intervals, we may apply Lemma 2.22 to obtain

$$\sum_i (\operatorname{diam} U_i)^t \geq \left(\frac{\lambda_{\min}{}^t}{M'}\right) \sum_{j=1}^{k} |I_j|^t > 0,$$

which completes the proof. $\qquad\qquad\qquad\qquad\qquad\qquad\qquad\quad \square$

Exercise 2.15. Given numbers $\lambda_i > 0$ such that $\sum_{i=1}^{k} \lambda_i < 1$, show that the function

$$P(t) = \log\left(\sum_{i=1}^{k} \lambda_i^t\right)$$

is real analytic, decreasing and convex for $t \in \mathbb{R}$. Find a condition on $\lambda_1, \ldots, \lambda_k$ that guarantees that $P(t)$ is strictly convex. Describe the asymptotic behaviour of $P(t)$ as $t \to \pm\infty$.

Exercise 2.16. Given a number $\alpha \in [0,1]$, construct a subset $Z \subset [0,1]$ whose Hausdorff dimension is α.

Exercise 2.17. Construct an uncountable subset of the unit interval $[0,1]$ whose Hausdorff dimension is zero.

c. Moran constructions. Moran's theorem applies not only to geometrically constructed Cantor sets on the line, but to the general class of constructions described below, an example of which is shown in Figure 1.20.

Definition 2.24. A *Moran construction* is a geometric construction that defines a limiting set $C \subset \mathbb{R}^d$ as follows:

(1) Begin with $k > 1$ basic sets $\Delta_1, \ldots, \Delta_k \subset \mathbb{R}^d$, each of which is the closure of its interior, and ratio coefficients $\lambda_i > 0$ such that $\sum_{i=1}^{k} \lambda_i < 1$.

(2) Via an iterative procedure, construct disjoint basic sets $\Delta_{w_1 \ldots w_n}$ such that $\Delta_{w_1 \ldots w_n}$ is similar to $\Delta_{w_1 \ldots w_{n-1}}$, with scaling factor given by the ratio coefficient λ_{w_n}; that is, $\Delta_{w_1 \ldots w_{n-1}}$ is the image of $\Delta_{w_1 \ldots w_n}$ under a similarity transformation with scaling factor λ_{w_n}.

(3) Place the basic sets $\Delta_{w_1 \ldots w_n}$ in the appropriate sets $\Delta_{w_1 \ldots w_{n-1}}$ in any position such that they are disjoint.

(4) Define C as the limiting set $C = \bigcap_{n \geq 1} \bigcup_{w_1 \ldots w_n} \Delta_{w_1 \ldots w_n}$.

Just as is the case for geometrically constructed Cantor sets on the line, the Hausdorff dimension of a Moran construction is given by Moran's formula (2.14). The proof in the higher dimensional setting $(d \geq 2)$ is more or less a carbon copy of the one just given, but the reader is strongly encouraged to work through the details, as this argument is easily the most intricate we have come across so far.

Remark. This result also holds in some cases where the requirement that the basic sets $\Delta_{w_1 \ldots w_n}$ are disjoint is replaced with the weaker assumption that their interiors are disjoint. Sierpiński gaskets are good examples of this situation.

Moran's theorem also extends to a slightly more general class of constructions in a different way. In the definition above, the basic sets at the first step of the construction can be chosen arbitrarily, together with their diameters, provided the basic sets at subsequent steps are homothetic images with the appropriate ratio coefficients. In fact, since it is the asymptotic behaviour which determines Hausdorff dimension, the same result holds if the shapes and sizes of basic sets are chosen completely arbitrarily for *any* finite number of steps, after which the similarity rules and ratio coefficients take over.

Exercise 2.18. Use Moran's theorem to verify the values you guessed in Exercise 2.3 for the Hausdorff dimension of the Sierpiński gasket and the von Koch curve.

Exercise 2.19. Compute the Hausdorff dimension of the subset $E \subset [0,1]$ whose decimal expansions do not contain the digit 5.

Exercise 2.20. Given $\alpha \in [0,\infty]$, find a compact metric space X and a set $Z \subset X$ such that $\dim_H Z = \alpha$.

d. Dynamical constructions and iterated function systems. Moran constructions are not the only way to obtain a fractal set. Although a dynamically defined Cantor set on the line is the result of a Moran construction if the corresponding map is piecewise linear, a repeller for a non-linear map cannot generally be obtained from a Moran construction, and so is not dealt with at all by Theorem 2.21. To study such sets, we will need to introduce techniques from measure theory, which we will do in the next chapter.

In the meantime, we observe that even without introducing non-linearity into the picture, things become more subtle when we work in higher dimensions. So instead of a map f which is defined piecewise on a number of intervals $I_j \subset [0,1]$, and which maps each one linearly onto the entire unit interval, we consider a map f which is defined piecewise on a number of rectangles $R_j \subset R = [0,1]^2$, and which maps each one linearly onto the entire unit square R.

Rather than attempting to understand f by drawing its graph and using a cobweb diagram, which would require four dimensions (one more than the average human being can reliably visualise), we look at the process by which the repeller of f is constructed. If we write f_j for the restriction of the map f to the rectangle R_j, then f_j is a homeomorphism from R_j to R, which expands distances; thus R_j is the image of R under the inverse map f_j^{-1}, which contracts distances.

Following the same idea as in Lecture 3, write $R_{w_1 \ldots w_n}$ for the set of all points $\mathbf{x} \in R$ such that $f^{i-1}(\mathbf{x}) \in R_{w_i}$ for all $1 \leq i \leq n$. Writing $g_j = f_j^{-1}$ for the inverse maps, we have the following analogue of (1.7):

$$(2.18) \qquad R_{w_1 \ldots w_n} = g_{w_n} \circ g_{w_{n-1}} \circ \cdots \circ g_{w_2} \circ g_{w_1}(R).$$

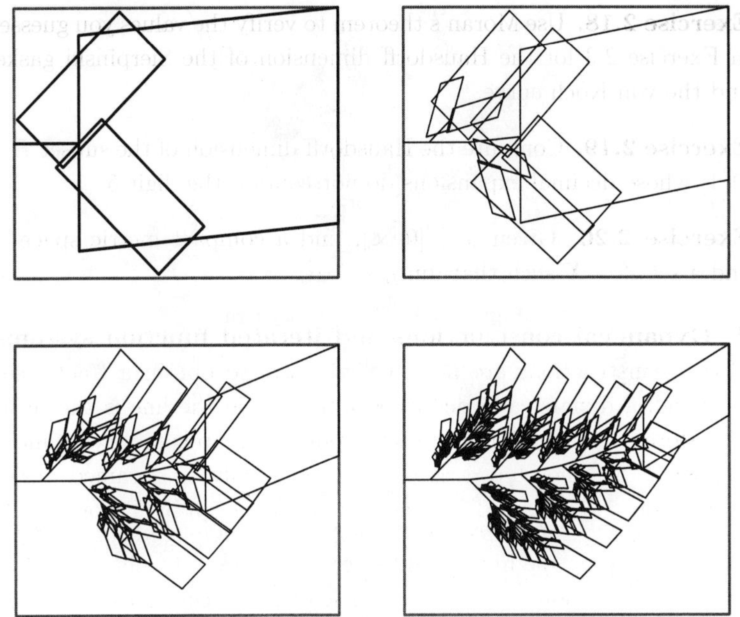

Figure 2.4. An iterated function system with overlapping images.

If f is expanding ($d(f(x), f(y)) > d(x,y)$ for every x, y), then each g_j is contracting ($d(g_j(x), g_j(y)) < d(x,y)$), and so the collection of maps $\{g_1, \ldots, g_k\}$ is an example of the following general object.

Definition 2.25. An *iterated function system (IFS)* is a finite collection of continuous contracting one-to-one maps $g_j \colon R \to R$, where $R \subset \mathbb{R}^d$ is closed. Defining basic sets $R_{w_1 \ldots w_n}$ by (2.18), the *limiting set* of the iterated function system $\{g_1, \ldots, g_k\}$ is

$$C = \bigcap_{n \geq 1} \bigcup_{(w_1, \ldots, w_n)} R_{w_1 \ldots w_n}.$$

Exercise 2.21. Obtain the middle-third Cantor set, the Sierpiński gasket, and the von Koch curve as limiting sets of iterated function systems.

By placing various hypotheses on the maps g_j, we can restrict to smaller classes of iterated functions systems, about which more can be

Figure 2.5. The Barnsley fern.

proved. For example, if R is compact and the maps g_j are contracting similarity transformations whose images are disjoint $(g_i(R) \cap g_j(R) = \emptyset$ for $i \neq j)$, then the limiting set C can once again be obtained via a Moran construction. The ratio coefficients of the construction are the scaling factors of the maps g_j, and we may compute the Hausdorff dimension of C using Moran's theorem.

If we drop the assumption of conformality, however, and merely require the maps g_j to be linear and contracting, then we no longer have a Moran construction (see Figure 2.6), and new tools are required to compute the Hausdorff dimension.

As long as the images $g_j(R)$ are disjoint, the limiting set of the iterated function system may be found as the repeller of a piecewise defined Markov map with domain $R_1 \cup \cdots \cup R_k$; if the images overlap, then we are in new territory. We will not discuss this case further, beyond giving one striking example. The first panel of Figure 2.4 shows an iterated function scheme with four linear contractions g_1, g_2, g_3, and g_4; the remaining three panels show the unions of the images $g_{w_n} \circ \cdots \circ g_{w_1}(R)$ taken over all sequences $(w_1, \ldots, w_n)$, for $n = 2$, $n = 4$, and $n = 6$.

The limiting set C of this particular IFS is called the *Barnsley fern*. A stochastically-generated approximation to C is shown in Figure 2.5; this example illustrates the capability of fractal constructions to produce strikingly realistic images.

Lecture 11

a. Box dimension: another way of measuring dimension. In the definition of the Hausdorff function $m(Z, \alpha)$, we considered covers whose sets have diameter less than or equal to ε; within a particular cover, we might find sets on many different scales, some of which could have diameter much smaller than ε.

An alternate approach is to restrict our attention to covers by sets on the same (small) scale; thus we denote by $\mathcal{D}'(Z, \varepsilon)$ the collection of all countable open covers $\mathcal{U}$ of Z such that $\operatorname{diam} U = \varepsilon$ for every $U \in \mathcal{U}$. We then define a set function

$$(2.19) \qquad r(Z, \alpha, \varepsilon) = \inf_{\mathcal{D}'(Z, \varepsilon)} \sum_i (\operatorname{diam} U_i)^\alpha.$$

This differs from the definition (2.2) of $m(Z, \alpha, \varepsilon)$ in only a single symbol; $\leq$ is replaced by $=$ in the definition of the collection of covers. Nevertheless, the effect of this change is quite drastic; in the first place, the argument that $m(Z, \alpha, \varepsilon)$ depends monotonically on ε does not apply to $r(Z, \alpha, \varepsilon)$, since the collections of admissible covers for two different values of ε are disjoint!

As a result of this change, we have no *a priori* guarantee that the limit of $r(Z, \alpha, \varepsilon)$ as $\varepsilon \to 0$ exists; indeed, there are many examples for which it does not. To deal with this difficulty, we need the concept of *upper and lower limits*.

Definition 2.26. Given a sequence $(x_n) \subset \mathbb{R}$, recall that a point $x \in \mathbb{R}$ is an accumulation point of (x_n) if there exists a subsequence (x_{n_k}) which converges to x. The *lower limit* of (x_n) is

$$\varliminf_{n \to \infty} x_n = \inf\{x \mid x \text{ is an accumulation point of } (x_n)\},$$

and the *upper limit* is

$$\varlimsup_{n \to \infty} x_n = \sup\{x \mid x \text{ is an accumulation point of } (x_n)\}.$$

The lower and upper limits are sometimes denoted by $\liminf$ and $\limsup$, respectively, and one may hear the terms *infimum (supremum) limit*, or possibly *limit inferior (superior)*.

Example 2.27. Define two sequences by $x_n = 1/n$ and $y_n = 1-1/n$, and interweave them:

$$z_{2n-1} = x_n, \qquad z_{2n} = y_n.$$

Then the sequence (z_n) does not converge, but it has subsequences which do. The set of accumulation points of (z_n) is $\{0,1\}$, and so

$$\varliminf_{n\to\infty} z_n = 0, \qquad \varlimsup_{n\to\infty} z_n = 1.$$

Exercise 2.22. Show that the lower and upper limits may equivalently be defined by

$$\varliminf_{n\to\infty} x_n = \lim_{n\to\infty} \left(\inf_{m\geq n} x_m \right),$$

$$\varlimsup_{n\to\infty} x_n = \lim_{n\to\infty} \left(\sup_{m\geq n} x_m \right).$$

Furthermore, show that $x = \varliminf_n x_n$ if and only if both of the following conditions hold:

(1) For every $\varepsilon > 0$, there exists N such that $x_n \geq x - \varepsilon$ for every $n \geq N$.

(2) There exists a subsequence (x_{n_k}) of (x_n) which converges to x.

Similarly, show that $x = \varlimsup_n x_n$ if and only if:

(1) For every $\varepsilon > 0$, there exists N such that $x_n \leq x + \varepsilon$ for every $n \geq N$.

(2) There exists a subsequence (x_{n_k}) of (x_n) which converges to x.

An immediate consequence of the definition is that $\varliminf x_n = \varlimsup x_n$ if and only if $\lim x_n$ itself exists, in which case it is equal to the common value and is the only accumulation point.

We have given the definition of the lower and upper limits for a discrete index (n), but it goes through equally well in the case of a continuous index (such as ε). Thus returning to the function $r(Z, \alpha, \varepsilon)$, where $Z \subset \mathbb{R}^d$, $\alpha \geq 0$, and $\varepsilon > 0$, we define

$$\underline{r}(Z, \alpha) = \varliminf_{\varepsilon \to 0} r(Z, \alpha, \varepsilon),$$

$$\overline{r}(Z, \alpha) = \varlimsup_{\varepsilon \to 0} r(Z, \alpha, \varepsilon).$$

We have the following partial analogue of Proposition 2.1:

Proposition 2.28. *The set functions $\underline{r}(\cdot, \alpha)$ and $\overline{r}(\cdot, \alpha)$ satisfy the following properties.*

(1) Normalisation: $\underline{r}(\emptyset, \alpha) = \overline{r}(\emptyset, \alpha) = 0$ for all $\alpha > 0$.

(2) Monotonicity: $\underline{r}(Z_1, \alpha) \leq \underline{r}(Z_2, \alpha)$ and $\overline{r}(Z_1, \alpha) \leq \overline{r}(Z_2, \alpha)$ whenever $Z_1 \subset Z_2$.

Proof. Follows immediately from the definitions. $\qquad\qquad\qquad\square$

Conspicuously absent from Proposition 2.28 is the subadditivity property which held for $m(\cdot, \alpha)$. The proof of that property relied on the construction of an ε-cover as a union of covers of arbitrarily small diameter; because the definition of $r(Z, \alpha, \varepsilon)$ does not allow us to use sets of diameter less than ε, the proof does not go through here. While we will see in the next lecture that $\overline{r}(Z, \alpha, \varepsilon)$ is subadditive for *finite* collections, no such result holds for countable collections, or for $\underline{r}(Z, \alpha, \varepsilon)$. The consequences of this will become apparent shortly.

As functions of α, both $\underline{r}(Z, \alpha)$ and $\overline{r}(Z, \alpha)$ have similar properties to $m(Z, \alpha)$; there are critical values $\underline{\alpha}_C$ and $\overline{\alpha}_C$ below which the value of the function is ∞, and above which it is 0. Just as the critical value of $m(Z, \alpha)$ determines the Hausdorff dimension of Z, the critical values of $\underline{r}$ and $\overline{r}$ are also dimensional quantities, known as the *lower box dimension* and *upper box dimension*, respectively,[6] and denoted $\underline{\dim}_B Z$ and $\overline{\dim}_B Z$. As with α_C, we have

$$\underline{\dim}_B Z = \underline{\alpha}_C = \inf\{\alpha > 0 \mid \underline{r}(Z, \alpha) = 0\}$$
$$= \sup\{\alpha > 0 \mid \underline{r}(Z, \alpha) = \infty\},$$
$$\overline{\dim}_B Z = \overline{\alpha}_C = \inf\{\alpha > 0 \mid \overline{r}(Z, \alpha) = 0\}$$
$$= \sup\{\alpha > 0 \mid \overline{r}(Z, \alpha) = \infty\}.$$

Exercise 2.23. Compute the lower and upper box dimensions of the following sets:

(a) $A = \{0, 1, \frac{1}{2}, \frac{1}{3}, \frac{1}{4}, \dots\}$.

(b) $B = \{0, 1, \frac{1}{4}, \frac{1}{9}, \frac{1}{16}, \dots\}$.

---

[6]In the literature, one may also find the box dimensions referred to as *box counting dimensions, entropy dimensions,* or *capacities.*

b. Properties of box dimension. As an immediate consequence of Proposition 2.28, we have the following analogue of Proposition 2.8:

Proposition 2.29. *The upper and lower box dimensions have the following basic properties.*

(1) Normalisation: $\underline{\dim}_B \emptyset = \overline{\dim}_B \emptyset = 0$.

(2) Monotonicity: $\underline{\dim}_B Z_1 \leq \underline{\dim}_B Z_2$ and $\overline{\dim}_B Z_1 \leq \overline{\dim}_B Z_2$ whenever $Z_1 \subset Z_2$.

(3) If $\{Z_i\}$ is any countable collection of subsets of $\mathbb{R}^d$, then

$$\underline{\dim}_B \left(\bigcup Z_i \right) \geq \sup_i \left(\underline{\dim}_B Z_i \right),$$

$$\overline{\dim}_B \left(\bigcup Z_i \right) \geq \sup_i \left(\overline{\dim}_B Z_i \right).$$

Property (3) follows immediately from property (2), and is weaker than its analogue in Proposition 2.8 because of the failure of countable subadditivity for the lower and upper box dimensions; in Example 2.36, we will see that the inequality may become strict.

First, though, let us address the definition (2.19) of $r(Z, \alpha, \varepsilon)$, which played a key role in the description of the lower and upper box dimensions. Since we restrict our attention to covers in which *every* set has diameter ε, every term in the sum $\sum_i (\operatorname{diam} U_i)^\alpha$ is the same! It seems rather silly, then, to continue writing it as a sum, and indeed (2.19) is equivalent to

$$r(Z, \alpha, \varepsilon) = \inf_{\mathcal{D}'(Z,\varepsilon)} \varepsilon^\alpha N(\mathcal{U}),$$

where $N(\mathcal{U})$ denotes the number of elements in the cover $\mathcal{U}$. Thus we write

$$N(Z, \varepsilon) = \inf_{\mathcal{D}'(Z,\varepsilon)} N(\mathcal{U})$$

for the minimal number of elements in a cover of Z by open sets of diameter ε and obtain

$$(2.20) \qquad r(Z, \alpha, \varepsilon) = \varepsilon^\alpha N(Z, \varepsilon).$$

Observe that $N(Z, \varepsilon)$ is finite if the set Z is bounded, and that there may exist different covers $\mathcal{U}$ with $N(\mathcal{U}) = N(Z, \varepsilon)$, so that there is not necessarily a unique optimal cover.

Exercise 2.24. Show that the lower and upper box dimensions may be characterised by

$$\underline{\dim}_B Z = \underline{\alpha}_C = \lim_{\varepsilon \to 0} \frac{\log N(Z,\varepsilon)}{\log(1/\varepsilon)}, \tag{2.21}$$

$$\overline{\dim}_B Z = \overline{\alpha}_C = \overline{\lim_{\varepsilon \to 0}} \frac{\log N(Z,\varepsilon)}{\log(1/\varepsilon)}. \tag{2.22}$$

Exercise 2.25. Show that the lower and upper box dimensions of a set $Z \subset \mathbb{R}^d$ can also be computed as

$$\underline{\dim}_B Z = \varliminf_{\varepsilon \to 0} \frac{\log L(Z,\varepsilon)}{-\log \varepsilon}, \quad \overline{\dim}_B Z = \varlimsup_{\varepsilon \to 0} \frac{\log L(Z,\varepsilon)}{-\log \varepsilon},$$

where $L(Z,\varepsilon)$ is largest number of *disjoint* balls of radius ε centred at points in Z.

Exercise 2.26. Given $Z \subset \mathbb{R}^d$, consider sequences $n_k \to \infty$ and $\varepsilon_k \to 0$. Assume that for every k the set Z can be covered by n_k balls of radius ε_k. Show that

$$\underline{\dim}_B Z \le \varliminf_{k \to \infty} \frac{\log n_k}{-\log \varepsilon_k}.$$

If in addition, there exists $0 < c < 1$ such that $\varepsilon_{k+1} \ge c\varepsilon_k$ for all k, then show that

$$\overline{\dim}_B Z \le \varlimsup_{k \to \infty} \frac{\log n_k}{-\log \varepsilon_k}.$$

Proposition 2.7 showed that defining $m(Z, \alpha, \varepsilon)$ in terms of open *balls* rather than open *sets* does not change the critical value α_C, and hence leads to an equivalent definition of the Hausdorff dimension. The same is true for the lower and upper box dimensions, and the proof goes through verbatim; hence we may also use $N_B(Z, \varepsilon)$, the smallest cardinality of a cover of Z by balls of radius ε, in (2.21) and (2.22).

Because $N(Z, \varepsilon)$ and $N_B(Z, \varepsilon)$ must be finite in order for the definition of lower and upper box dimension to make any sense, we restrict our attention to *bounded* subsets of $\mathbb{R}^d$. Furthermore, we need only consider *compact* subsets Z, thanks to the following fact.

Proposition 2.30. *The box dimension (lower or upper) of a set Z is the same as the box dimension (lower or upper) of its closure.*

Proof. Recall that the closure $\overline{Z}$ of a set $Z \subset \mathbb{R}^d$ is just the union of Z and its accumulation points. If $\{B(x_i, \varepsilon/2)\}$ covers Z, then $\{B(x_i, \varepsilon)\}$ covers $\overline{Z}$, and so

$$N_B(\overline{Z}, 2\varepsilon) \leq N_B(Z, \varepsilon) \leq N_B(\overline{Z}, \varepsilon),$$

from which (2.21) and (2.22) show that $\underline{\dim}_B Z = \underline{\dim}_B \overline{Z}$, and also $\overline{\dim}_B Z = \overline{\dim}_B \overline{Z}$. $\square$

As a consequence of Proposition 2.30, it suffices to consider subsets of $\mathbb{R}^d$ which are both closed and bounded, hence compact.

As before, we want to understand how the box dimensions behave under continuous maps or changes of metric. The most important result is the following analogue of Proposition 2.17.

Proposition 2.31. *If $f \colon \mathbb{R}^d \to \mathbb{R}^d$ is Lipschitz, then $\underline{\dim}_B f(Z) \leq \underline{\dim}_B Z$ and $\overline{\dim}_B f(Z) \leq \overline{\dim}_B Z$ for every compact $Z \subset \mathbb{R}^d$.*

Proof. If $\mathcal{U} = \{U_i\}$ is a cover of Z by open sets of diameter ε, then $\operatorname{diam} f(U_i) \leq L \operatorname{diam} U_i$, where L is a Lipschitz constant for f. By "fattening up" the sets $f(U_i)$ by an appropriate amount, we can turn $f(\mathcal{U}) = \{f(U_i)\}$ into an open cover of $f(Z)$ by sets of diameter precisely $L\varepsilon$; it follows that $N(f(Z), L\varepsilon) \leq N(Z, \varepsilon)$, and so

$$\frac{\log N(f(Z), L\varepsilon)}{\log L + \log(1/L\varepsilon)} \leq \frac{\log N(Z, \varepsilon)}{\log(1/\varepsilon)}.$$

Taking lower and upper limits as $\varepsilon \to 0$ gives the result. $\square$

It follows that if f is bi-Lipschitz, then Z and $f(Z)$ have the same box dimensions. In particular, the lower and upper box dimensions do not change if we pass to a strongly equivalent metric.

Finally, we remark that as with the Hausdorff dimension, the box dimensions may be defined in any separable metric space, not just $\mathbb{R}^d$.

Lecture 12

a. Relationships between the various dimensions. Let us examine the relationship between the Hausdorff dimension and the two box dimensions. It follows immediately from the definitions that

$$m(Z, \alpha) \leq \underline{r}(Z, \alpha) \leq \overline{r}(Z, \alpha)$$

for any $Z \subset \mathbb{R}^d$ and $\alpha > 0$, and thus we have the relations

$$(2.23) \qquad \dim_H Z \leq \underline{\dim}_B Z \leq \overline{\dim}_B Z.$$

Exercise 2.27. Show that $\underline{\dim}_B[0,1] = \overline{\dim}_B[0,1] = 1$.

One of our goals will be to establish conditions on Z under which the quantities in (2.23) all coincide. When this occurs, we may refer to the common value as the *fractal dimension* without fear of ambiguity. We will see that the three quantities agree for a wide range of examples, including some rather complicated sets. We will also see relatively simple examples of sets Z for which the inequalities in (2.23) become strict. The challenge is to develop some criteria which may let us know what sort of behaviour to expect for a particular Z.

For a very important family of sets, those resulting from Moran constructions, all three quantities coincide.

Theorem 2.32. *Let C be the limit set of a Moran construction. Then* $\dim_H C = \underline{\dim}_B C = \overline{\dim}_B C$.

Proof. Thanks to (2.23), it suffices to show that $\overline{\dim}_B C \leq \dim_H C$. By Moran's theorem (Theorem 2.21), $\dim_H C$ is the unique solution of (2.14), and so we show that $\overline{\dim}_B C \leq t$, where $\sum_i \lambda_i^t = 1$.

Thus we want to show that $\overline{r}(C,t) < \infty$ by bounding $r(C,t,\varepsilon)$ from above; this is accomplished by producing a suitable cover of C by open sets of diameter ε, as follows. Given $x \in C$, let $n(x)$ be the unique integer such that

$$\left| I_{w_1 \ldots w_{n(x)-1}} \right| \geq \varepsilon > \left| I_{w_1 \ldots w_{n(x)}} \right|;$$

the existence of $n(x)$ is guaranteed by the fact that $|I_{w_1 \ldots w_n}| \to 0$ as $n \to \infty$. Thus for each x we may choose an open interval $U(x)$ of length ε which contains $I_{w_1 \ldots w_{n(x)}}$. Because C is compact, it can be covered by a finite collection of basic intervals $\{I_{w_1 \ldots w_{n(x_i)}}\}_{i=1}^N$. Without loss of generality, we may take this collection to be disjoint; then for the open cover $\mathcal{U} = \{U(x_i)\}_{i=1}^N$, we have

$$\sum_i (\operatorname{diam} U(x_i))^t \leq \frac{1}{\lambda_{\min}^t} \sum_i \left| I_{w_1 \ldots w_{n(x_i)}} \right|^t$$

$$(2.24) \qquad \qquad = \frac{1}{\lambda_{\min}^t} \left(|I_1|^t + \cdots + |I_k|^t \right),$$

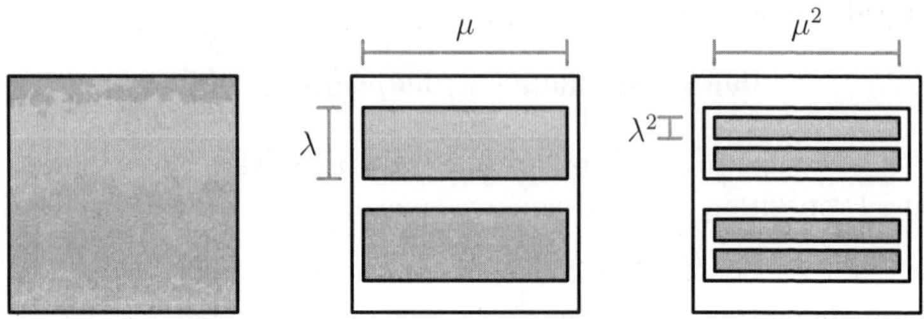

Figure 2.6. A Cantor construction in $\mathbb{R}^2$ which is not Moran.

where the last equality follows by the same calculation as in the proof of Theorem 2.21.

Writing K for the last expression in (2.24), we see that $r(C, t, \varepsilon) \leq K$ for every $\varepsilon > 0$; since K does not depend on ε, this gives $\overline{r}(C, t) \leq K < \infty$. This implies that $\overline{\dim}_B C \leq t = \dim_H C$, and the desired result follows. $\qquad\square$

Exercise 2.28. Let I_1, I_2, and I_3 be three disjoint subintervals of $[0, 1]$, and let $f\colon I_1 \cup I_2 \cup I_3 \to [0, 1]$ be the corresponding one-dimensional linear full-branched Markov map. Compute the Hausdorff and box dimensions of the repeller for f.

Exercise 2.29. Compute the Hausdorff dimension and lower and upper box dimensions of the Cantor set obtained via the following geometric construction in $\mathbb{R}^2$:

Starting with the unit square in $\mathbb{R}^2$, choose two disjoint rectangles with sides λ and μ (where $0 < \lambda < \frac{1}{2} < \mu < 1$), which are spaced one exactly above the other. At the next step of the construction inside each of these rectangles choose two disjoint rectangles with sides λ^2 and μ^2, which are spaced one exactly above the other. Continue in the same fashion (see Figure 2.6).

It is possible to extend the result of Theorem 2.32 to certain other examples, as the following exercise shows.

Exercise 2.30. Let C be the middle-third Cantor set and let C' be obtained from C by adding a single point inside each complementary interval. Show that

$$\dim_H C' = \underline{\dim}_B C' = \overline{\dim}_B C' = \dim_H C.$$

The competing ideas of dimension do not always agree, however; using Proposition 2.30, we can give an example of a set for which the inequality in (2.23) becomes strict.

Example 2.33. Let $Z = \mathbb{Q} \cap [0, 1]$ be the set of all rational numbers in the unit interval. Then Z is countable, which implies that $\dim_H Z = 0$ by Corollary 2.9. However, Z is dense in $[0, 1]$, and so

$$\underline{\dim}_B Z = \underline{\dim}_B [0, 1],$$
$$\overline{\dim}_B Z = \overline{\dim}_B [0, 1].$$

By Exercise 2.27, this implies that $\underline{\dim}_B Z = \overline{\dim}_B Z = 1$.

There are many other examples of sets for which the Hausdorff and box dimensions do not agree, and for which in addition the two box dimensions do not agree. Indeed, the coincidence of all three quantities is somehow a special case, which usually occurs only if the set Z arises from a geometric construction or a dynamical system.

Before giving further negative results in the form of counterexamples, we give one more positive result, one more relationship between the different sorts of dimensions which always holds. For general subsets $Z \subset \mathbb{R}^d$, even though the inequalities in (2.23) may become strict, it is still possible to prove the following (rather deep) theorem, due to Lev Pontryagin and Lev Shnirel'man [**PS32**], which is a stronger version of Hausdorff's theorem (Theorem 2.19). As with the Hausdorff dimension, the lower and upper box dimensions depend on the choice of metric; it follows from Proposition 2.31 that passing to a strongly equivalent metric preserves these dimensions, but they may change if the new metric is merely equivalent.

Theorem 2.34 (Pontryagin–Shnirel'man theorem). *Given a set $Z \subset \mathbb{R}^d$, the topological dimension $\dim Z$ and the lower box dimensions $\underline{\dim}_B^\rho Z$ are related by the following variational principle:*

$$(2.25) \qquad \dim Z = \inf_{\rho \sim d} \underline{\dim}_B^\rho Z.$$

That is, the topological dimension is the infimum of the possible lower box dimensions, taken over all metrics ρ which are equivalent to the standard metric d.

No analogue of this result holds for the upper box dimension, and so this theorem is in some sense the best result possible. Pontryagin and Shnirel'man were the first to introduce the concept of lower box dimension, which they called the *metric order*, a piece of terminology which has fallen by the wayside. Perhaps because no such result holds for the upper box dimension, they did not consider that quantity.

b. A counterexample. We now construct an example which shows that the three different dimensional quantities may all take different values for relatively simple subsets of $[0, 1]$.

Theorem 2.35. *Given any $0 < \alpha \leq \beta < 1$, there exists a countable closed set $A \subset [0, 1]$ such that $\dim_H A = 0$, $\underline{\dim}_B A = \alpha$, and $\overline{\dim}_B A = \beta$.*

Proof. Consider the sequence $a_n = e^{-n} \to 0$. We will construct a set $A \subset [0, 1]$ as an increasing sequence beginning at 0; the first few terms will be separated by a gap of length a_1, the next few by a gap of length a_2, and so on. That is, the set A will be the sequence

$$(2.26) \quad \left\{ 0, a_1, 2a_1, \ldots, b_1 a_1, \right.$$

$$b_1 a_1 + a_2, b_1 a_1 + 2a_2, \ldots, b_1 a_1 + b_2 a_2,$$

$$\vdots$$

$$\left(\sum_{k=1}^{n} b_k a_k \right) + a_{n+1}, \ldots, \left(\sum_{k=1}^{n} b_k a_k \right) + b_{n+1} a_{n+1},$$

$$\left. \left(\sum_{k=1}^{n+1} b_k a_k \right) + a_{n+2}, \ldots \right\}$$

together with its limit point, where (b_n) is a sequence of non-negative integers which we will choose so as to obtain the desired result for the lower and upper box dimensions. We write the endpoints between sequences of differently spaced points as

$$T_n = \sum_{k=1}^{n} b_k a_k,$$

and see that $\lim_{n\to\infty} T_n = T$, the limit point of A.

The fact that $\dim_H A = 0$ follows immediately from the fact that A is countable, and so it remains to choose (b_n) so as to guarantee $\underline{\dim}_B A = \alpha$ and $\overline{\dim}_B A = \beta$. The key properties of our sequence (b_n) will be as follows:

(1) (b_n) tends to infinity monotonically as $n \to \infty$.

(2) $\sum_{n=1}^{\infty} a_n b_n < 1$.

(3) The exponential growth rate of the partial sums $S_n = \sum_{k=1}^{n} b_k$ is given by

$$\varliminf_{n\to\infty} \frac{1}{n} \log S_n = \alpha, \qquad \varlimsup_{n\to\infty} \frac{1}{n} \log S_n = \beta.$$

(4) The "tail" $[T_n, T]$ of the set A is not too long: there exists a constant C such that

$$\frac{T - T_n}{a_n S_n} \le C$$

for all n.

The summability property (2) guarantees that A is bounded. The significance of the partial sums S_n is that they let us estimate $N(\varepsilon)$, the cardinality of an optimal cover by open sets of diameter ε. Indeed, given $\varepsilon > 0$, we choose $n = n(\varepsilon)$ such that

$$e^{-(n+1)} < \varepsilon \le e^{-n},$$

and observe that since the first S_n points in the sequence defining A are all separated by a distance of at least e^{-n}, we must have $N(A, \varepsilon) \ge S_n$.

Furthermore, the number of intervals of length ε required to cover the entire interval $[T_n, T]$ is at most

$$\frac{T - T_n}{e^{-(n+1)}} = \frac{1}{e} \frac{T - T_n}{a_n} \le \frac{C}{e} S_n,$$

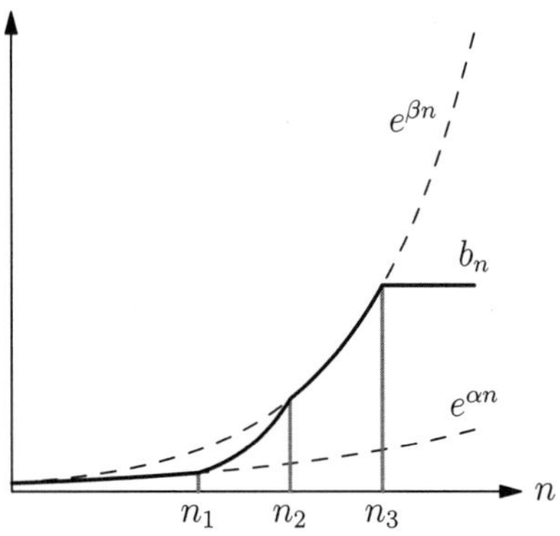

Figure 2.7. Building a sequence (b_n).

using property (4). This shows that $N(A, \varepsilon) \leq (1 + C/e)S_n$, and hence

$$\frac{\log S_n}{n+1} \leq \frac{\log N(A, \varepsilon)}{\log(1/\varepsilon)} \leq \frac{\log(1 + C/\varepsilon) + \log S_n}{n}.$$

The result for the lower and upper box dimensions then follows immediately from property (3); thus it remains only to produce a sequence with these properties. To this end, we follow a four-step recursive procedure, illustrated in Figure 2.7. At first, b_n is just the integer part of $e^{\alpha n}$, and so S_n also grows at the same rate as $e^{\alpha n}$. It follows that the quantity $(\log S_n)/n$ converges to α as n grows, and so we may choose n_1 such that

$$(2.27) \qquad\qquad \left| \frac{1}{n_1} \log S_{n_1} - \alpha \right| < \frac{1}{2}.$$

Now we would like to let b_n follow the function $e^{\beta n}$ for a while, to approximate the desired upper limit, but if we jump directly to the graph of $e^{\beta n}$ at n_1, we will find that the sequence we eventually produce fails property (4), and so we must be slightly more careful. Thus for $n > n_1$, we let b_n grow exponentially, with $b_{n+1} = M b_n$ for some fixed $e^{\beta} < M < e$, until it reaches the upper function at n_2, at which point we have b_n follow $e^{\beta n}$ until

$$(2.28) \qquad\qquad \left| \frac{1}{n_3} \log S_{n_3} - \beta \right| < \frac{1}{2}.$$

Finally, for $n > n_3$, we leave b_n constant until it is once again equal to $e^{\alpha n}$ at n_4 (which is somewhere off the right edge of the graph in Figure 2.7). Then we iterate all four steps of this procedure, replacing the bound $1/2$ in (2.27) and (2.28) with $(1/2)^k$ at the kth iteration.

We must now verify that the sequence (b_n) so constructed has properties (1)–(4). Monotonicity and divergence to ∞ are immediate from the definition (see also Figure 2.7), which establishes (1).

To see (2), observe that $b_n \leq e^{\beta n}$ and $a_n = e^{-n}$, hence

$$\sum_{n=1}^{\infty} a_n b_n \leq \sum_{n=1}^{\infty} e^{(\beta-1)n} = \frac{e^{\beta-1}}{1 - e^{\beta-1}} < \infty,$$

where we use the fact that $\beta < 1$ and hence $e^{\beta-1} < 1$.

For (3), we need the estimates (2.27) and (2.28), or rather, their generalisations for large k. The first of these gives

$$\left| \frac{1}{n_{4k+1}} \log S_{n_{4k+1}} - \alpha \right| < \left(\frac{1}{2} \right)^{k+1},$$

and hence

$$\varliminf_{n \to \infty} \frac{1}{n} \log S_n \leq \lim_{k \to \infty} \frac{1}{n_{4k+1}} \log S_{n_{4k+1}} = \alpha.$$

Furthermore, we have

$$S_n = \sum_{k=1}^{n} b_k \geq \sum_{k=1}^{n} e^{\alpha n} = \frac{e^{\alpha(n+1)} - 1}{e^{\alpha} - 1},$$

and so

$$\log S_n \geq \log(e^{\alpha(n+1)} - 1) - \log(e^{\alpha} - 1),$$

which gives

$$\frac{1}{n} \log S_n \geq \frac{1}{n} \log(e^{\alpha(n+1)} - 1) - \frac{1}{n} \log(e^{\alpha} - 1) \to \alpha.$$

This establishes the first half of (3), and the second half is proved similarly.

Finally, using the fact that $S_n \geq b_n$ and $b_{n+1}/b_n \leq M$, we have

$$\frac{T - T_n}{a_n S_n} \leq \sum_{k=n+1}^{\infty} \frac{e^{-k} b_k}{e^{-n} b_n} \leq \sum_{k=n+1}^{\infty} \left(\frac{M}{e} \right)^{k-n} = \sum_{k=1}^{\infty} \left(\frac{M}{e} \right)^{k} < \infty,$$

where the sum converges to some finite constant C by our choice of M. Thus property (4) holds as well.

Having shown that the sequence $\{b_n\}$ satisfies the properties (1)–(4), it follows from the previous discussion that the set A given in (2.26) has the dimensions claimed. $\qquad\qquad\square$

In some sense, this is the simplest possible counterexample to equality in (2.23); any simpler set is just finite, and then all three quantities are immediately 0.

Note that Theorem 2.35 does *not* provide a set A with $\dim_H A = 0$, $0 < \underline{\dim}_B A < 1$, and $\overline{\dim}_B A = 1$; this case, which requires a different construction, is left as an exercise for the reader.

Exercise 2.31. Find the Hausdorff and box dimensions of the following sets in $\mathbb{R}^2$:

(a) $\{(\frac{1}{2^p}, \frac{1}{2^q}) \mid p, q \in \mathbb{N}\}$.

(b) $\{(\frac{1}{p^2}, \frac{1}{q^2}) \mid p, q \in \mathbb{N}\}$.

(c) $\{(\frac{1}{n}, \frac{1}{2^n}) \mid n \in \mathbb{N}\}$.

c. Stability and subadditivity. An important property of the dimensional quantities we have seen so far is their behaviour under taking countable unions. We saw in Proposition 2.8 that $\dim_H \left(\bigcup_i Z_i\right) = \sup_i \dim_H Z_i$, while the best that Proposition 2.29 could offer for the lower and upper box dimensions was an inequality. Indeed, the example of the rational numbers in the unit interval showed that the box dimensions can increase when we take a countable union; taking Z_i to be singleton sets which exhaust $\mathbb{Q} \cap [0, 1]$, we saw that

$$\underline{\dim}_B \left(\bigcup_i Z_i\right) = 1 > 0 = \sup_i \underline{\dim}_B Z_i,$$

$$\overline{\dim}_B \left(\bigcup_i Z_i\right) = 1 > 0 = \sup_i \overline{\dim}_B Z_i.$$

The reason for the discrepancy between the two sorts of dimensions is the fact that while $m(Z, \alpha)$ is countably subadditive (Property (3) in Proposition 2.1), neither $\underline{r}(Z, \alpha)$ nor $\overline{r}(Z, \alpha)$ has this property. The best we can hope for in this case is *finite* subadditivity:

Exercise 2.32. Show that for any finite collection of sets $\{Z_i\}_{i=1}^k$ in $\mathbb{R}^d$ and any $\alpha \geq 0$, we have

$$(2.29) \qquad \overline{r}\left(\bigcup_{i=1}^k Z_i\right) \leq \sum_{i=1}^k \overline{r}(Z_i, \alpha),$$

and that consequently

$$(2.30) \qquad \overline{\dim}_B\left(\bigcup_{i=1}^k Z_i\right) = \max\left\{\overline{\dim}_B Z_i \mid i = 1, \ldots, k\right\}.$$

Even this weakened property only holds for the *upper* box dimension; for the set function $\underline{r}(Z, \alpha)$ associated to the lower box dimension, we have no subadditivity property at all, as the following example shows.

Example 2.36. Fixing $0 < \alpha < \beta < 1$, we may use the construction in the proof of Theorem 2.35 to find $Z_1 \subset [0,1]$ and $Z_2 \subset [2,3]$ such that $\dim_H Z_1 = \dim_H Z_2 = 0$, $\underline{\dim}_B Z_1 = \underline{\dim}_B Z_2 = \alpha$, and $\overline{\dim}_B Z_1 = \overline{\dim}_B Z_2 = \beta$. We wish to modify the construction slightly to ensure that $\underline{\dim}_B(Z_1 \cup Z_2) > \alpha$.

The key idea in the construction was for b_n to follow first one exponential curve and then the other, transferring at the appropriate indices n_i. In our present case, we wish to start b_n^1 (the sequence defining Z_1) on the lower exponential curve, and b_n^2 (defining Z_2) on the upper curve; furthermore, we define a *single* sequence n_i of indices for *both* b_n^1 and b_n^2 by requiring that the estimates on $S_{n_i}^1$ and $S_{n_i}^2$ are both within the range given by (2.27) and (2.28).

Defining Z_1 and Z_2 in this way, we see that if $\varepsilon > 0$ is such that $(\log N(Z_1, \varepsilon))/(-\log \varepsilon)$ is near α, then $(\log N(Z_2, \varepsilon))/(-\log \varepsilon)$ will be near β, and vice versa. Since $N(Z_1 \cup Z_2, \varepsilon) \geq N(Z_1, \varepsilon) + N(Z_2, \varepsilon)$, it can then be shown that $(\log N(Z_1 \cup Z_2, \varepsilon))/(-\log \varepsilon)$ is bounded away from α, and hence $\underline{\dim}_B Z_1 \cup Z_2 > \alpha = \underline{\dim}_B Z_1 = \underline{\dim}_B Z_2$.

This is the second case we have seen in which the lower and upper box dimensions behave in fundamentally different ways. The first was the Pontryagin–Shnirel'man theorem for the lower box dimension, which had no analogue for the upper box dimension. These two episodes illustrate the fact that the three different dimensional quantities—Hausdorff dimension, lower box dimension, and upper box dimension—are all qualitatively distinct and operate according to different rules. And this is not yet the whole story; there are other "dimensions" waiting in the wings. All the quantities we have discussed so far have been *global* quantities, which depend on the entire set or space we are interested in. Eventually, in Chapter 4, we will introduce dimensional quantities of a *local* nature, which may vary from point to point within the object we are studying. We will also introduce dimensional quantities which characterise not only the spatial structure of a set, but the time evolution of a dynamical system.

Chapter 3

Measures: Definitions and Examples

Lecture 13

a. A little bit of measure theory. As we have seen, computing the Hausdorff dimension of a set can be very difficult, even if the set is geometrically quite regular, as was the case for the Cantor sets we have considered so far. For sets whose self-similarity is not quite so regular, the situation becomes even worse; for example, consider the non-linear map shown in Figure 1.23. This map generates a repelling Cantor set C through the same construction that we carried out for the linear map in Figure 1.12, and we may ask what the Hausdorff dimension of C is. Thus we must study a Cantor construction in which the ratio coefficients are no longer constant but may change at each step of the iteration. This occurs because the map is no longer linear, and so how much contraction (in the construction) or expansion (in the map) occurs at each step now depends on which point we consider, not just on which interval it is in.

How does the Hausdorff dimension respond to this change in the construction? The problems which arise at this stage are much more difficult than those we encountered in proving Theorem 2.21, and we will need new tools to deal with them.

A key idea will be to somehow sidestep the fact that ratio coefficients may vary by studying "asymptotic ratio coefficients" which give the average rate of contraction over a large number of steps of the construction. If this average rate converges, we can still hope to say something; however, at some point it will become necessary to distinguish between "bad" points of C, which we want to ignore because the average rate does not converge, and "good" points, which we can deal with, and we will need to show that there are in some sense "more" of the latter.

In order to make all this precise, we need to expand our toolkit to include the idea of a *measure*. Without further ado, then, we have the following definition.

Definition 3.1. Let X be any set (which will be called our *space*), and let $\mathcal{A}$ be a collection of subsets of X. $\mathcal{A}$ is called an *algebra* if

(1) $\emptyset, X \in \mathcal{A}$.

(2) $X \setminus A \in \mathcal{A}$ whenever $A \in \mathcal{A}$.

(3) $A_1 \cup A_2 \in \mathcal{A}$ whenever $A_1, A_2 \in \mathcal{A}$.

Property (3) immediately implies that $\mathcal{A}$ is closed under finite unions: $\bigcup_{i=1}^{n} A_i \in \mathcal{A}$ whenever $A_1, \ldots, A_n \in \mathcal{A}$. If in addition $\mathcal{A}$ is closed under *countable* unions, that is, if

(4) $\bigcup_{n=1}^{\infty} A_i \in \mathcal{A}$ whenever $A_n \in \mathcal{A}$ for every $n \in \mathbb{N}$,

then $\mathcal{A}$ is a σ-*algebra*. We will refer to the elements of $\mathcal{A}$ as *measurable sets*. A set function $m \colon \mathcal{A} \to [0, \infty]$, which assigns to each measurable set a non-negative number (or possibly ∞), is a *measure* if it satisfies the following properties.

(1) $m(\emptyset) = 0$.

(2) *Monotonicity:* $m(A_1) \leq m(A_2)$ whenever $A_1 \subset A_2$ and both are measurable.

(3) *Countable additivity:* $m(\bigcup_i) = \sum_i m(A_i)$ whenever $\{A_i\} \subset \mathcal{A}$ is a countable collection of disjoint measurable sets.

Property (3) is sometimes referred to as σ-*additivity*. The triple $(X, \mathcal{A}, m)$ is known as a *measure space*.

A finite measure μ (that is, one for which $\mu(X) < \infty$) may be referred to as a *mass distribution*, since we may think of it as describing how a specific amount of mass is distributed over a space. If the measure is normalised so that $\mu(X) = 1$, we often call μ a *probability measure*, since in this case subsets of X may be thought of as events in a probabilistic process.

Before moving on to meatier applications, we give a few very basic examples of measure spaces.

Example 3.2. Let X be any set, and let $\mathcal{A} = \{\emptyset, X\}$. Then $\mathcal{A}$ is a σ-algebra, and we may define a measure m by setting $m(\emptyset) = 0$ and assigning an arbitrary value to $m(X)$. This, of course, is a completely trivial example.

Example 3.3. Let X be any set, and let $\mathcal{A} = 2^X$ be the power set of X; that is, the collection of all subsets of X. $\mathcal{A}$ is again a σ-algebra (in fact, it is the largest possible σ-algebra on X), and we may define a measure ν by

$$\nu(A) = \begin{cases} \mathrm{card}(A) & A \text{ finite,} \\ \infty & \text{otherwise.} \end{cases}$$

So if A is a finite set, ν counts the number of points in A; otherwise, it gives ∞. This is known as the *counting measure* on X.

Example 3.4. Let X and $\mathcal{A}$ be as in the previous example, and fix $x \in X$. Define a measure δ_x by

$$\delta_x(A) = \begin{cases} 1 & x \in A, \\ 0 & x \notin A. \end{cases}$$

Thus δ_x simply measures whether or not A contains the point x. This is known as the *point measure* sitting on x.

A measure gives us several things. First and foremost, it allows us to quantify the size of subsets of X, to say which are large and which are small. For the counting measure ν, a set is large according to how many points it contains, while for the point measure δ_x, a set is large if and only if it contains the point x, regardless of how many other points it contains.

Consequently, given a function $\varphi \colon X \to \mathbb{R}$, a measure μ assigns weights to the parts of X on which φ takes various values, and hence allows us to meaningfully speak of the *average value* of φ via the process of *Lebesgue integration*, a detailed description of which is beyond the scope of this book.

Because a given space X may carry many different measures, this average value depends on the measure μ (the method of averaging) as well as the function φ (the thing which is averaged). For example, the average value of φ with respect to the point measure δ_x is just $\varphi(x)$, since the measure gives no weight to the value of the function elsewhere.

This second interpretation is the beginning of the functional analytic approach to measure theory, which views measures in terms of their interaction with functions, as sets of weights to be used for averaging. The first interpretation, in terms of sets, is more important for our purposes; of particular importance is the fact that a given measure μ tells us which sets may be disregarded as being negligible as far as μ is concerned.

Definition 3.5. Given a measure space $(X, \mathcal{A}, \mu)$, a set $E \in \mathcal{A}$ is a *null set* for μ if $\mu(E) = 0$. Given some property P which may be true or false at each point of X, we say that P is true *almost everywhere* with respect to μ (or more concisely, μ-a.e.) if the set of points on which it fails is a null set for μ.

The idea that a property may be true almost everywhere, without necessarily being true everywhere, is one of the most important tools in measure theory; as we will see, many of the properties in which we are interested fall into this category. It turns out that if we are interested in measure-theoretic properties, then we are not giving up too much by letting the property fail on a null set; since null sets have zero measure, what happens on them is unimportant from a measure-theoretic point of view.

For the counting measure ν, the only null set is the empty set; hence a property holds ν-a.e. if and only if it holds everywhere. In contrast, the point measure δ_x has many more null sets; indeed, every set which does not contain x is a null set for δ_x. Thus a property holds

δ_x-a.e. if and only if it holds at the single point x, whether or not it holds anywhere else; this is because if it holds at x, then the set of all points y at which it fails is contained in $X \setminus \{x\}$, and hence is a null set, and vice versa.

To illustrate this contrast, consider two functions $\psi, \varphi \colon X \to \mathbb{R}$. These functions are equal ν-almost everywhere if and only if they are in fact equal everywhere, but they are equal δ_x-almost everywhere if and only if $\psi(x) = \varphi(x)$, which is a very different condition.

The notion of "almost everywhere" for the counting measure and for point measures is quite straightforward (albeit very different between the two examples). We will see a more sophisticated incarnation of this notion in the next section, when we discuss Lebesgue measure.

b. Lebesgue measure and outer measures. We now introduce a more complicated example of a measure space—Lebesgue measure on $\mathbb{R}^d$, which we first mentioned in the proof of Theorem 2.20. In the case $d = 1$, which we will consider first, this generalises the idea of "length" to apply to a broader class of sets than merely intervals. For $d = 2$, it generalises area; for $d = 3$, it generalises volume; and for $d \geq 4$, it generalises d-dimensional volume.

The full construction of Lebesgue measure is one of the primary parts of measure theory, and a complete treatment of all the details requires most of a graduate-level course, so our discussion here will necessarily be somewhat abbreviated, and we will omit proofs.[1]

In order to construct a measure space $(X, \mathcal{A}, m)$, we must do two things. First, we must produce a σ-algebra $\mathcal{A}$, in which we would like to include as many sets as possible, to make our measure as useful as possible. Second, we must figure out how to define a set function m which satisfies the three properties of a measure. In particular, it is far from clear how to guarantee that whatever function m we construct is σ-additive, especially if the collection $\mathcal{A}$ of sets for which this must be checked is very large.

---

[1] The interested reader is referred to *Measure Theory* by Paul Halmos [**Hal78**], or *Real Analysis* by H. L. Royden [**Roy88**], for a complete exposition.

There is a standard procedure in measure theory, developed by Carathéodory, which deals with both these challenges at once; for this we need the notion of *outer measure*.

Definition 3.6. A set function $m^* : 2^X \to [0, \infty]$ is an *outer measure* if it satisfies

(1) $m^*(\emptyset) = 0$.

(2) $m^*(A_1) \leq m^*(A_2)$ whenever $A_1 \subset A_2$.

(3) $m^* \left(\bigcup_i A_i \right) \leq \sum_i m^*(A_i)$ for any countable collection of sets A_i.

The first two properties are exactly those which we required for a measure; however, the third property of a measure, countable additivity, has been replaced here with the weaker property of countable *subadditivity*. In that regard, the notion of outer measure is weaker than the notion of measure; however, we require the outer measure to be defined for *every* subset of X.

In fact, we have already seen an example of an outer measure; the three properties above are exactly what we proved in Proposition 2.1 for the set function $m(\cdot, \alpha)$. Thus Proposition 2.1 may be rephrased as the statement that $m(\cdot, \alpha)$ is an outer measure on $\mathbb{R}^d$; from now on, we will write it as $m_H(\cdot, \alpha)$ so as to stress the origin of this particular outer measure, and to avoid confusion with our notation m for a measure.

Once we have an outer measure, there is a canonical way to produce both a σ-algebra $\mathcal{A}$ and a measure m. The key step is the following definition.

Definition 3.7. Given an outer measure $m^* : 2^X \to [0, \infty]$, we say that $E \subset X$ is *measurable* if for every $A \subset X$, we have

$$(3.1) \qquad m^*(A) = m^*(A \cap E) + m^*(A \cap (X \setminus E)).$$

As we need to gain σ-additivity in order to have a measure, this is a reasonable definition to make; after all, (3.1) is just a very particular case of finite additivity. Indeed, it turns out to be enough; this justifies our reuse of the word "measurable", which we earlier reserved for elements of the σ-algebra on which a measure is defined.

Theorem 3.8. *Let $\mathcal{A}$ be the collection of measurable subsets of X for an outer measure m^*, and let m be the restriction of m^* to $\mathcal{A}$; that is, $m(A) = m^*(A)$ for $A \in \mathcal{A}$. Then*

(1) $\mathcal{A}$ is a σ-algebra.

(2) m is a measure.

The procedure of passing from an outer measure to a measure and a σ-algebra of measurable sets is a completely general one, which works for any (X, m^*).

Restricting our attention now to the case $X = \mathbb{R}$, we follow the above recipe by first defining an outer measure on $\mathbb{R}$ as follows:

$$(3.2) \qquad m^*(A) = \inf\left\{ \sum_k \ell(I_k) \,\Big|\, \bigcup_k I_k \supset A, \ I_k \in \mathcal{C} \text{ for all } k \right\},$$

where $\mathcal{C}$ is the class of all open intervals and $\ell \colon \mathcal{C} \to [0, \infty]$ is the set function $\ell(I) = \operatorname{diam}(I)$ (which we denoted earlier by $|I|$). This is a particular example of a *Carathéodory construction*, an important procedure for building an outer measure, and hence a measure, using a set function on some family of subsets of X; we will examine this procedure in more detail in the next lecture.

Observe that (3.2) is very reminiscent of (2.3), which defined $m_H(A, 1)$. The only difference between the two is that the latter requires that we consider covers whose diameter becomes arbitrarily small, whereas m^* is not concerned with the size of the elements of the cover. In fact, this makes no difference to the actual value of the set function.

Proposition 3.9. $m^*(A) = m_H(A, 1)$ *for every $A \subset \mathbb{R}$.*

Proof. It follows from the definitions that $m^*(A) \leq m_H(A, 1, \varepsilon)$ for every $\varepsilon > 0$, since the infimum in the latter is taken over a smaller collection of covers, and hence $m^*(A) \leq m_H(A, 1)$.

To prove the reverse inequality, fix $\gamma > 0$, and observe that if $m_H(A, 1) < \infty$, then there exists $\varepsilon > 0$ such that

$$m_H(A, 1) \leq m_H(A, 1, \varepsilon) + \gamma,$$

and that there exists a cover $\{I_k\}$ of A by open intervals such that

$$m^*(A) \geq \sum_k \ell(I_k) - \gamma.$$

If the interval I_k has length greater than ε, we can cover it with intervals $J_{k,i}$ of length less than ε in such a way that

$$\sum_i \ell(J_{k,i}) \leq \ell(I_k) + \frac{\gamma}{2^k}.$$

It follows that

$$m^*(A) \geq \sum_{k,i} \ell(J_{k,i}) - 2\gamma \geq m_H(A, 1, \varepsilon) - 2\gamma \geq m_H(A, 1) - 3\gamma,$$

and since $\gamma > 0$ was arbitrary, the result follows. The case $m_H(A, 1) = \infty$ is similar. $\qquad\qquad\square$

Corollary 3.10. *m^* is an outer measure.*

It follows that m^* defines a σ-algebra $\mathcal{A} \subset 2^{\mathbb{R}}$ and a measure Leb: $\mathcal{A} \to [0, \infty]$; this is one-dimensional Lebesgue measure. One may check that every interval I (whether open, closed, or neither) is Lebesgue measurable and that $m(I) = \ell(I)$, so that m really is an extension of length.

Indeed, most of the sets we usually encounter are measurable; $\mathcal{A}$ contains all open sets, all closed sets, and all countable sets, along with a great deal more. Thus the limit set of any Moran construction is measurable, being closed.

Exercise 3.1. Show that any set $E \subset \mathbb{R}$ with $\dim_H E < 1$ has Lebesgue measure zero, and deduce that Lebesgue-a.e. real number is irrational.

Using Theorem 2.21 and Exercise 3.1, we see that limit sets of Moran constructions have Lebesgue measure zero. If we construct the limit set as the repeller of a map f, this means that Lebesgue-a.e. point x has a trajectory which eventually leaves the domain of f.

Lebesgue measure can be defined not just on the real line, but on any $\mathbb{R}^d$. Simply replace the open intervals in the above procedure by open balls, and the length function ℓ with d-dimensional volume. One may once again prove that $m^*(A) = m_H(A, d)$, although the argument is slightly more difficult. It then follows that m^* is an outer measure, and the resulting measure is d-dimensional Lebesgue measure, which agrees with the usual idea of d-dimensional volume for familar geometric shapes such as balls and cubes.

c. Hausdorff measures. The fact that $m_H(\cdot, \alpha)$ is an outer measure is not restricted to integer values of α, but holds for any $\alpha > 0$. We call this the *Hausdorff outer measure*, and the measure it induces is called *Hausdorff measure*. In this manner, we obtain quite a large collection of measures sitting on $\mathbb{R}^d$; in fact, a one-parameter family of them, indexed by α.

Consider the middle-third Cantor set $C \subset [0,1]$. What is its Hausdorff measure? We know from the definition of Hausdorff dimension that

$$(3.3) \qquad m_H(C, \alpha) = \begin{cases} \infty & \text{if } \alpha < \dim_H C, \\ 0 & \text{if } \alpha > \dim_H C; \end{cases}$$

the result $\mathrm{Leb}(C) = m_H(C, 1) = 0$ is just a particular case of this more general fact. It follows from Moran's equation (2.14) that the critical value is $\alpha_C = \log 2 / \log 3$.

Exercise 3.2. Complete the proof of Proposition 2.6 by proving that $m_H(C, \alpha_C) = 1$.

The result of Exercise 3.2 is valid for any limit set of a Moran construction. The key fact is the relationship (2.14) between the ratio coefficients, which lets us move between different levels of the construction without changing the potential of a cover by basic intervals. Like Goldilocks, we find that other values of α are either too big or too small for $m_H(C, \alpha)$ to measure C properly, thanks to (3.3), but that α_C is "just right". For this particular choice of α, we get a Hausdorff measure which sees the size of the Cantor set as non-zero and finite, and which will be of great utility to us later on.

In the meantime, one final property of the Hausdorff measures is worth noting. Both the Hausdorff measures and the outer measures they are induced from are *translation invariant*; that is, $m_H(A, \alpha) = m_H(A + \mathbf{v}, \alpha)$ for every $A \subset \mathbb{R}^d$, $\mathbf{v} \in \mathbb{R}^d$, and $\alpha > 0$, where $A + \mathbf{v}$ is the image of A under a translation by the vector $\mathbf{v}$. In particular, this is true for Lebesgue measure. Note that for example, the point measure δ_x defined in the previous lecture is not translation invariant.

Lecture 14

a. Choosing a "good" outer measure. The Carathéodory construction of an outer measure described in (3.2) is actually quite a common one. In the construction of the Lebesgue measure, we took $\mathcal{C}$ to be the class of open intervals and ℓ to be the length function $\ell((a,b)) = b - a$. What would happen if we chose a different set function ℓ? We can still define m^*, determine which sets are measurable, and obtain a set function m—but will it be a measure?

We examine the possible outcomes by considering three candidate set functions on $\mathcal{C}$:

$$\ell_1((a,b)) = e^{b-a},$$
$$\ell_2((a,b)) = (b-a)^2,$$
$$\ell_3((a,b)) = \sqrt{b-a}.$$

The proof that the function m^* defined in (3.2) is monotonic and σ-subadditive does not rely on the particular form of ℓ, and so these properties hold whatever set function we begin with. However, in order to have $m^*(\emptyset) = 0$, we must be able to find sets $U \in \mathcal{C}$ for which $\ell(U)$ is arbitrarily small; or if $\emptyset \in \mathcal{C}$ (which it is in this case, since $(a,a) = \emptyset$), we must have $\ell(\emptyset) = 0$. Hence if we take ℓ_1 as our set function, m^* will not be an outer measure.

Thus we must demand that either $\ell(U)$ take arbitrarily small values on $\mathcal{C}$ or $\ell(\emptyset) = 0$; both ℓ_2 and ℓ_3 satisfy this requirement. It follows that these set functions will define outer measures m_2^* and m_3^*, respectively.

Exercise 3.3. Show that $m_2^*(A) = 0$ for every $A \subset \mathbb{R}$; in particular, show that every subset of X is measurable, but $m_2^*(A) \neq \ell(A)$ for $A \in \mathcal{C}$.

Given that the whole point of introducing an outer measure using ℓ was to extend the definition of ℓ beyond the elements of $\mathcal{C}$, the result of Exercise 3.3 is rather undesirable; m_2^* does not agree with ℓ on the class $\mathcal{C}$ of intervals!

To avoid this behaviour, we consider only set functions ℓ such that the outer measure they induce agrees with ℓ on the members of $\mathcal{C}$. It is not difficult to show that ℓ_3 has this property; however, there is now another problem. In order for the result of the Carathéodory construction to be a genuine extension of the initial set function, the σ-algebra of measurable sets should contain the collection $\mathcal{C}$ of intervals, but this is not the case for ℓ_3. Indeed, given $a < c < b$, one immediately sees that

$$m_3^*((a,b)) = \sqrt{b-a} \neq \sqrt{c-a} + \sqrt{b-c} = m_3^*((a,c]) + m_3^*((c,b)),$$

and so intervals are non-measurable!

The upshot of all this is that if we want the measure induced by ℓ to be "sensible"—that is, if we want it to agree with ℓ on intervals, and if we want intervals to be measurable—then we must choose ℓ more carefully. The following turns out to suffice: expand $\mathcal{C}$ to include *all* intervals, whether open, closed, or half-closed and half-open, and require that ℓ be countably additive on $\mathcal{C}$. That is, if $I_1, I_2, \ldots$ are disjoint intervals such that $\bigcup_i I_i = I$ is also an interval, then

$$\ell(I) = \sum_i \ell(I_i).$$

It may be shown (although we do not do so) that if ℓ has this property, then it induces an outer measure m^*, and hence a measure m on a σ-algebra $\mathcal{A}$, such that $\mathcal{C} \subset \mathcal{A}$ and $m(I) = \ell(I)$ for all $I \in \mathcal{C}$.

However, the requirement that ℓ be countably additive is a very restrictive condition; indeed, if we include the further requirement that ℓ be translation invariant, the only possibility for ℓ is a constant multiple of the length function, leading to Lebesgue measure or a scalar multiple thereof.

b. Bernoulli measures on symbolic space. We now move to another context and discuss measures on the symbolic space Σ_k^+. As before, we follow the Carathéodory approach; first we define a set function ℓ on a collection $\mathcal{C}$ of relatively simple sets, then we use ℓ

to define an outer measure m^* and a measure m as in the previous section. But what should our set $\mathcal{C}$ be?

In the previous section, we considered the set of intervals in $\mathbb{R}$; these sets arise naturally from the metric structure of $\mathbb{R}$ as the balls of various radii. The analogous sets in Σ_k^+ are the cylinder sets, and indeed we can define a measure m on Σ_k^+ by specifying a countably additive set function ℓ on the collection $\mathcal{C}$ of cylinders given by (1.19).

To define a countably additive set function ℓ on cylinders, we first choose non-negative numbers $p_1, \ldots, p_k \geq 0$ with the property that $p_1 + \cdots + p_k = 1$. Then we let ℓ be given by

$$(3.4) \qquad m^*(C_{w_1 \ldots w_n}) = p_{w_1} p_{w_2} \cdots p_{w_n}.$$

Example 3.11. If $k = 2$, the condition on the numbers p_i reduces to $p_1 + p_2 = 1$, and we often write $p = p_1$, $q = p_2 = 1 - p$. Then

$$(3.5) \qquad m^*(C_{w_1 \ldots w_n}) = p^{a_n(w)} q^{n - a_n(w)},$$

where $a_n(w)$ is the number of times the symbol 1 appears in the sequence $w_1, \ldots, w_n$.

If we repeatedly toss a weighted coin such that the probability of heads appearing on any given toss is p, and the probability of tails is q, then $\ell(C_{w_1 \ldots w_n})$ is the probability that the first n tosses give the result $w_1, \ldots, w_n$, where $w_j = 1$ denotes heads and $w_j = 2$ denotes tails.

Now we want to build a measure m on Σ_k^+ from the set function ℓ. In light of the discussion in the previous section, we hope to obtain a measure for which cylinders are measurable, and which agrees with the set function on cylinders. As a first step, we show that m^* is additive on the class of cylinders; that is, if $C_1, \ldots, C_n$ are cylinders whose union $C_1 \cup \cdots \cup C_n$ is also a cylinder, then

$$(3.6) \qquad \ell(C_1 \cup \cdots \cup C_n) = \ell(C_1) + \cdots + \ell(C_n).$$

This in turn follows from the formula

$$(3.7) \qquad \ell(C_{w_1 \ldots w_n}) = \ell(C_{w_1 \ldots w_n 1}) + \cdots + \ell(C_{w_1 \ldots w_n k}),$$

which is a direct consequence of (3.4). The proof that (3.7) implies (3.6) is exactly the same argument that we used in the proof of Moran's theorem, with $p_1, \ldots, p_k$ taking the place of $\lambda_1^t, \ldots, \lambda_k^t$.

The set function $\ell\colon \mathcal{C} \to \mathbb{R}$ defines an outer measure m^* on Σ_k^+ as in (3.2), which in turn defines a σ-algebra $\mathcal{A}$ and a measure m by Theorem 3.8. Using the additivity of ℓ, it can be shown (although we do not do so), that the considerations discussed at the beginning of this lecture are satisfied: cylinders are measurable and m agrees with ℓ on cylinders.

The measure m which is built from the set function ℓ in (3.4) is called a *Bernoulli measure*. In fact, there are many such measures, since we may choose any positive parameters $p_1, \ldots, p_k$ which sum to 1. Such a set of parameters is called a *probability vector*, and we will often write $\mathbf{p} = (p_1, \ldots, p_k)$.

Bernoulli measures all have the following interesting property:

Proposition 3.12. *Let m be a Bernoulli measure on Σ_k^+, and let $C_{w_1\ldots w_n}$ be any n-cylinder. Then*

$$(3.8) \qquad m(\sigma^{-1}(C_{w_1\ldots w_n})) = m(C_{w_1\ldots w_n}).$$

Proof. Recall that the preimage of the cylinder $C_{w_1\ldots w_n}$ is

$$\sigma^{-1}(C_{w_1\ldots w_n}) = \{w \in \Sigma_k^+ \mid \sigma(w) \in C_{w_1\ldots w_n}\}$$
$$= C_{1w_1\ldots w_n} \cup \cdots \cup C_{kw_1\ldots w_n}.$$

Then it follows from (3.4) that

$$m(\sigma^{-1}(C_{w_1\ldots w_n})) = p_1 m(C_{w_1\ldots w_n}) + \cdots + p_k m(C_{w_1\ldots w_n})$$
$$= (p_1 + \cdots + p_k) m(C_{w_1\ldots w_n})$$
$$= m(C_{w_1\ldots w_n}). \qquad \square$$

It follows from Proposition 3.12 that $m(\sigma^{-1}E) = m(E)$ for *any* measurable set $E \subset \Sigma_k^+$. We say that m is *invariant* with respect to σ; we may also call such a measure *shift-invariant*.

c. Measures on Cantor sets. Let C be a Cantor set with coding map $h\colon \Sigma_k^+ \to C$, and recall the conjugacy between the shift σ and the map $f\colon C \to C$, which is illustrated in the commutative diagram (1.11). We may use this conjugacy to transfer a measure m on Σ_k^+ to a measure μ on C, as follows.

The measurable subsets of C will be precisely those sets A whose preimages $h^{-1}(A)$ are measurable; we may define the measure of such sets as

$$(3.9) \qquad \mu(A) = m(h^{-1}(A)).$$

After having built one finite measure on C, the Hausdorff measure $m_H(\cdot, t)$, we now have a whole family of finite measures on C which come from the Bernoulli measures on Σ_k^+. The Hausdorff measure corresponds to the particular choice $p_i = \lambda_i^t$; all the other measures are new.

All of these measures are f-invariant; this follows since

$$\mu(f^{-1}(A)) = m(h^{-1}f^{-1}(A)) = m(\sigma^{-1}h^{-1}(A)) = m(h^{-1}(A)) = \mu(A).$$

Thus we have one class of f-invariant measures on the Cantor set—the Bernoulli measures. We could consider many other classes of measures as well; the point measures are an easy set of examples. However, as we are interested in studying the dynamics of f, only the invariant measures are really important, and a point measure δ_x is only invariant if x is a fixed point for f. A more interesting set of examples comes from tampering a little more with the imaginary coins (or dice) we used to define Bernoulli measure, and allowing them to have some memory....

d. Markov measures. The coin flips described in Example 3.11 are independent events, as the outcome of a given flip does not depend on the outcomes which came before it. But what if this was not the case? What if the process had some sort of memory, so that the future could depend on the past?

Example 3.13. Suppose that, as in Example 3.11, we toss a coin repeatedly and record $w_j = 1$ for heads and $w_j = 2$ for tails. We use a fair coin; however, we tamper with the procedure a little bit, as follows. If, immediately after we record a result, we flip the coin and get a result which is the same as what we just wrote down, then we flip the coin again and do not record that result (let's say we just don't like getting the same thing twice in a row). In this case, we record whatever happens on the second toss, regardless of whether it agrees or disagrees with the previous result.

What effect does this have on the probabilities? If $w_n = 1$, so that the last result was heads, then only way that we will record $w_{n+1} = 1$, another heads, is if we actually flip heads *twice* in a row, and the coin is stubborn enough to override our bias towards change. Thus w_{n+1} will be 1 with probability 1/4, and 2 with probability 3/4. If, on the other hand, $w_n = 2$, then the probabilities are reversed. The system remembers what happened just before, and this affects the probabilities.

A process such as this, which exhibits finite memory is known as a *Markov process*. Just as a sequence of independent events (a Bernoulli process) is modeled by a Bernoulli measure, a Markov process is modeled by a *Markov measure* on Σ_k^+. To construct such a measure, we fix a probability vector $\boldsymbol{\pi} = (\pi_1, \ldots, \pi_k)$ and a $k \times k$ *stochastic matrix*; that is, a matrix $P = (p_{ij})$ with non-negative entries such that for every i,

$$(3.10) \qquad \sum_{j=1}^{k} p_{ij} = 1.$$

We also require that the probability vector $\boldsymbol{\pi}$ be *stationary*; that is, that $\pi P = \pi$, so that π is a left eigenvector of P with eigenvalue 1. In terms of the entries, this amounts to

$$(3.11) \qquad \sum_{j=1}^{k} \pi_i p_{ij} = \pi_j$$

for every $i = 1, \ldots, k$.

Now we construct an additive set function on cylinders by

$$(3.12) \qquad \ell(C_{w_1 \ldots w_n}) = \pi_{w_1} p_{w_1 w_2} p_{w_2 w_3} \cdots p_{w_{n-1} w_n},$$

and follow the usual Carathéodory construction to obtain a measure m on Σ_k^+. This is the *Markov measure* associated to the probability vector $\boldsymbol{\pi}$ and the stochastic matrix P.

Exercise 3.4. Use the property (3.10) of a stochastic matrix P to show that the set function defined in (3.12) is additive on cylinders.

Exercise 3.5. Show that the Markov measure generated by π and P is shift-invariant.

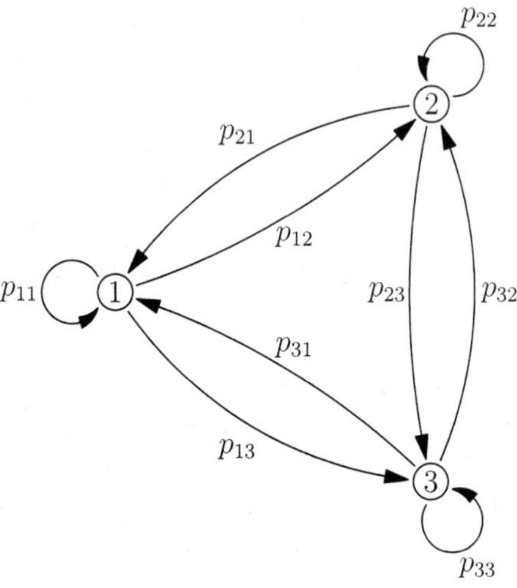

Figure 3.1. A Markov measure on Σ_3^+.

One interpretation of (3.12) is as follows. Consider a process which may give any one of k results at a given time $n \in \mathbb{N}$, and write w_n for the result observed. Then each sequence $w \in \Sigma_k^+$ represents a particular instance of the process, in which the results $w_1, w_2, \ldots$ were observed. As was the case with Bernoulli measures, the Markov measure $m(C_{w_1 \ldots w_n})$ of a cylinder gives the probability that the first n results are $w_1, \ldots, w_n$.

Alternatively, we may think of w_n as representing the state of a system at time n. An element π_i of the probability vector gives the probability of beginning in state i, or indeed of being in state i at an arbitrary time (if we have no other information), and an entry p_{ij} of the stochastic matrix P gives the probability of going from state i to state j at any given time step; that is, it gives the conditional probability of being in state j at time $n + 1$, given that the state at time n is i.

We may represent this graphically, as shown in Figure 3.1 for the case $k = 3$. The vertices of the graph represent the states of the system, and to the edge from vertex i to vertex j is associated a *transition probability* p_{ij}. The condition that the matrix P be stochastic may be rephrased as the condition that the outgoing transition probabilities from any given vertex sum to 1; that is, if we are at vertex i now, then at the next time step, we will be *somewhere*. We may be at vertex i again (with probability p_{ii}), but we will not simply vanish from the graph.

In this interpretation, the symbolic space Σ_k^+ may be thought of as the set of all possible paths along edges of the graph, and the measure of a cylinder $C_{w_1\ldots w_n}$ is the probability that a randomly chosen itinerary on the graph begins with $w_1, \ldots, w_n$.

Now what happens if one of the transition probabilities is 0? If $p_{ij} = 0$, then the probability of going from vertex i to vertex j is 0, and so we may as well erase this edge from the graph. What does this do to the set of all possible paths, which was in one-to-one correspondence with symbolic space Σ_k^+?

We will address this question in the next lecture, after introducing another definition from measure theory.

Lecture 15

a. The support of a measure. Fix a point $x \in \mathbb{R}^d$, and consider the point measure δ_x on $\mathbb{R}^d$ defined in Exercise 3.4. If $E \subset \mathbb{R}^d$ does not contain x, then as far as the measure is concerned, the points in E may be neglected without losing anything important, since $\delta_x(E) = 0$. Thus from the measure theoretic point of view, the measure space $(\mathbb{R}^d, \delta_x)$ is the same as the measure space $(\{x\}, \delta_x)$, since the only difference between the two is a set of zero measure (the complement of $\{x\}$).

In the same vein, given any topological space X and a measure m on X, there is a canonical way to decompose X into two parts, one of which may be discarded, since it has measure zero with respect to m, and the other of which carries all the information about the measure, without betraying the topology of X. The latter set is called the *support* of m, and is the smallest closed set of full measure:

$$(3.13) \qquad \operatorname{supp} m = \{x \in X \mid m(E) > 0 \text{ for any } E \ni x, \ E \text{ open}\}.$$

The measure m sits on the support $\operatorname{supp} m$ in the following sense: an open set has positive measure if it intersects the support non-trivially and zero measure otherwise.

Example 3.14. Consider Lebesgue measure Leb on the unit interval $[0,1]$. Given $a < b$, we have $\operatorname{Leb}((a,b)) = b - a > 0$, so open intervals have positive measure. Since open sets are countable unions of open intervals, we see that all open sets have positive measure, hence $\operatorname{supp} \operatorname{Leb} = [0,1]$.

Example 3.15. For the point measure δ_x, an open set has positive measure if and only if it contains x. Thus $\operatorname{supp} \delta_x = \{x\}$.

Example 3.16. Let X be the symbolic space Σ_k^+. If m is the Bernoulli measure given by a probability vector with positive entries, we see from (3.4) that cylinders have positive measure; since any open set is a union of cylinders, it follows that the support of m is the entire space Σ_k^+.

Similarly, if m is a Markov measure for which all entries of both the probability vector and the stochastic matrix are positive, then all cylinders have positive measure by (3.12), and we again have $\operatorname{supp} m = \Sigma_k^+$.

Exercise 3.6. Let m be the Bernoulli measure on Σ_3^+ given by weights (p_1, p_2, p_3), and suppose that $p_1 = 0$. Describe $\operatorname{supp} m$.

Returning to our question from the previous section, we may ask what $\operatorname{supp} m$ looks like if m is a Markov measure for which one or more of the entries p_{ij} are equal to 0. In this case, we see that the measure of a cylinder $C_{w_1 \ldots w_n}$ is positive precisely when all the entries $p_{w_j w_{j+1}}$ are positive for $j = 1, \ldots, n-1$. In particular, if any of the entries in the stochastic matrix vanish, then some cylinders have zero measure. Since cylinders are open, this means that the support of m is *not* the whole space Σ_k^+.

To describe $\operatorname{supp} m$ for an arbitrary Markov measure m, we introduce a $k \times k$ matrix whose entries are either 0 or 1, which keeps track of which entries of P vanish; define a *transition matrix* $A = (a_{ij})$ by

$$a_{ij} = \begin{cases} 0 & p_{ij} = 0, \\ 1 & p_{ij} > 0. \end{cases}$$

A is *primitive* if there exists n such that every entry of A^n is positive; for technical reasons, we will *only* consider primitive transition matrices.

The transition matrix A records which transitions $w_j \to w_{j+1}$ have a non-zero probability of occurring. We say that a sequence $w \in \Sigma_k^+$ is *admissible* if $a_{w_j w_{j+1}} = 1$ for all j; that is, if we only follow edges in the graph which carry a non-zero probability.

Exercise 3.7. Using the fact that A is primitive, show that for every $m \geq n$ and $i, i' = 1, \ldots, k$, there exists an admissible word w with $w_1 = i$ and $w_{m+1} = i'$.

Note that a cylinder has positive measure precisely when it is defined by an admissible sequence, and so we consider the set of admissible sequences

$$(3.14) \qquad \Sigma_A^+ = \{w \in \Sigma_k^+ \mid a_{w_j w_{j+1}} = 1 \text{ for all } j \in \mathbb{N}\}.$$

If $E \subset \Sigma_k^+$ is open and intersects Σ_A^+ non-trivially, there exists a cylinder $C_{w_1 \ldots w_n} \subset E \cap \Sigma_A^+$, and we see that $m(C_{w_1 \ldots w_n}) > 0$ since $p_{w_j w_{j+1}} > 0$ for all $j = 1, \ldots, n - 1$; it follows that $m(E) \geq m(C_{w_1 \ldots w_n}) > 0$, and thus $\operatorname{supp} m = \Sigma_A^+$.

In Lecture 14(c), we saw how to use the coding map h to go from a measure m on symbolic space Σ_k^+ to a measure $\mu = m \circ h^{-1}$ on a Cantor set $C \subset [0, 1]$. If m is a Markov measure whose support is not the whole space, then some cylinders in Σ_k^+ have zero measure; hence their images, which are open sets in C, have zero measure as well.

In particular, we have $\operatorname{supp} \mu \subsetneq C$. We obtain $\operatorname{supp} \mu$ by removing from C precisely those points whose codings are not admissible. Thus at each step of the construction we remove any basic interval which would introduce an inadmissible coding. The result of this *Markov construction* is another Cantor set, which we will denote C_A.

Example 3.17. Let C be a Cantor set modelled on Σ_2^+, and consider a Markov measure μ on C for which $p_{11} = 0$ and for which no other entries of P or π vanish. Then the associated transition matrix is $A = \left(\begin{smallmatrix} 0 & 1 \\ 1 & 1 \end{smallmatrix}\right)$, and the corresponding Markov construction is shown in Figure 3.2. At the second step, I_{11} is erased, since every $x \in I_{11}$ has a coding which begins with $1, 1, \ldots$, and hence is not admissible, so

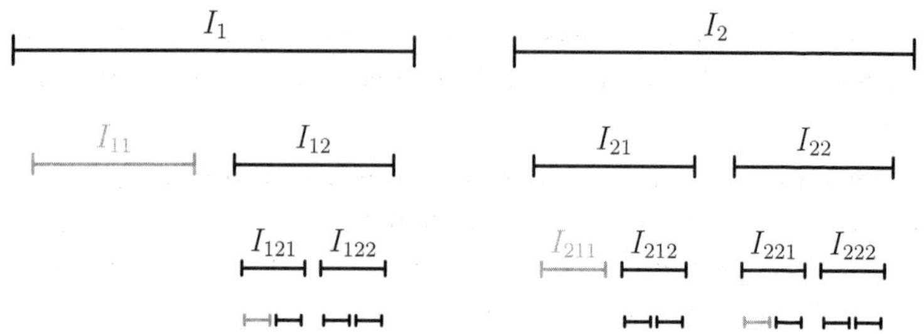

Figure 3.2. A Markov construction.

$x \notin \operatorname{supp} \mu$. Similarly, I_{211} must be erased at the third step, and I_{1211} and I_{2211} are taken away at the fourth.

This gives us another class of constructions to consider: Given a Markov measure μ, we would like to characterise $C_A = \operatorname{supp} \mu$ by determining its Hausdorff dimension. Because we have deleted basic intervals at each step of the construction, we can no longer apply Moran's theorem, and the question becomes much more complicated. We will need a new set of tools, which will be developed in the next chapter.

b. Subshifts of finite type: One-dimensional Markov maps. *A priori*, the Markov constructions described in the previous section seem rather artificial and contrived; it may not be obvious just why such constructions would be important or where they would appear.

In fact, they are tremendously important in dynamics; as a very simple example, consider a piecewise linear map of the interval $[0, 1]$ of the sort shown in Figure 3.3.

This map generalises the sort of map we saw in Figure 1.17; in both cases, we begin with disjoint intervals $I_1, \ldots, I_k$ and define a map $f \colon I_1 \cup \cdots \cup I_k \to [0, 1]$. Before, we demanded that f map each interval I_i homeomorphically to the entire interval $[0, 1]$, whereas now we demand rather less.

Definition 3.18. A *one-dimensional Markov map* is a map $f \colon I_1 \cup \cdots \cup I_k \to [0, 1]$, where the I_j are disjoint intervals in $[0, 1]$, which satisfies the following two conditions.

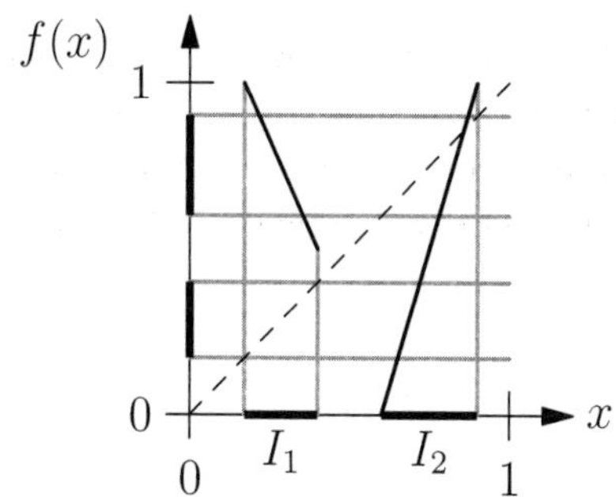

Figure 3.3. A map which is modelled by a subshift.

(1) The restriction of f to each I_i is a homeomorphism onto its image.

(2) Each image $f(I_i)$ contains every I_j it intersects; that is, if i, j are such that $f(I_i) \cap I_j \neq \emptyset$, then $I_j \subset f(I_i)$.

As before, we may consider the set of points whose trajectories remain within the domain of definition of f, and obtain a repelling Cantor set C_A, where the transition matrix A is given by

$$(3.15) \qquad a_{ij} = \begin{cases} 1 & \text{if } f(I_i) \cap I_j \neq \emptyset, \\ 0 & \text{if } f(I_i) \cap I_j = \emptyset. \end{cases}$$

We also have a coding map which conjugates $f \colon C_A \to C_A$ to a symbolic model. In this case, however, the symbolic model is not the entire space Σ_k^+, but rather a subset Σ_A^+. For the map shown in Figure 3.3, we see that points in I_2 can be mapped to either I_1 or I_2, but points in I_1 can only be mapped to I_2. Thus the transition matrix is $A = \left(\begin{smallmatrix} 0 & 1 \\ 1 & 1 \end{smallmatrix} \right)$.

It is obvious that if $w \in \Sigma_A^+$ is an admissible sequence, then its image $\sigma(w)$ under the shift will be admissible as well. Thus $\sigma(\Sigma_A^+) \subset \Sigma_A^+$, and so Σ_A^+ is invariant under the action of σ. We may consider the restriction of the shift σ to the domain Σ_A^+, which is called a *subshift of finite type*; for the sake of clarity, the shift $\sigma \colon \Sigma_k^+ \to \Sigma_k^+$ is often referred to as the *full shift*.

Subshifts of finite type form a very important class of models in dynamics, whose applications stretch far beyond the simple interval maps we have mentioned so far. Together with the Markov measures sitting on them, they provide a powerful tool with which to study questions in dimension theory and to examine the stochastic properties of certain broad classes of dynamical systems, which are much more general than the one-dimensional Markov maps we have studied so far.

Chapter 4

Measures and Dimensions

Lecture 16

a. The Uniform Mass Distribution Principle: Using measures to determine dimension. Now that we have added a little bit of measure theory to our toolkit, we return to the problem of computing the Hausdorff dimension of a repeller C for an arbitrary one-dimensional Markov map f, which may not be linear or full-branched. Thanks to Moran's theorem, we already understand the case where f *is* linear and full-branched. Because this is the simplest case, we will return to it frequently in order to illustrate the techniques we develop, but these techniques are applicable in more general settings.

We begin with the following theorem, which shows that the existence of finite measures with particular scaling properties has consequences for the Hausdorff dimension of sets with positive measure.

Theorem 4.1 (Uniform Mass Distribution Principle). *Suppose that μ is a finite measure on $\mathbb{R}^d$ and that there exist $\alpha, K, \delta > 0$ such that for every ball with radius $r \leq \delta$, we have*

$$\mu(B(x, r)) \leq K r^\alpha. \tag{4.1}$$

Then if $E \subset \mathbb{R}^d$ is measurable and $\mu(E) > 0$, we have $\dim_H E \geq \alpha$.

Proof. For any $0 < \varepsilon < \delta$ and any cover $\mathcal{U} = \{U_i\}$ of E by balls of diameter less than ε, we have

$$\sum_i (\operatorname{diam} U_i)^\alpha \geq \sum_i \frac{\mu(U_i)}{K} \geq \frac{\mu\left(\bigcup_i U_i\right)}{K} \geq \frac{\mu(E)}{K} > 0,$$

and hence $m_H(Z, \alpha) \geq \mu(E)/K > 0$, so $\dim_H E \geq \alpha$. $\qquad\square$

Theorem 4.1 gives a lower bound for the Hausdorff dimension, which is typically the more difficult bound to obtain. Indeed, finding a measure μ which satisfies the hypothesis of the theorem can be quite hard (or even impossible), and so we will soon want to weaken the assumption on μ. First, though, we consider the particular case of a Moran construction in $[0, 1]$ with ratio coefficients $\lambda_1, \ldots, \lambda_k > 0$, $\sum_i \lambda_i < 1$, and describe a measure μ for which the Uniform Mass Distribution Principle may be applied.

As we showed in the proof of Moran's theorem, if $\mathcal{V}(r)$ denotes the collection of basic intervals whose diameter is between $r\lambda_{\min}$ and $r/\lambda_{\min}$, then each ball of radius r intersects at most M' elements of $\mathcal{V}(r)$, where the constant M' is independent of r. In particular, $B(x, r)$ can be covered by at most M' elements of $\mathcal{V}(r)$.

We shall show that (4.1) holds for basic intervals with $\alpha = t$, where t is the root of Moran's equation (2.14). Then for each element I of $\mathcal{V}(r)$, we will have $\mu(I) \leq K(\operatorname{diam} I)^\alpha \leq K'r^\alpha$, where $K' = K/\lambda_{\min}{}^\alpha$, and hence $\mu(B(x, r)) \leq M'K'r^\alpha$.

Now for a Bernoulli measure μ given by a probability vector $(p_1, \ldots, p_k)$, we have

$$\mu(I_{w_1 \ldots w_n}) = p_{w_1} \cdots p_{w_n},$$
$$(\operatorname{diam} I_{w_1 \ldots w_n})^t = \lambda_{w_1}^t \cdots \lambda_{w_n}^t \left(|I_{w_1}|^t / \lambda_{w_1}^t \right).$$

Thus in order to have $\mu(I_{w_1 \ldots w_n})/(\operatorname{diam} I_{w_1 \ldots w_n})^t$ bounded by a constant independent of $(w_1, \ldots, w_n)$, we take $p_i = \lambda_i^t$. This defines a Bernoulli measure μ which satisfies (4.1), and so the Uniform Mass Distribution Principle gives the lower bound $\dim_H C \geq t$, since the set C has measure $\mu(C) = 1 > 0$.

Of course in this case we already knew the answer, but this illustrates the principle of using measures to obtain bounds on dimensional quantities which might otherwise be difficult to gain information about. Certainly the calculations in the proof of Theorem 4.1 are simpler than those involved in computing the potential associated to a cover by basic intervals, and then showing that we can move between various levels of the construction without changing the potential. Furthermore, the present argument shows that *any* subset of C which is given positive measure by the Bernoulli measure μ must have Hausdorff dimension equal to α, which we could not have deduced directly from Moran's theorem.

Exercise 4.1. Let C be the middle-third Cantor set, and compute the Hausdorff dimensions of the following sets:

(a) The "Cantor tartan" $E = \{(x, y) \in \mathbb{R}^2 \mid \text{either } x \in C \text{ or } y \in C\}$.

(b) The set $F = \{(x, y) \in \mathbb{R}^2 \mid x \in C \text{ and } 0 \leq y \leq x^2\}$.

b. Pointwise dimension and the Non-uniform Mass Distribution Principle. Not every set in which we may be interested is so obliging as to admit a measure satisfying the conditions of the Uniform Mass Distribution Principle. That result applies to certain Bernoulli measures on limit sets of Moran constructions, and also to Lebesgue measure on $\mathbb{R}^d$; however, further progress beyond these two rather restrictive cases requires us to introduce a weaker condition on the measure, which still gives us information about the dimension of certain sets.

In the statement of Theorem 4.1, the constants α, K, δ appearing in (4.1) were to be independent of both scale and position; that is, they could not depend on either the centre x or the radius r of the ball $U = B(x, r)$. We now consider a slightly more general condition, in which the constants are still independent of r but may vary between different locations in $\mathbb{R}^d$, i.e., may depend on x.

Definition 4.2. The *pointwise dimension of μ at x* is the limit

$$(4.2) \qquad d_\mu(x) = \lim_{r \to 0} \frac{\log \mu(B(x, r))}{\log r},$$

if the limit exists.

Of course, the limit may not exist at every point x, and so the pointwise dimension may not be defined for all $x \in \mathbb{R}^d$; furthermore, even when it does exist, it may vary from one point to another.

We emphasise that the pointwise dimension is a property of the *measure* μ, rather than of any particular set. This is in contrast to previous dimensional quantities we have seen, which were all properties of sets. Furthermore, it is a local property rather than a global one, being defined in terms of neighbourhoods of a single point.

The existence of the pointwise dimension at a point x leads to the following estimates. For every $\varepsilon > 0$, there exists $\delta > 0$ such that for every $0 < r < \delta$,

$$d_\mu(x) - \varepsilon \leq \frac{\log \mu(B(x,r))}{\log r} \leq d_\mu(x) + \varepsilon.$$

Since $r < 1$, the function $t \mapsto r^t$ is decreasing in t, and so

$$r^{d_\mu(x)-\varepsilon} \geq \mu(B(x,r)) \geq r^{d_\mu(x)+\varepsilon},$$

which gives us bounds reminiscent of (4.1). However, the scale δ for which this estimate holds may vary from point to point, and $d_\mu(x)$ itself may also vary and may not exist everywhere. Despite this, we will eventually see that the pointwise dimension provides a useful tool for gaining information about the dimension of a set by considering an appropriate measure.

Example 4.3. Consider a piecewise linear map $f \colon I_1 \cup I_2 \to [0,1]$ as in Figure 1.12, where $|I_1| = |I_2| = \lambda < 1/2$. The repeller of f is a Cantor set C modelled on Σ_2^+, and both ratio coefficients in the construction of C are equal to λ.

Let m be a Bernoulli measure on Σ_2^+ with probability vector (p, q), where $p, q > 0$, $p + q = 1$, and let μ be the corresponding Bernoulli measure on C. Given $x \in C$, what is the pointwise dimension $d_\mu(x)$?

Say $x = h(w)$; the pointwise dimension is determined by the measure of the balls $B(x,r)$ centred at x, and for appropriate values of r, these are just the basic intervals $I_{w_1 \dots w_n}$. We have $|I_{w_1 \dots w_n}| = \lambda^n$, and so $\log r = n \log \lambda$ (up to some bounded difference which will vanish in the limit). Furthermore, $\mu(I_{w_1 \dots w_n}) = p^{a_n(w)} q^{n - a_n(w)}$, where $a_n(w)$ is the number of times the symbol 1 appears in the string $(w_1, \dots, w_n)$.

In the simplest case, we have $p = q = 1/2$, and so $\mu(I_{w_1 \ldots w_n}) = (1/2)^n$. Thus the pointwise dimension is given by

$$(4.3) \qquad d_\mu(x) = \lim_{n \to \infty} \frac{\log(\mu(I_{w_1 \ldots w_n}))}{n \log \lambda} = \frac{\log 2}{-\log \lambda} = \dim_H C.$$

Observe that in the example above, the Cantor set C is just the support of the measure μ. This suggests some connection between the pointwise dimension of a measure and the Hausdorff dimension of its support. However, the relationship is not always this simple. For most choices of p and q, the pointwise dimension does not exist everywhere, and does not always equal $\dim_H C$ where it does exist. Indeed, consider the points $x = h(1, 1, 1, \ldots)$ and $y = h(2, 2, 2, \ldots)$ in C. An easy calculation shows that $d_\mu(x) = \log p / \log \lambda$ and $d_\mu(y) = \log q / \log \lambda$; thus the pointwise dimension at x does not equal the pointwise dimension at y except in the special case $p = q = 1/2$.

The true relationship between the pointwise dimension of a measure and the Hausdorff dimension of its support is somewhat more subtle. Not only may the pointwise dimension vary from point to point, but there may be points at which the limit in (4.2) does not even exist; thus we consider instead the lower and upper limits, which always exist. These are referred to as the *lower and upper pointwise dimensions* of the measure μ at the point x:

$$\underline{d}_\mu(x) = \varliminf_{r \to 0} \frac{\log \mu(B(x, r))}{\log r},$$

$$\overline{d}_\mu(x) = \varlimsup_{r \to 0} \frac{\log \mu(B(x, r))}{\log r}.$$

We have $\underline{d}_\mu(x) \le \overline{d}_\mu(x)$ for any measure μ and any point x; the two coincide if and only if the limit in (4.2) exists, in which case their common value is the pointwise dimension.

The following result generalises Theorem 4.1:

Theorem 4.4 (Non-uniform Mass Distribution Principle). *Suppose that μ is a finite measure on $\mathbb{R}^d$, that $E \subset \mathbb{R}^d$ has positive measure ($\mu(E) > 0$), and that there exists $\alpha > 0$ such that*

$$(4.4) \qquad\qquad\qquad \underline{d}_\mu(x) \ge \alpha$$

for μ-almost every $x \in E$. Then $\dim_H E \ge \alpha$.

Proof. Given $\varepsilon > 0$, we show that $m_H(E, \alpha - \varepsilon) > 0$, as follows. For almost every point $x \in E$, we have

$$\alpha \leq \underline{d}_\mu(x) = \lim_{r \to 0} \frac{\log \mu(B(x,r))}{\log r},$$

and so there exists $\delta > 0$ such that for all $0 < r < \delta$,

$$\alpha - \varepsilon \leq \frac{\log \mu(B(x,r))}{\log r},$$

which leads to the inequality

$$(4.5) \qquad\qquad r^{\alpha - \varepsilon} \geq \mu(B(x,r)).$$

We would like to argue that by picking δ small enough, we can use (4.5) to proceed exactly as in the proof of Theorem 4.1; however, since δ may depend on x, we must be slightly more careful.

Given $n \in \mathbb{N}$, let E_n be the set of points $x \in E$ such that (4.5) holds for all $0 < r < 1/n$. Because (4.4) holds for almost every $x \in C$, we have $\mu\left(\bigcup_{n \geq 1} E_n\right) = 1$, and so there exists n such that $\mu(E_n) > 0$.

It suffices to show that $m_H(E, \alpha - \varepsilon) \geq m_H(E_n, \alpha - \varepsilon) > 0$. Indeed, for any open cover $\{B(x_i, r_i)\}$ of E_n by balls of radius $r_i \leq \delta_n$, we have

$$\sum_i \operatorname{diam}(B(x_i, r_i))^{\alpha - \varepsilon} = 2^{\alpha - \varepsilon} \sum_i r_i^{\alpha - \varepsilon} \geq 2^{\alpha - \varepsilon} \sum_i \mu(B(x_i, r_i))$$

$$\geq 2^{\alpha - \varepsilon} \mu\left(\bigcup_i B(x_i, r_i)\right) \geq 2^{\alpha - \varepsilon} \mu(E_n).$$

It follows that $m_H(E_n, \alpha - \varepsilon) \geq 2^{\alpha - \varepsilon} \mu(E_n) > 0$, and so $\dim_H E_n \geq \alpha - \varepsilon$. By the monotonicity property of Hausdorff dimension, this implies that $\dim_H E \geq \alpha - \varepsilon$, and since $\varepsilon > 0$ is arbitrary, we have $\dim_H E \geq \alpha$. $\qquad\square$

Lecture 17

a. Variable pointwise dimension. We must emphasise the crucial fact that the Non-uniform Mass Distribution Principle does not require the bound (4.4) to hold at points x which lie outside of E (the set we are interested in), but only at points in E (even then, only at μ-almost every point, and it even suffices to have it only on a set of positive μ-measure). This is vital to applications, because the pointwise dimension of a measure typically varies from point to point, and so we may well be interested in cases in which the bound fails on a fairly large set outside of E.

To illustrate this phenomenon, let us return to the setting of Example 4.3, a Cantor repeller C in the interval for a piecewise linear map f as in Figure 1.12, where the two basic intervals I_1 and I_2 have equal length λ. Consider the Bernoulli measure μ on C which is generated by the probability vector (p, q), so that $\mu(I_1) = p$, $\mu(I_2) = q$. We want to compute the pointwise dimension of μ at a point $x \in C$, and relate this to the Hausdorff dimension of C, which as we already know from Moran's theorem, is $\dim_H C = -\log 2/\log \lambda$.

To begin with, recall that we have a one-to-one correspondence between elements of the Cantor set C and elements of the symbolic space Σ_2^+, and so we consider the sequence $w = h^{-1}(x)$ which gives the coding of the point x.

In order to compute the ratio in (4.2), we may replace $B(x, r)$ with $I_{w_1 \ldots w_n}$, the basic interval containing x, and r with $|I_{w_1 \ldots w_n}| = \lambda^n$, since the error term between the two ratios vanishes in the limit. From the definition of a Bernoulli measure, we have $\mu(I_{w_1 \ldots w_n}) = p^{a_n(w)} q^{n-a_n(w)}$, where $a_n(w)$ is the number of times the symbol 1 appears in the sequence $w_1, \ldots, w_n$.

Proceeding naïvely and ignoring any questions regarding existence of limits, we see that

$$
\begin{aligned}
d_\mu(x) &= \lim_{n \to \infty} \frac{\log \mu(I_{w_1 \ldots w_n})}{\log |I_{w_1 \ldots w_n}|} = \lim_{n \to \infty} \frac{\log(p^{a_n(w)} q^{n-a_n(w)})}{\log \lambda^n} \\
&= \lim_{n \to \infty} \frac{a_n(w) \log p + (n - a_n(w)) \log q}{n \log \lambda} \\
&= \frac{\left(\lim_{n \to \infty} \frac{a_n(w)}{n}\right) \log p + \left(1 - \lim_{n \to \infty} \frac{a_n(w)}{n}\right) \log q}{\log \lambda}.
\end{aligned}
$$

$$(4.6)$$

Thus everything hinges on the value (and existence) of the limit

$$(4.7) \qquad \alpha(x) = \lim_{n \to \infty} \frac{a_n(h^{-1}(x))}{n}.$$

The ratio $a_n(w)/n$ is simply the proportion of ones in the first n entries of w, and so the limit (if it exists) is the asymptotic frequency of ones. But what is this value, and when does it exist?

Consider first the simple case $p = q = 1/2$. Then we may think of μ as giving the probabilities of particular sequences of outcomes of a random process, where at each step, we have an equal probability of choosing a 1 or a 2 for the sequence, as if we were flipping a fair coin. Intuitively, we expect such a process to yield an approximately equal number of ones and twos in the long run, and so we expect to see the ratio $a_n(w)/n$ converge to $1/2$ as n goes to infinity.

Of course, we can construct particular sequences for which this is not the case; for $w = (1, 1, 1, \dots)$, we have $a_n(w) = n$ for all n, and so $\alpha(h(w)) = 1$. Similarly, $w = (2, 2, 2, \dots)$ gives a limiting value of 0, and in fact, given any $\alpha_0 \in [0, 1]$, it is not hard to construct an $x \in C$ for which $\alpha(x) = \alpha_0$; there are also many points for which the limit does not exist. So we can only expect the limit to be equal to $1/2$ for certain "good" points x, which we hope are in some way typical. But typical in what sense?

Let G be the set of "good" points, and B be the set of "bad" points; that is, $G = \{x \in C \mid \alpha(x) = 1/2\}$ and $B = C \setminus G$. We just saw that B is non-empty; indeed, it is dense in C. To see this, consider a point $x \in C$ with periodic coding $(w_1, w_2, \dots, w_N, w_1, w_2, \dots)$; then w has an asymptotic frequency of ones which is equal to $a_N(w)/N$. For most such points, this is not equal to $1/2$, and such points are dense in C. Thus topologically, B is fairly large.

From a measure-theoretic point of view, however, B is negligible; one can show (see Proposition 4.5 below) that $\alpha(x)$ exists and is equal to $1/2$ μ-almost everywhere, and so B is a null set.

Now we can complete the calculation in (4.6) for the case $p = q = 1/2$ to obtain

$$d_\mu(x) = \frac{\log 2}{-\log \lambda}$$

for μ-almost every $x \in C$. In and of itself, this is not very helpful; after all, we already saw in (4.3) that this is true for *every* $x \in C$ and that the common value is $\dim_H C$. However, that result did not give us any information on $d_\mu(x)$ for other values of p and q, while (4.6) does, using the following result, which corresponds to the *strong law of large numbers* from probability theory.

Proposition 4.5. *Let μ be the Bernoulli measure on C with probability vector (p_1, p_2). Then for μ-a.e. x,*

$$\alpha(x) = \lim_{n \to \infty} \frac{a_n(h^{-1}(x))}{n} = p_1.$$

Proof. For each $\varepsilon > 0$ and $n \in \mathbb{N}$, consider the set

$$Z_{\varepsilon,n} = \left\{ x \in C \;\middle|\; \left| \frac{a_n(h^{-1}(x))}{n} - p_1 \right| \geq \varepsilon \right\}.$$

The set B of "bad" points in C is the set of all points x for which $a_n(h^{-1}(x))/n \not\to p_1$; thus $x \in C$ is in B if and only if there exists $\varepsilon > 0$ such that for every $N \in \mathbb{N}$, there exists $n \geq N$ such that $x \in Z_{\varepsilon,n}$. That is,

$$B = \bigcup_{\varepsilon > 0} \bigcap_{N \in \mathbb{N}} \bigcup_{n \geq N} Z_{\varepsilon,n}.$$

We want to show that the set of "bad" points is a null set, that is, $\mu(B) = 0$. To this end, we first estimate the measure of $Z_{\varepsilon,n}$.

$Z_{\varepsilon,n}$ consists of all the points $x = h(w)$ for which

$$\left| \frac{a_n(w)}{n} - p_1 \right| \geq \varepsilon.$$

Since this inequality depends only on the first n symbols in w, we observe that if $x = h(w) \in Z_{\varepsilon,n}$, then $I_{w_1 \ldots w_n} \subset Z_{\varepsilon,n}$. Let $\mathcal{W}_{\varepsilon,n}$ denote the set of all n-tuples $w_1 \ldots w_n$ such that $I_{w_1 \ldots w_n} \subset Z_{\varepsilon,n}$. Then we have the following computation:

$$\mu(Z_{\varepsilon,n}) = \sum_{w_1 \ldots w_n \in \mathcal{W}_{\varepsilon,n}} \mu(I_{w_1 \ldots w_n})$$

$$= \sum_{w_1 \ldots w_n \in \mathcal{W}_{\varepsilon,n}} p_{w_1} \cdots p_{w_n}$$

$$\leq \frac{1}{\varepsilon^2} \sum_{w_1 \ldots w_n \in \mathcal{W}_{\varepsilon,n}} \left(\frac{a_n(w)}{n} - p_1 \right)^2 p_{w_1} \cdots p_{w_n}$$

$$\leq \frac{1}{n^2 \varepsilon^2} \sum_{w_1 \ldots w_n} (a_n(w) - np_1)^2 p_{w_1} \cdots p_{w_n},$$

where the final sum is taken over all n-tuples $w_1 \ldots w_n$, and where for each n-tuple we choose an arbitrary $w \in I_{w_1 \ldots w_n}$ to evaluate $a_n(w)$. Observe that because a_n only depends on the first n symbols in w, it does not matter which w we choose.

We proceed by expanding the square to get

$$(a_n(w) - np_1)^2 = a_n(w)^2 - 2np_1 a_n(w) + n^2 p_1^2$$
$$= A_1(w) + A_2(w) + A_3(w),$$

where $A_1(w) = a_n(w)^2$, $A_2(w) = -2np_1 a_n(w)$, and $A_3(w) = n^2 p_1^2$. Thus the inequality becomes

$$\mu(Z_{\varepsilon,n}) \leq \frac{1}{n^2 \varepsilon^2} \sum_{w_1 \ldots w_n} (A_1(w) + A_2(w) + A_3(w)) p_{w_1} \cdots p_{w_n},$$

and it remains to evaluate the sums $S_i = \sum_{w_1 \ldots w_n} A_i(w) p_{w_1} \cdots p_{w_n}$ for $i = 1, 2, 3$. We compute each of these separately. The third is the easiest:

$$S_3 = \sum_{w_1 \ldots w_n} n^2 p_1^2 p_{w_1} \cdots p_{w_n} = n^2 p_1^2,$$

using the fact that $p_1 + p_2 = 1$. To compute S_1 and S_2, we write $\delta(w_j) = 1$ if $w_j = 1$ and $\delta(w_j) = 0$ otherwise, and observe that $a_n(w) = \delta(w_1) + \cdots + \delta(w_n)$. Thus

$$S_2 = \sum_{w_1 \ldots w_n} -2np_1 a_n(w) p_{w_1} \cdots p_{w_n}$$
$$= -2np_1 \sum_{w_1 \ldots w_n} \sum_{j=1}^{n} \delta(w_j) p_{w_1} \cdots p_{w_n}$$
$$= -2np_1 \sum_{j=1}^{n} \sum_{w_j} \delta(w_j) p_{w_1}$$
$$= -2n^2 p_1^2.$$

Finally, we have $\delta(w_j)^2 = \delta(w_j)$, and so

$$a_n(w)^2 = \left(\sum_{j=1}^{n} \delta(w_j) \right) + \left(\sum_{i \neq j} \delta(w_i) \delta(w_j) \right).$$

It follows that

$$S_1 = \sum_{w_1 \ldots w_n} a_n(w)^2 p_{w_1} \cdots p_{w_n}$$

$$= \sum_{w_1 \ldots w_n} \left[\left(\sum_{j=1}^{n} \delta(w_j) \right) + \left(\sum_{i \neq j} \delta(w_i)\delta(w_j) \right) \right] p_{w_1} \cdots p_{w_n}$$

$$= \left(\sum_{j=1}^{n} \sum_{w_j} \delta(w_j) \right) + \left(\sum_{i \neq j} \sum_{w_i w_j} \delta(w_i)\delta(w_j) p_{w_i} p_{w_j} \right)$$

$$= np_1 + n(n-1)p_1^2.$$

Putting it all together, we have

$$\mu(Z_{\varepsilon,n}) \leq \frac{1}{n^2 \varepsilon^2}(S_1 + S_2 + S_3)$$

$$\text{(4.8)} \qquad = \frac{1}{n^2 \varepsilon^2}\left(np_1 + n(n-1)p_1^2 - 2n^2 p_1^2 + n^2 p_1^2 \right)$$

$$= \frac{p_1 - p_1^2}{n\varepsilon^2}.$$

We now use this estimate to show that $\mu(B) = 0$. Naïvely, we may try to estimate $\mu\left(\bigcup_{n \geq N} Z_{\varepsilon,n} \right)$ by summing the estimate (4.8) over all $n \geq N$; however, this sum diverges, and so this is of no use to us.

The way forward is to observe that since the ratio $a_n(w)/n$ cannot vary too quickly as n changes, the sets $Z_{\varepsilon,n}$ for nearby values of n may be compared to each other by allowing ε to vary slightly. Indeed, given $x \in Z_{2\varepsilon,n}$, one of the two following inequalities holds (here $w = h^{-1}(x)$):

$$\text{(4.9)} \qquad \begin{aligned} a_n(w) &\geq p_1 n + 2\varepsilon n, \\ a_n(w) &\leq p_1 n - 2\varepsilon n. \end{aligned}$$

Now given $n \leq m \leq (1 + \varepsilon)n$, we have

$$a_n(w) \leq a_m(w) \leq a_n(w) + \varepsilon n.$$

Taking $\varepsilon > 0$ small enough so that $1 > p_1 + \varepsilon$, the first inequality in (4.9) implies

$$a_m(w) \geq a_n(w) > p_1 n + (p_1 + \varepsilon)\varepsilon n + \varepsilon n \geq p_1 m + \varepsilon m,$$

and provided $p_1 - \varepsilon > 0$, the second implies

$$a_m(w) \le a_n(w) + \varepsilon n \le (p_1 - \varepsilon)n \le (p_1 - \varepsilon)m.$$

Taking this all together, we see that

$$Z_{2\varepsilon,n} \subset Z_{\varepsilon,m}$$

for every $n \le m \le (1+\varepsilon)n$. Thus

$$\bigcup_{n \ge N} Z_{2\varepsilon,n} \subset \bigcup_{k \ge 0} Z_{\varepsilon,n_k},$$

where $n_k = \lfloor (1+\varepsilon)^k N \rfloor$, and using (4.8), we see that

$$\mu\left(\bigcup_{n \ge N} Z_{2\varepsilon,n}\right) \le \sum_{k=0}^{\infty} \mu(Z_{\varepsilon,n_k})$$

$$\le \frac{p_1 - p_1^2}{N\varepsilon^2} \sum_{k=0}^{\infty} (1+\varepsilon)^{-k} \le \frac{p_1 - p_1^2}{N\varepsilon^2} \frac{1+\varepsilon}{\varepsilon}.$$

It follows that

$$\mu\left(\bigcap_{N \in \mathbb{N}} \bigcup_{n \ge N} Z_{2\varepsilon,n}\right) = 0$$

for every $\varepsilon > 0$, which implies $\mu(B) = 0$, and we are done. $\qquad \square$

The result of Proposition 4.5 reveals the following important property of Bernoulli measures.

Definition 4.6. Given a dynamical system $f\colon X \to X$, an invariant measure μ is said to be *ergodic* if any invariant subset $E \subset X$ is either a null set ($\mu(E) = 0$) or a set of full measure ($\mu(X \setminus E) = 0$).

Ergodic measures are irreducible in the sense that they do not recognise any non-trivial invariant subsets of X. Furthermore, they are building blocks for all invariant measures: if a measure μ is not ergodic, the space can be decomposed into two invariant subsets each of positive measure (and thus non-negligible). If the restriction of μ to these subsets is not ergodic, we can further decompose them; continuing with this process (which can take uncountably many steps!), we obtain the decomposition of μ into ergodic components.

A crucial property that distinguishes ergodic measures from non-ergodic ones is the following: given a measurable set $A \subset X$ (which may or may not be invariant), we have for a.e. point x (which may or may not lie in A) that the trajectory originating at x visits A with an asymptotic frequency equal to the measure of A. More precisely, for a.e. x the following limit exists:

$$\lim_{n \to \infty} \frac{N(x, A, n)}{n} = \mu(A),$$

where $N(x, A, n)$ is the number of integers $0 \leq k < n$ for which $f^k(x) \in A$.

This property is actually equivalent to ergodicity (a fact which we do not prove). Proposition 4.5 establishes this property for Bernoulli measures in the case where A is a basic set I_j, and can in fact be used to establish it for arbitrary measurable sets; thus Bernoulli measures are ergodic.

b. Hausdorff dimension of exact dimensional measures. We write μ_p for the Bernoulli measure on Σ_2^+ with probability vector $(p, 1 - p)$. Applying Proposition 4.5 to (4.6), we see that

$$(4.10) \qquad d_{\mu_p}(x) = \frac{p \log p + (1 - p) \log(1 - p)}{\log \lambda}$$

for μ_p-almost every $x \in C$. Thus the pointwise dimension $d_{\mu_p}(x)$ exists and is constant μ_p-a.e.

Definition 4.7. Let μ be a finite measure on $\mathbb{R}^d$. If there exists $\alpha \in \mathbb{R}$ such that $d_\mu(x) = \alpha$ for μ-a.e. $x \in \mathbb{R}^d$, we say that μ is *exact dimensional*. We refer to the constant value α as the *Hausdorff dimension of the measure*, and write $\dim_H \mu = \alpha$.

It follows from Theorem 4.4 that if μ is an exact dimensional measure and $Z \subset \mathbb{R}^d$ is such that $\mu(Z) = 1$, then $\dim_H Z \geq \dim_H \mu$. Thus if we rewrite (4.10) for a Bernoulli measure μ_p on C as

$$(4.11) \qquad \dim_H \mu_p = \frac{p \log p + (1 - p) \log(1 - p)}{\log \lambda},$$

we obtain the lower bound $\dim_H C \geq \dim_H \mu_p$. This bound is tight in the following sense: one can check that the quantity $\dim_H \mu_p$ takes its maximum value when $p = q = 1/2$, and that in this case we have $\dim_H C = \dim_H \mu_p$.

Thus the Bernoulli measure $\mu_{1/2}$ corresponding to $p = q = 1/2$ differs from the other Bernoulli measures μ_p for $p \neq q$ in two ways:

(1) The pointwise dimension $d_{\mu_{1/2}}(x)$ exists and is constant *everywhere* on C, not just almost everywhere.

(2) The Hausdorff dimension of $\mu_{1/2}$ is equal to the Hausdorff dimension of C. Such a measure, whose Hausdorff dimension is equal to the Hausdorff dimension of the space, is called a *measure of maximal dimension*.

These two facts are related, as a consequence of the following result, which mirrors the Non-uniform Mass Distribution Principle, but gives an *upper* bound for the Hausdorff dimension of a set.

Theorem 4.8. *Suppose μ is a finite measure on $\mathbb{R}^d$, and that there exists $\alpha > 0$ such that*

$$\tag{4.12} \underline{d}_\mu(x) \leq \alpha$$

*for **every** $x \in Z \subset \mathbb{R}^d$. Then $\dim_H Z \leq \alpha$.*

Proof. We show that $m_B(Z, \alpha') < \infty$ for every $\alpha' > \alpha$; thus for every such α', we must exhibit $C > 0$ such that for every $\varepsilon > 0$, there exists an ε-cover $\{B(x_i, r_i)\}$ with $\sum_i r_i^{\alpha'} \leq C$.

We require the following geometric lemma (which we do not prove):

Lemma 4.9 (Besicovitch covering lemma). *Consider Euclidean space $\mathbb{R}^d$. There exists a constant K, depending only on the dimension d, such that for every set $Z \subset \mathbb{R}^d$ and every open cover of Z of the form $\mathcal{U} = \{B(x, r(x)) \mid x \in Z\}$, where $r \colon Z \to (0, \infty)$ is an arbitrary function, there exists a countable subcover $\mathcal{U}' = \{B(x_i, r(x_i) \mid i \in \mathbb{N}\}$ with multiplicity bounded by K (that is, every point of Z is contained in at most K elements of $\mathcal{U}'$).*

Given $\alpha' > \alpha$ and $\varepsilon > 0$, we use Lemma 4.9 to obtain a "good" cover of Z, as follows. For every $x \in Z$, we have

$$\varliminf_{r \to 0} \frac{\log \mu(B(x, r))}{\log r} < \alpha',$$

and thus there exists $0 < r(x) < \varepsilon$ such that $\log \mu(B(x, r(x)))/\log r < \alpha'$, or equivalently,

$$\mu(B(x, r(x))) > r(x)^{\alpha'}.$$

Now by Lemma 4.9, there exists a cover $\mathcal{U}' = \{B(x_i, r(x_i))\}$ with multiplicity bounded by K, and hence

$$\sum_i r(x_i)^{\alpha'} < \sum_i \mu(B(x_i, r(x_i))) \leq K\mu(\mathbb{R}^d) < \infty.$$

The bound only depends on n and μ, thus since $\varepsilon > 0$ and $\alpha' > \alpha$ were arbitrary, we have the desired result. $\square$

Corollary 4.10. *If μ is a finite measure on $\mathbb{R}^d$, and $Z \subset \mathbb{R}^d$ is such that $\mu(Z) > 0$ and $\underline{d}_\mu(x) = \alpha$ for all $x \in Z$, then $\dim_H Z = \alpha$.*

Using Theorems 4.4 and 4.8, it is possible in certain cases to use information about the pointwise dimension of certain measures to not only estimate the Hausdorff dimension of a given set, but to compute it precisely. We have just seen that this is the case for a repeller of a one-dimensional full-branched Markov map with constant slope. Eventually, we will address the case where the slope may vary and where the map may not be full-branched.

Remark. Our definition of $\dim_H \mu$ differs from the usual one. The usual approach is to define the Hausdorff dimension of a measure by

$$\dim_H \mu = \inf\{\dim_H Z \mid \mu(Z) = 1\},$$

and then show that if μ is exact dimensional, then $d_\mu(x) = \dim_H \mu$ at μ-a.e. x. Thus for exact dimensional measures (the only context in which our definition applies), our definition agrees with the usual one: one inequality is a consequence of the Non-uniform Mass Distribution Principle, which immediately implies $\dim_H \mu \leq \inf\{\dim_H Z \mid \mu(Z) = 1\}$, and the other follows from Theorem 4.8, since we may take $Z = \{x \mid d_\mu(x) = \dim_H \mu\}$ and obtain $\mu(Z) = 1$, $\dim_H Z = \dim_H \mu$.

c. Pointwise dimension of Hausdorff measures. Since we now have all the tools in place, we digress briefly to give a result on the pointwise dimension of the Hausdorff measures $m_H(\cdot, \alpha)$, which we will need in Chapter 6 when we consider the Hausdorff dimension of direct products of sets.

Proposition 4.11. *Let $\alpha \geq 0$, and suppose that $Z \subset \mathbb{R}^d$ is such that $0 < m_H(Z, \alpha) < \infty$. Define a measure μ by $\mu(E) = m_H(E \cap Z, \alpha)$. Then we have $\underline{d}_\mu(x) = \alpha$ for μ-a.e. $x \in Z$.*

Proof. Consider the two sets

$$Z^- = \{x \in Z \mid \underline{d}_\mu(x) < \alpha\},$$
$$Z^+ = \{x \in Z \mid \underline{d}_\mu(x) > \alpha\};$$

we will show that $\mu(Z^-) = \mu(Z^+) = 0$. To this end, let $Z_n^- = \{x \in Z \mid \underline{d}_\mu(x) \leq \alpha - 1/n\}$, and observe that $Z^- = \bigcup_n Z_n^-$. Since μ is a finite measure, we may apply Theorem 4.8 to Z_n^- and obtain $\dim_H Z_n^- \leq \alpha - 1/n < \alpha$. It follows that $\mu(Z_n^-) = m_H(Z_n^-, \alpha) = 0$. Taking the union over all $n \in \mathbb{N}$ and using countable subadditivity, we see that

$$\mu(Z^-) \leq \sum_n \mu(Z_n^-) = 0.$$

Similarly, let $Z_n^+ = \{x \in Z \mid \underline{d}_\mu(x) \geq \alpha + 1/n\}$, so that $Z^+ = \bigcup_n Z_n^+$. If $\mu(Z_n^+) \geq 0$, then the Non-uniform Mass Distribution Principle gives

$$\dim_H Z_n^+ \geq \alpha + \frac{1}{n} > \alpha = \dim_H Z,$$

a contradiction since $Z_n^+ \subset Z$. It follows that $\mu(Z_n^+) = 0$ for every n, and so $\mu(Z^+) = 0$. This completes the proof, since Z^- and Z^+ contain all points x such that $\underline{d}_\mu(x) \neq \alpha$. $\qquad\square$

Remark. Proposition 4.11 gives a partial converse to Corollary 4.10 by showing the existence of a measure μ supported on Z with $\underline{d}_\mu(x) = \dim_H Z$ almost everywhere, under the assumption that the Hausdorff measure of Z is positive and finite. In fact, the assumption that $m_H(Z, \alpha)$ is finite can be dropped: given Z and α such that $m_H(Z, \alpha) = \infty$, one may show that there exists $Z' \subset Z$ such that $0 < m_H(Z', \alpha) < \infty$, although we do not prove this here (see [**Fal03**]).

Lecture 18

a. Local entropy. In the previous lecture, we studied the dimensional properties of various measures on the symbolic space Σ_k^+ and a Cantor repeller C, with the goal of gaining information about the geometry of C. This analysis was carried out in terms of the cylinder sets $C_{w_1 \ldots w_n} \subset \Sigma_k^+$ and their counterparts, the basic sets $I_{w_1 \ldots w_n}$, which appeared because they are the sets $B(x, r)$ in the definition of pointwise dimension.

However, the cylinder sets have a dual characterisation, and this static definition (which makes no mention of the dynamics of either σ or f) is only one side of their identity. If we fix x, then as r decreases, the cylinder set corresponding to $B(x, r)$ becomes smaller and smaller; we pass from C_{w_1} to $C_{w_1 w_2}$ to $C_{w_1 w_2 w_3}$, and so on. Recall that n is the number of iterations it takes before the image of $C_{w_1 \ldots w_n}$ is the whole space:

$$\sigma^n(C_{w_1 \ldots w_n}) = \Sigma_k^+.$$

Thus we may characterise the cylinder set $C_{w_1 \ldots w_n}$ as the set of all points which follow the orbit of w (to within a small error term) for the first $n - 1$ iterates:

$$C_{w_1 \ldots w_n} = \{v \in \Sigma_k^+ \mid d_a(\sigma^j(v), \sigma^j(w)) < 1/a \text{ for all } 0 \leq j < n\},$$

where we use the metric d_a from (1.18), with $a > 2$. With this characterisation, refining the cylinder sets involves increasing the length of time n for which we require the orbits to be close, rather than decreasing the radius r in $B(x, r)$.

Thus instead of refining statically, in space, we are now refining dynamically, in time. It also makes sense to involve the dynamics of this system, which we did not do in the previous lecture, because the most important measures we have introduced so far, the Bernoulli and Markov measures on Σ_k^+ and C, are dynamically significant; they are invariant under the relevant dynamics, whether those are given by the shift σ or a one-dimensional Markov map f.

Keeping in mind the paradigm of dealing with time rather than space, the preceding considerations may be profitably generalised to the context of an arbitrary map f acting on a metric space (X, d).

Definition 4.12. Given $x \in X$, $n \in \mathbb{N}$, and $\delta > 0$, the *Bowen ball* centred at x of radius δ and order n is the set

$$(4.13) \quad B_f(x, n, \delta) = \{y \in X \mid d(f^j(x), f^j(y)) < \delta \text{ for all } 0 \leq j \leq n\}.$$

If the map f is chaotic (whichever precise definition we care to use), we expect small errors to get magnified by repeated iteration of f, and hence a point $y \in B(x, \delta) \setminus \{x\}$ should have an orbit which is eventually ejected from $B(f^j(x), \delta)$. Consequently, we expect the Bowen balls $B_f(x, n, \delta)$ to get smaller and smaller as n increases; this dynamic refinement parallels the static refinement that occurs for $B(x, r)$ as r decreases.

If X is symbolic space, then the Bowen balls $B_\sigma(x, n, \delta)$ coincide with the regular balls $B(x, r)$ (for appropriate values of n, δ, and r), as the cylinder sets play both roles. However, the two concepts are usually distinct. All the dimensional quantities we have defined so far—$d_\mu(x)$, $\dim_H \mu$, and $\dim_H Z$ (along with the box dimensions)— have been defined using the balls $B(x, r)$. Each of these dimensional quantities can also be defined using the Bowen balls $B_f(x, n, \delta)$, and in general we obtain new information, with a different interpretation, by doing so.

We begin with the pointwise dimension $d_\mu(x)$, which measured the exponential rate of decay of $\mu(B(x, r))$, and was heuristically defined as the quantity for which

$$\mu(B(x, r)) \approx r^{d_\mu(x)}.$$

If μ is the Bernoulli measure given by $(1/2, 1/2)$ on the symbolic space Σ_2^+, then the Bowen balls are cylinder sets, and we have, for $\delta = 1/a$,

$$\mu(B_\sigma(w, n, \delta)) = \mu(C_{w_1 \dots w_{n+1}}) = 2^{-(n+1)}.$$

We want to write this in terms of n (the quantity which varies as we refine dynamically) and a scaling parameter, without reference to anything else that depends on the map, the measure, or the point. Thus we observe that

$$2^{-(n+1)} \approx e^{-n \log 2},$$

and in general, we search for a scaling parameter α such that

$$\mu(B_f(x, n, \delta)) \approx e^{-n\alpha}.$$

Definition 4.13. The *local entropy* of the map f and the measure μ at the point x is

$$h_{\mu, f}(x) = \lim_{\delta \to 0} \lim_{n \to \infty} -\frac{1}{n} \log \mu(B_f(x, n, \delta)),$$

if the limit exists.

Swapping $B(x, r)$ for $B_f(x, n, \delta)$, and declaring the "radius" of the latter to be e^{-n}, this is very nearly a carbon copy of the definition of pointwise dimension.

Example 4.14. Let μ be a Bernoulli measure on Σ_k^+ with probability vector $\mathbf{p} = (p_1, \ldots, p_k)$. Given $\delta > 0$, let $N \in \mathbb{N}$ be the unique N such that $a^{-N} \leq \delta < a^{-(N-1)}$; thus for every $w \in \Sigma_k^+$ and $n \in \mathbb{N}$,

$$B_\sigma(w, n, \delta) = C_{w_1 \ldots w_{N+n}}.$$

It follows that

$$\mu(B_\sigma(w, n, \delta)) = \mu(C_{w_1 \ldots w_{N+n}}) = p_1^{a_{N+n}^1(w)} \cdots p_k^{a_{N+n}^k(w)},$$

where $a_m^i(w)$ denotes the number of times the symbol i appears in the itinerary $w_1, \ldots, w_m$. Then

$$-\frac{1}{n} \log \mu(B_\sigma(w, n, \delta)) = -\sum_{i=1}^{k} \frac{a_{N+n}^i(w)}{n} \log p_i,$$

and using the result of Proposition 4.5 (which holds for all k, not just $k = 2$), we have

$$h_{\mu, f}(w) = \lim_{\delta \to 0} \lim_{n \to \infty} -\sum_{i=1}^{k} \frac{a_{N+n}^i(w)}{N+n} \frac{N+n}{n} \log p_i$$

(4.14)

$$= -\sum_{i=1}^{k} p_i \log p_i$$

for μ-a.e. $w \in \Sigma_k^+$.

Thanks to Proposition 4.15 below, which shows that local entropy is an invariant of topological conjugacy, the same result holds if μ is a Bernoulli measure on a Cantor repeller C which is modelled on Σ_k^+. In both cases, we observe that if the entries of $\mathbf{p}$ are not all equal, then there will be points w at which the local entropy takes some other value, just as was the case for pointwise dimension.

Proposition 4.15. *Let (X, d) and (Y, ρ) be compact metric spaces, let $f \colon X \to X$ and $g \colon Y \to Y$ be continuous maps, and let $\phi \colon X \to Y$ be a homeomorphism such that $g \circ \phi = \phi \circ f$. Let μ be a g-invariant* *measure on Y, and let μ^* be the f-invariant measure on X given by $\mu^*(E) = \mu(\phi(E))$. Then the local entropies are related by*

$$(4.15) \qquad\qquad h_{\mu^*,f}(x) = h_{\mu,g}(\phi(x)).$$

Proof. Because X is compact, the continuous map $\phi \colon X \to Y$ is actually uniformly continuous. Thus for every $\varepsilon > 0$, there exists $\delta = \delta(\varepsilon) > 0$ such that $d(x, x') < \delta$ implies $d(\phi(x), \phi(x')) < \varepsilon$. This may be restated as the following inclusion, which holds for all $x \in X$:

$$\phi(B(x, \delta)) \subset B(\phi(x), \varepsilon).$$

This inclusion holds at x, $f(x)$, $f^2(x)$, etc., and so the Bowen balls are related by

$$\phi(B_f(x, n, \delta)) \subset B_g(\phi(x), n, \varepsilon)$$

for every $n \in \mathbb{N}$. In particular, this implies that

$$\mu^*(B_f(x, n, \delta)) \leq \mu(B_g(\phi(x), n\varepsilon)),$$

which gives

$$\lim_{n\to\infty} -\frac{1}{n} \log \mu^*(B_f(x, n, \delta)) \geq \lim_{n\to\infty} -\frac{1}{n} \log \mu(B_g(\phi(x), n, \varepsilon)).$$

Since $\delta \to 0$ as $\varepsilon \to 0$, taking the limit as $\varepsilon \to 0$ gives

$$h_{\mu^*,f}(x) \geq h_{\mu,g}(\phi(x)).$$

The proof of the other inequality is similar. $\qquad\qquad\qquad\square$

Remark. In order to prove the analogous result to Proposition 4.15 for the pointwise dimensions of μ^* and μ, we would need the homeomorphism ϕ to be bi-Lipschitz. Here that assumption is replaced by the statement that ϕ conjugates the dynamics of f and g.

b. Kolmogorov–Sinai entropy. Although the local entropy may vary from point to point and may fail to exist at some points, there are many important cases in which it exists and is constant almost everywhere, as in Example 4.14. This happens whenever the measure μ is ergodic, and we refer to the constant value as the *entropy of the measure μ* under the transformation f, and denote it by $h(\mu, f)$. This is the analogue of $\dim_H \mu$; we obtain the dimension of the measure by doing our refining in space, while we obtain the entropy by doing our refining in time.

Remark. The entropy $h(\mu, f)$ was first introduced by Andrei Kolmogorov in 1958 and in a somewhat more general form by Yakov Sinai in 1959; it is often called the *Kolmogorov–Sinai entropy*, although the terms *metric entropy* and *measure theoretic entropy* are also used. It is one of the most important characteristics in dynamics, and our discussion of it in this book does little more than scratch the surface.

As with the Hausdorff dimension of a measure, our definition of $h(\mu, f)$ as the almost everywhere value of a locally defined quantity is not the standard one, although it agrees with the standard definition in every case where our definition applies. There is a trade-off here: we have chosen to highlight the dimensional nature of the entropy at the cost of obscuring its significance as a measure of information, which echoes its origins in information theory. We content ourselves here with saying that heuristically, the entropy $h(\mu, f)$ may be thought of as the *average rate at which information is gained* as a typical trajectory of the system is observed.[1]

Recalling the result of Example 4.14, we may now rewrite (4.11) as the statement that for a Bernoulli measure μ on a Cantor set C which is the maximal invariant set for a piecewise linear one-dimensional full-branched Markov map with slope $1/\lambda$, the dimension and entropy are related by

$$(4.16) \qquad \dim_H \mu = \frac{h(\mu, f)}{-\log \lambda}.$$

This was proved in the particular case where the number of branches is $k = 2$, but the same computations show that it holds for any value of k; for general k, the Kolmogorov–Sinai entropy of the Bernoulli measure with probability vector $\mathbf{p}$ is given by (4.14) as $h(\mu, f) = -\sum_{i=1}^{k} p_i \log p_i$.

c. Topological entropy. So far, we have examined two new dimensional quantities obtained by replacing statically defined balls $B(x, r)$ with dynamically defined balls $B_f(x, n, \delta)$ in our earlier definitions. Both of these characterise a measure μ: at a local level, we have the

---

[1] The standard definition of entropy may be found in Peter Walters' book *An Introduction to Ergodic Theory* [**Wal75**]; an overview of the history of entropy in its various guises is given in Anatole Katok's survey paper *Fifty years of entropy in dynamics* [**Kat07**].

pointwise dimension $d_\mu(x)$ and the local entropy $h_{\mu,f}(x)$, while at a global level, we have the Hausdorff dimension (of a measure) $\dim_H \mu$ and the Kolmogorov–Sinai entropy $h(\mu, f)$.

Along similar lines, we now outline the definition of *topological entropy*, a dynamically defined quantity analogous to the Hausdorff dimension of a set. The reader is urged to compare each step with the corresponding step in the definition of Hausdorff dimension to get a feel for the general procedure; once again, we stress that if we call e^{-n} the "radius" of the Bowen ball $B_f(x, n, \delta)$, then the two definitions are nearly word-for-word identical.

Definition 4.16. Fix a separable metric space X, a continuous map $f\colon X \to X$, and a set $Z \subset X$. For every $N \in \mathbb{N}$ and $\delta > 0$, denote by $\mathcal{P}(Z, N, \delta)$ the collection of all countable covers of Z by Bowen balls $B_f(x, n, \delta)$ for which $x \in Z$ and $n \geq N$. Given $\alpha \geq 0$, define a set function $m_h(Z, \alpha, \delta)$ by

$$m_h(Z, \alpha, \delta) = \lim_{N \to \infty} \inf_{\mathcal{P}(Z,N,\delta)} \sum_i e^{-n_i \alpha}.$$

It may be shown that $m_h(Z, \cdot, \delta)$ has a jump-like graph with the form shown in Figure 2.1 for $m_H(Z, \cdot)$; thus we consider the critical parameter value

$$h_{\text{top}}(Z, f, \delta) = \sup\{\alpha \geq 0 \mid m_h(Z, \alpha, \delta) = \infty\}$$
$$= \inf\{\alpha \geq 0 \mid m_h(Z, \alpha, \delta) = 0\},$$

and define the *topological entropy* of the map f on the set Z by

$$h_{\text{top}}(Z, f) = \lim_{\delta \to 0} h_{\text{top}}(Z, f, \delta),$$

where one can show that the limit exists.

Remark. Once again, we are obliged to point out that this is not the usual way of defining the topological entropy. The more common definition actually mirrors the definition of box dimension: writing $N_f(Z, n, \delta)$ for the minimal number of Bowen balls $B_f(x, n, \delta)$ it takes to cover Z, one defines

$$h_{\text{top}}(Z, f) = \lim_{\delta \to 0} \lim_{n \to \infty} \frac{1}{n} \log N_f(Z, n, \delta).$$

This characterises the topological entropy as the growth rate (with respect to n) of the number of Bowen balls of order *exactly* n it takes to cover Z. The same quantity is also often expressed, in terms reminiscent of Exercise 2.25, as the growth rate of the maximal number of disjoint Bowen balls centred in Z, or equivalently, the number of orbit segments of length n which can be distinguished at a given finite scale.

In the case where Z is compact and f-invariant, our definition agrees with the more common one; however they can (and do) differ when Z is non-compact and/or non-invariant. In this case, the definition we have given is often the more appropriate tool, just as Hausdorff dimension is often a better tool than box dimension when the quantities differ.

Exercise 4.2. Let $\sigma \colon \Sigma_k^+ \to \Sigma_k^+$ be the full shift on k symbols, and show that $h_{\text{top}}(\Sigma_k^+, \sigma) = \log k$.

It can be shown that the topological entropy is an invariant of topological conjugacy; that is, if two maps are conjugated by a homeomorphism, then they have the same topological entropy (recall we saw a similar result for the local entropy). Thus if C is a repeller for a one-dimensional full-branched Markov map f, we have $h_{\text{top}}(C, f) = \log k$. In the particular case where f has constant slope $1/\lambda$, we obtain the following relationship between Hausdorff dimension and topological entropy, paralleling (4.16):

$$(4.17) \qquad \dim_H C = \frac{h_{\text{top}}(C, f)}{-\log \lambda}.$$

In the previous lecture, we obtained the left-hand side of (4.17) by maximising $\dim_H \mu$ over all Bernoulli measures. Comparing (4.16) and (4.17), we see that we can also get $h_{\text{top}}(C, f)$ by maximising $h(\mu, f)$ over all Bernoulli measures.

In fact, the analogies between dimension and entropy run very deep. There are analogues of Theorems 4.4 and 4.8 available for topological entropy, which relate it to local entropy, and hence to Kolmogorov–Sinai entropy. The proofs are similar to the proofs in the case of dimension; as we are not particularly concerned with these results here, we omit the details, referring the interested reader to [**Pes98**] for a full exposition of these matters.

Let us mention, however, that the idea of obtaining $h_{\text{top}}(C, f)$ by maximising $h(\mu, f)$ in fact works quite generally, thanks to the following *variational principle* (which we do not prove).

Theorem 4.17. *Let X be a compact metric space and $f\colon X \to X$ a continuous map. Then*

$$(4.18) \qquad h_{\text{top}}(X, f) = \sup\{h(\mu, f) \mid \mu \text{ is ergodic and } f\text{-invariant}\}.$$

We will not make use of this result in its full generality; however, we will refer to one specific case beyond the example of the full shift mentioned above. If X is a subshift of finite type or a Markov construction modelled on such a subshift, then it turns out that the supremum in (4.18) can be taken over all Markov measures μ supported on X (like Bernoulli measures, Markov measures are ergodic, provided their associated transition matrix is primitive). We will soon see how to compute the entropy for such measures, and this will give us a tool with which to compute the topological entropy of a subshift.

Exercise 4.3. A measure μ which achieves the supremum in (4.18) is called a *measure of maximal entropy*. Show that the measure of maximal entropy for the full shift on k symbols is the Bernoulli measure with probability vector $\left(\frac{1}{k}, \ldots, \frac{1}{k}\right)$.

Lecture 19

a. Entropy of Markov measures. We now return to the Markov constructions introduced in Lecture 15 and use our recently developed tools to gain more knowledge about the Hausdorff dimension of the resulting Cantor sets.

Let f be a one-dimensional Markov map, as in Figure 3.2, and suppose for the time being that f is piecewise linear, with $f'(x) = 1/\lambda$ everywhere in the domain of f. Then f determines a $k \times k$ transition matrix A and a repelling Cantor set C_A, and the dynamics of f on C_A are modelled by the subshift of finite type Σ_A^+ comprising all admissible sequences with respect to A.

In order to estimate $\dim_H C_A$ using the Non-uniform Mass Distribution Principle, we need a measure μ which gives positive weight to C_A.

Exercise 4.4. Show that if there exist i, j such that $a_{ij} = 0$, then $\mu(C_A) = 0$ for every Bernoulli measure μ.

Exercise 4.5. Let μ be a Markov measure on Σ_k^+ with stationary probability vector $\pi = (\pi_i)$ and stochastic matrix $P = (p_{ij})$, and let A be an arbitrary $k \times k$ transition matrix. Show that if there exist i, j such that $p_{ij} > 0$ and $a_{ij} = 0$, then $\mu(C_A) = 0$.

As a consequence of the previous two exercises, we must consider Markov measures supported on C_A in order to gain any information about $\dim_H C_A$. We begin by working in symbolic space.

Let μ be a Markov measure on Σ_k^+ with stationary probability vector $\pi = (\pi_i)$ and stochastic matrix $P = (p_{ij})$. Then as in Example 4.14, we have
$$B_\sigma(w, n, \delta) = C_{w_1 \ldots w_{N+n}},$$
where $a^{-N} \le \delta < a^{-(N-1)}$, and hence
$$\mu(B_\sigma(w, n, \delta)) = \pi_{w_1} p_{w_1 w_2} \cdots p_{w_{N+n-1} w_{N+n}}$$
$$= \pi_{w_1} \prod_{i=1}^{k} \prod_{j=1}^{k} p_{ij}^{a_{N+n}^{i,j}(w)},$$
where $a_m^{i,j}(w)$ denotes the number of indices $m' < m$ such that $w_{m'} = i$ and $w_{m'+1} = j$. Taking logarithms yields
$$\frac{1}{n} \log \mu(B_\sigma(w, n, \delta)) = \frac{\pi_{w_1}}{n} + \sum_{i=1}^{k} \sum_{j=1}^{k} \frac{a_{N+n}^{i,j}(w)}{n} \log p_{ij}.$$

A similar result to that in Proposition 4.5 shows that if the transition matrix A associated to the stochastic matrix P is primitive, then for μ-a.e. sequence $w \in \Sigma_k^+$, we have
$$(4.19) \qquad \lim_{m \to \infty} \frac{a_m^{i,j}(w)}{m} = \pi_i p_{ij},$$
and hence for such sequences w, the local entropy of μ is
$$h_{\mu,\sigma}(w) = -\sum_{i=1}^{k} \pi_i \sum_{j=1}^{k} p_{ij} \log p_{ij}.$$

Since this holds μ-a.e., we also have

$$(4.20) \qquad h(\mu, \sigma) = -\sum_{i=1}^{k} \pi_i \sum_{j=1}^{k} p_{ij} \log p_{ij}.$$

Exercise 4.6. Show that if f is a piecewise linear one-dimensional Markov map with constant expansion $|f'(x)| = 1/\lambda$, then the pointwise dimension and the local entropy exist at precisely the same points, and they are related by

$$d_\mu(x) = \frac{h_{\mu, f}(x)}{-\log \lambda}.$$

Example 4.18. Consider a Markov map as in Exercise 4.6 for which the transition matrix A is $\left(\begin{smallmatrix} 0 & 1 \\ 1 & 1 \end{smallmatrix}\right)$, as in Example 3.17. If P is the stochastic matrix for a Markov measure supported on C_A, then since $p_{11} = 0$ and the entries in each row of P must sum to 1, we have $p_{12} = 1$; thus the stochastic matrix has the form $P_a = \left(\begin{smallmatrix} 0 & 1 \\ a & 1-a \end{smallmatrix}\right)$, where $a = p_{21} \in [0, 1]$. If we interpret the entries of P as giving the probability of a transition from one state to another, then the fact that p_{11} vanishes means that every time we are in state 1, we must immediately return to state 2 at the next time step, with zero probability of remaining in state 1.

Given $a \in [0, 1]$, we wish to construct a Markov measure with stochastic matrix P_a. Thus we need a stationary probability vector $\pi = (p, q)$. After some straightforward calculations (which the reader is encouraged to carry out), the requirement that $\pi P_a = \pi$ forces us to take

$$\pi = \pi_a = \left(\frac{1}{1+a}, \frac{a}{1+a} \right).$$

Now we denote by μ_a the Markov measure with probability vector π_a and stochastic matrix P_a. Thus we have a one-parameter family of Markov measures μ_a for which $\mu_a(C_A) = 1$. The entropy of these measures is given by

$$h(\mu_a, f) = -\pi_1(p_{11} \log p_{11} + p_{12} \log p_{12})$$
$$- \pi_2(p_{21} \log p_{21} + p_{22} \log p_{22})$$
$$= -\frac{1}{1+a}(a \log a + (1-a) \log(1-a)).$$

Thus, by Exercise 4.6, the pointwise dimension is equal μ_a-almost everywhere to

$$\phi(a) = \frac{a \log a + (1-a) \log(1-a)}{(1+a) \log \lambda},$$

so $\dim_H \mu_a = \phi(a)$. As before, Theorem 4.4 implies that $\dim_H C_A \geq \phi(a)$ for all $a \in [0,1]$, and so we want to maximise ϕ in order to find the best bound, which will hopefully give us the exact value of $\dim_H C_A$.

By solving $\phi'(a) = 0$ for the critical value of a, we find that ϕ achieves its maximum at $a = a_0 = (3 - \sqrt{5})/2$, and hence

$$(4.21) \qquad \dim_H C_A \geq \phi(a_0) = \frac{\log(1 + \sqrt{5}) - \log 2}{- \log \lambda}.$$

A further computation (which we also omit) shows that $d_{\mu_{a_0}}(x) = \dim_H \mu_{a_0}$ at *every* point $x \in C_A$, and thus by Theorem 4.8, we have equality in (4.21). Thus μ_{a_0}, in addition to being the measure of maximal entropy, is also the measure of maximal dimension.

Exercise 4.7. Consider the limiting set C_A for the Markov geometric construction which starts from three disjoint intervals $I_1, I_2, I_3 \subset [0,1]$, and which has ratio coefficients $\lambda_1 = \lambda_2 = \lambda_3 = \lambda$ with $0 < \lambda < \frac{1}{3}$ and transition matrix

$$A = \begin{pmatrix} 0 & 1 & 0 \\ 1 & 1 & 0 \\ 0 & 0 & 1 \end{pmatrix}.$$

Use the technique described above to compute the Hausdorff dimension of C_A. (Note that A is not primitive.)

b. Hausdorff dimension of Markov constructions. We now describe a procedure for computing the measure of maximal entropy for an arbitrary subshift of finite type Σ_A^+. Using the relationship between dimension and entropy given by Exercise 4.6, we will then be able to determine the Hausdorff dimension of certain Markov constructions.

Let A be a $k \times k$ matrix of zeros and ones, and Σ_A^+ the associated subshift of finite type. Let χ be the largest positive eigenvalue[2]

---

[2] We use χ rather than the more common notation λ so as to avoid confusion with our use of λ to denote a ratio coefficient.

of A, and fix a left eigenvector $(u_1, \ldots, u_k)$ and a right eigenvector $(v_1, \ldots, v_k)$ for χ, both with positive entries, and normalised so that $\sum_{i=1}^{k} u_i v_i = 1$. Let μ be the Markov measure with probability vector

$$\pi_i = u_i v_i$$

and stochastic matrix

$$p_{ij} = \frac{a_{ij}}{\lambda} \frac{v_j}{v_i};$$

thus the weight which μ assigns each cylinder is given by

$$\mu(C_{w_1 \ldots w_n}) = \chi^{-n} u_{w_1} v_{w_n},$$

provided w is admissible. μ is called the *Parry measure* for the subshift Σ_A^+.

Since cylinders coincide with Bowen balls, we may proceed as in Example 4.14, and compute the local entropy as follows:

$$\begin{aligned}
h_{\mu,\sigma}(w) &= \lim_{n \to \infty} -\frac{1}{n} \log \mu(C_{w_1 \ldots w_n}) \\
&= \lim_{n \to \infty} -\frac{1}{n} \left(-n \log \chi + \log u_{w_1} + \log v_{w_n} \right) \\
&= \log \chi.
\end{aligned}$$

This holds for *every* $w \in \Sigma_A^+$. Thus, the analogues of Theorems 4.4 and 4.8 for entropies together imply that $h_{\text{top}}(\Sigma_A^+) = \log \chi$. The Parry measure μ is a measure of maximal entropy for Σ_A^+; in fact, one can show that it is the *unique* measure of maximal entropy.

Exercise 4.8. Find the Parry measure for the subshift in Example 4.18, and compare the results of this section with the results we obtained there.

The procedure described above allows us to compute a measure of maximal entropy for a subshift. Given a one-dimensional piecewise linear Markov map f with constant slope, and an invariant measure μ for f, Exercise 4.6 tells us that the pointwise dimension of the Parry measure exists and is constant everywhere, hence the Parry measure is the measure of maximal dimension as well, and we have the following theorem:

Theorem 4.19. *Let f be a piecewise linear Markov map of the interval (which may not be full-branched) with $|f'(x)| = \lambda^{-1}$ everywhere f*

is defined. Let A be the transition matrix associated to f and C_A the limiting Cantor set of the associated Markov construction. Let χ be the largest positive eigenvalue of the matrix A. Then

$$(4.22) \qquad \dim_H C_A = \frac{\log \chi}{-\log \lambda}.$$

Thus we have successfully computed the Hausdorff dimension of a Cantor set defined by a Markov construction *when the rate of expansion is constant*. This procedure does not, however, tell us what to do when the rate of expansion $|f'(x)|$ is no longer constant....

Lecture 20

a. Lyapunov exponents. Throughout the previous two lectures, we dealt only with examples where $|f'(x)| = 1/\lambda$ was constant on the domain of f; this had the effect that Bowen balls $B_f(x, n, \delta)$ were really just usual balls $B(x, \delta\lambda^{-n})$, and consequently each dimension was related to the corresponding entropy by a factor of $-\log \lambda = \log |f'(x)|$.

If we allow $|f'(x)|$ to vary from point to point, either by considering a linear map where the branches of f have different slopes or by letting f be non-linear, then things are more complicated. Because the various entropies—$h_{\mu,f}(x)$, $h(\mu, f)$, and $h_{\text{top}}(Z, f)$—are invariant under topological conjugacy, the results of the previous two lectures still tell us everything we want to know about them.

However, the various dimensions—$d_\mu(x)$, $\dim_H \mu$, and $\dim_H Z$—are defined not in terms of the dynamics of f, but in terms of the metric structure of X. Thus, if we wish to gain information about them via the dynamics of f (that is, via the entropies), as we did in the previous lecture, we must figure out what is to replace the factor of $\log |f'(x)|$, which is no longer constant.

Our initial goal will be to compare $B_f(x, n, \delta)$ and $B(x, r)$, and to use this to compare $h_{\mu,f}(x)$ and $d_\mu(x)$ for a given measure μ. To this end, consider a continuously differentiable map $f \colon E \to \mathbb{R}$, where $E \subset \mathbb{R}$ is the domain of definition of f, and let $x, y \in E$ be two points which are close together. We want to compare their trajectories, to see how the distance between $f^n(x)$ and $f^n(y)$ varies with n.

Taking the Taylor expansion of f around x gives

$$f(y) = f(x) + f'(x)(y - x) + o(y - x),$$

and so the distance between $f(x)$ and $f(y)$ is given by

$$d(f(x), f(y)) = |f(x) - f(y)| \approx |f'(x)|\, d(x, y),$$

where the error term is on the order of $(d(x, y))^2$. Thus the distance between the trajectories is multiplied by a factor of approximately $|f'(x)|$ when we pass from x and y to their images under f.

Passing to the second iterates, we have

$$\begin{aligned}
d(f^2(x), f^2(y)) &= d(f(f(x)), f(f(y))) \\
&\approx |f'(f(x))|\, d(f(x), f(y)) \\
&\approx |f'(f(x))f'(x)|\, d(x, y),
\end{aligned}$$

and in general, after n iterations the estimate is

$$(4.23) \qquad d(f^n(x), f^n(y)) \approx \left(\prod_{i=0}^{n-1} |f'(f^i(x))| \right) d(x, y).$$

The size of the error term in (4.23) depends on the distance between $f^{n-1}(x)$ and $f^{n-1}(y)$; if the trajectories become far enough apart, the error term becomes large, and the estimate is no longer valid. However, the closer together we choose x and y to be, the longer it takes for this to happen, and so we can make the estimate valid for as long as we like by choosing x and y close enough together.

We are interested, then, in the behaviour of the quantity $d_n(x) = \prod_{i=0}^{n-1} |f'(f^i(x))|$, which gives the amount of expansion in a neighbourhood of x after n iterations. In the case when f is piecewise linear as in Figure 1.17, the derivative f' takes two values:

$$(4.24) \qquad f'(x) = \begin{cases} \lambda_1^{-1} & x \in I_1, \\ \lambda_2^{-1} & x \in I_2. \end{cases}$$

Thus the rate of growth of $d_n(x)$ depends on whether the iterates $f^i(x)$ are in I_1 or in I_2. We would like to have some information on this rate of growth by finding some real number λ for which the product asymptotically behaves like $e^{\lambda n}$. This is made precise by the following definition.

Definition 4.20. The *Lyapunov exponent* of the map f at the point x is

$$(4.25) \qquad \lambda_f(x) = \lim_{n \to \infty} \frac{1}{n} \log \left(\prod_{i=0}^{n-1} |f'(f^i(x))| \right),$$

if the limit exists.

Remark. The Lyapunov exponent tells us how quickly the distance between two nearby points grows under repeated iterations of f. It may be thought of as the asymptotic rate of expansion of the map or, alternately, as the asymptotic rate of growth of a small error term. It was introduced by Aleksandr Lyapunov in 1892 to study the stability of solutions of differential equations [**Lya92**].

Proposition 4.21. *Let* $f \colon I_1 \cup I_2 \to [0, 1]$ *be piecewise linear, as in Figure 1.17, with* $|I_1| = \lambda_1$ *and* $|I_2| = \lambda_2$, *and let* μ *be a Bernoulli measure on* C *with probability vector* (p, q). *Then the Lyapunov exponent exists* μ-*almost everywhere, and* $\lambda_f(x) = -(p \log \lambda_1 + q \log \lambda_2)$.

Proof. The basic interval $I_{w_1 \dots w_n}$ consists of precisely those points x for which $f^{j-1}(x) \in I_{i_j}$ for each $j = 1, \dots, n$. Thus for $x \in I_{w_1 \dots w_n}$,

$$(4.26) \qquad d_n(x) = \prod_{i=0}^{n-1} |f'(f^i(x))| = \prod_{j=1}^{n} \lambda_{i_j}^{-1} = \lambda_1^{-a_n(w)} \lambda_2^{-(n-a_n(w))},$$

where $a_n(w)$ is the number of times the symbol 1 appears in the first n terms $w_1, w_2, \dots, w_n$. Recalling that $a_n(w)/n \to p$ for μ-almost every point x (Proposition 4.5), we may compute the Lyapunov exponent $\lambda(x)$ for such points:

$$\begin{aligned}
\lambda(x) &= \lim_{n \to \infty} \frac{1}{n} \log d_n(x) \\
&= \lim_{n \to \infty} \frac{1}{n} \log \left(\lambda_1^{-a_n(x)} \lambda_2^{-(n-a_n(x))} \right) \\
&= \lim_{n \to \infty} - \left(\frac{a_n(x)}{n} \log \lambda_1 + \frac{n - a_n(x)}{n} \log \lambda_2 \right) \\
&= -(p \log \lambda_1 + (1 - p) \log \lambda_2).
\end{aligned} \qquad \square$$

Definition 4.22. Given an ergodic measure μ, the Lyapunov exponent $\lambda_f(x)$ exists and is constant μ-a.e.; we refer to the almost everywhere value as the *Lyapunov exponent of* μ, and denote it by $\lambda(\mu, f)$.

We can also compute the pointwise dimension of a Bernoulli measure in the context of Proposition 4.21. The length of the basic interval $I_{w_1 \ldots w_n}$ depends on $a_n(w)$, and so in the calculation of the pointwise dimension $d_\mu(x)$, the denominator in (4.6) must be replaced by

$$
(4.27) \qquad
\begin{aligned}
\log |I_{w_1 \ldots w_n}| &= \log \left(\lambda_1^{a_n(w)} \lambda_2^{n - a_n(w)} \right) \\
&= a_n(w) \log \lambda_1 + (n - a_n(w)) \log \lambda_2.
\end{aligned}
$$

Using the result of Proposition 4.5, we have, for almost every $x \in C$,

$$
d_\mu(x) = \frac{p \log p + q \log q}{p \log \lambda_1 + q \log \lambda_2};
$$

we can rewrite this as

$$
(4.28) \qquad \dim_H \mu = \frac{h(\mu, f)}{\lambda(\mu, f)}.
$$

Remark. In fact, one can prove for a large class of one-dimensional maps (not just linear Markov maps) that some version of the following relationship holds:

$$
B_f(x, n, \delta) \approx B\left(x, \delta e^{-n \lambda_f(x)} \right).
$$

This allows us to relate the pointwise dimension, local entropy, and Lyapunov exponent by the following pointwise analogue of (4.28):

$$
d_\mu(x) = \frac{h_{\mu, f}(x)}{\lambda_f(x)}.
$$

This relationship has (4.28) as a corollary.

Now let f be a one-dimensional non-linear full-branched expanding Markov map, and let C be the maximal repelling Cantor set for f. Then the ratio coefficients change at each step in the construction of C, and so Moran's theorem is of no use to us in determining the Hausdorff dimension of C. However, (4.28) still holds in this case, and so by the Non-uniform Mass Distribution Principle, we have

$$
(4.29) \qquad \dim_H C \geq \frac{h(\mu, f)}{\lambda(\mu, f)}.
$$

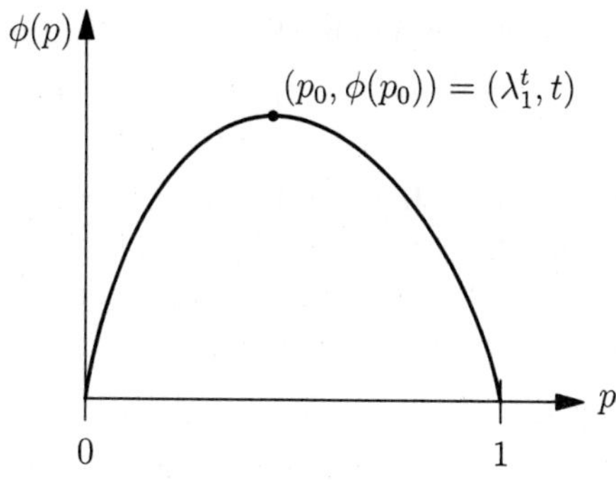

Figure 4.1. Hausdorff dimension of the Bernoulli measures μ_p.

That is, the Hausdorff dimension of the Cantor set is bounded below by the ratio of the entropy and the Lyapunov exponent for any Bernoulli measure μ; in fact, this holds for any ergodic measure.

Since this ratio depends on the particular choice of measure, what we really have is a whole family of lower bounds on $\dim_H C$; it is natural to ask if one can find an invariant measure μ for which equality is achieved, and the ratio between the entropy and the Lyapunov exponent gives exactly the Hausdorff dimension. In fact, this measure of maximal dimension can be shown to exist for any expanding one-dimensional Markov map (i.e., a one-dimensional Markov map for which $|f'(x)| > 1$ for every x), even in the non-linear case. This gives a general technique for finding the Hausdorff dimension of a dynamically defined Cantor set on the line, which more or less accomplishes the original task we set out to perform.

b. Fractals within fractals. To illustrate the general procedure, let us return to the particular case of a piecewise linear full-branched one-dimensional Markov map f on two intervals (whose lengths λ_1 and λ_2 may differ), and go hunting for a measure of maximal dimension, for which the pointwise dimension is constant *everywhere*, and the lower bound (4.29) on $\dim_H C$ actually gives equality.

We begin our search with the family of Bernoulli measures. Once again, write μ_p for the Bernoulli measure with probability vector

$(p, 1-p)$, and define a function $\phi\colon [0,1] \to \mathbb{R}$ by

$$\phi(p) = \dim_H \mu_p = \frac{h(\mu_p, f)}{\lambda(\mu_p, f)} = \frac{p \log p + (1-p) \log(1-p)}{p \log \lambda_1 + (1-p) \log \lambda_2};$$

the graph of a typical ϕ is shown in Figure 4.1. It follows from (4.28) and (4.29) that $\dim_H C \geq \phi(p)$ for all $p \in [0,1]$. In order to find the best bound, one solves for the critical point p_0 at which $\phi'(p_0) = 0$ and the function ϕ achieves its maximum.

Exercise 4.9. Show that the function ϕ achieves its maximum at the point $p_0 = \lambda_1^t$, where t is the unique solution of Moran's equation $\lambda_1^t + \lambda_2^t = 1$, and that $\phi(p_0) = t$.

The result of Exercise 4.9 shows that the bound $\dim_H C \geq \phi(p_0)$ is optimal, since we have

$$\dim_H C = t = \phi(p_0) = \sup_{p \in [0,1]} \phi(p).$$

The probability vector associated to the Bernoulli measure μ_{p_0} is $(p_0, 1-p_0) = (\lambda_1^t, \lambda_2^t)$, and in addition to being the measure of maximal dimension, μ_{p_0} turns out to be the Hausdorff measure $m_H(\cdot, t)$.

Let us step back a moment to take stock of all this, and survey the picture we have just painted. We begin with a Cantor set C, whose structure we hope to understand by examining various invariant measures for the map f. The Lebesgue measure of C is 0, so the measures which we use are very different from Lebesgue measure; to use the language of measure theory, they are *singular*, since they give positive measure to sets of Lebesgue measure zero. To each $p \in [0,1]$ we can associate a Bernoulli measure μ_p; upon discarding from C a null set for μ_p, we are left with a set $C_p \subset C$ of full measure—that is, $\mu_p(C_p) = 1$—on which the pointwise dimension $d_{\mu_p}(x)$ is constant and equal to $h_{\mu_p}(f)/\lambda_{\mu_p}(f)$.

The fact that pointwise dimension exists and is constant everywhere on C_p means that the measure of a small ball whose centre is in C_p scales as

$$(4.30) \qquad\qquad \mu_p(B(x,r)) \approx r^{d_{\mu_p}} = r^{\phi(p)}.$$

Thus μ_p exhibits a sort of measure-theoretic self-similarity on C_p, which is an analogue of the geometric self-similarity possessed by certain very regular fractals, such as the middle-third Cantor set and the Sierpiński gasket, where the geometric structure appears the same at every scale. Such geometric regularity is not present in the Cantor sets generated by non-linear expanding maps, but (4.30) displays the measure-theoretic regularity of μ_p across all scales.

One may also recall the relationship between existence of the pointwise dimension at x and the convergence of the ratio $a_n(w)/n$, which measures the relative amount of time the orbit of x spends in I_1. While this ratio approaches many different limits depending on x (and sometimes fails to converge at all), the limit exists and is equal to p for every $x \in C_p$. Thus C_p also has a very regular structure in terms of how the trajectories of points in C_p apportion their time between I_1 and I_2.

These are some of the ways in which the set C_p is a better fractal set than C itself, since all the points in C_p have more or less the same "average" behaviour. For a given Bernoulli measure μ_p, C_p is the set of all points which are "typical" with respect to μ_p, and so C_p depends on the choice of p. In fact, if $\lambda_1 \neq \lambda_2$, then one can show that these sets are pairwise disjoint; $C_p \cap C_{p'} = \emptyset$ for $p \neq p'$. Thus $\{C_p \mid 0 \leq p \leq 1\}$ gives a whole *family* of "fractals within fractals", illustrating the deeply complicated *multifractal* structure of the Cantor set C. If we let C' be the set containing all points of C which do not lie in any of the sets C_p, then we can write the following *multifractal decomposition* of C:

$$C = \left(\bigcup_{p \in [0,1]} C_p \right) \cup C'.$$

Each C_p is f-invariant and supports the measure μ_p in the sense that $\mu_p(C_p) = 1$. This is the beginning of what is known as *multifractal analysis* of dynamical systems, in which the function $\phi(p)$, which is known as the *dimension spectrum for pointwise dimensions*,[3] plays an important role.

---

[3]In fact, the dimension spectrum as obtained from $\phi(p)$ by an affine change of coordinates, but the two functions carry the same information.

Chapter 5

Discrete-Time Systems: The FitzHugh–Nagumo Model

Lecture 21

a. The FitzHugh–Nagumo model for neurons. And now, as they say, for something completely different. Setting aside for the time being our discussion of Cantor sets, measures, entropy, dimension, and all such manner of things, we turn our attention to a model from biology, which attempts to describe the propagation of an impulse through a neuron, and which seems to have absolutely nothing to do with the subject matter at hand.

In the course of our examination of this model, we will study the dynamics of a two-dimensional map and discover various mechanisms which lead to chaotic behavior and which are associated with the presence of fractal sets with a complicated geometric structure, such as hyperbolic attractors, horseshoes, and homoclinic tangles. We will study and compute some dimensional characteristics of these fractal sets using the machinery developed in the previous chapters, and we will see how they relate to the dynamics of the map. In fact, we will deal not just with a single two-dimensional map, but with a family of such maps, depending on a parameter, and we will see some dramatic

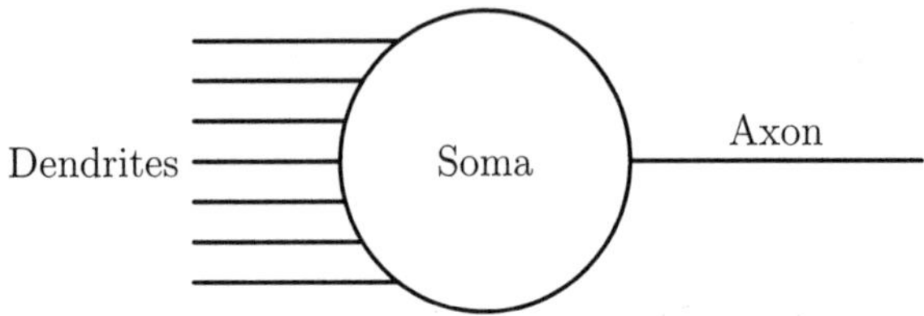

Figure 5.1. A schematic diagram of a neuron.

changes (bifurcations) in the dynamics of the map as the parameter varies.

First, though, we introduce the model, and thus we must speak biologically for a spell.

The (functional) structure of a neuron is shown in Figure 5.1. The neuron receives signals along its *dendrites*, of which it may have up to several hundred, and after processing these signals in the *soma*, it sends a single output along its *axon*.

An axon can be thought of as a tube whose boundary is selectively permeable to various ions. This permeability is mediated via a multitude of *ion channels*, each of which may be either open or closed; when open, each channel permits one species of ion (the most important being sodium and potassium) to pass from one side of the boundary to the other, altering the relative concentrations of each type of ion between the interior of the axon and the external environment. If these concentrations differ, then there is a voltage difference across the boundary of the tube, between the interior and exterior.

This voltage difference (called the *action potential* or sometimes the *transmembrane potential*) can vary in time; it affects the opening and closing of ion channels, and hence the rate at which charge flows across the boundary, which feeds back to influence the voltage difference itself. The voltage difference at one location on the axon also affects the time evolution of the voltage difference at nearby locations; in this manner, the action potential can propagate along the length of the axon, which is how signals are transmitted from one neuron to another.

In order to understand the propagation of an impulse through a neuron, we want a model which describes two things:

(1) The time evolution of the action potential at a given site on the axon, without regard to its value elsewhere.

(2) Diffusion of the action potential down the length of the axon, and the consequent propagation of an electrical signal from one neuron to another.

A naïve approach would be to directly apply the fundamental laws of physics which govern the interaction and motion of single particles (i.e., molecules), and to keep track of absolutely everything that goes on in the neuron. This is readily seen to be a preposterous line of attack due to the sheer scale of the system, which has many levels of structure lying between the macroscopic description we are interested in and the microscopic level of elementary particles at which the fundamental laws apply. With billions upon billions of ions moving in each axon, it is foolhardy to try to track each individual particle.

Having ruled out the bottom-up approach, we resort to a top-down *phenomenological* approach; that is, we use our knowledge about the functional structure of the system in conjunction with empirical observations to build a model which is consistent with the fundamental theory, but is not directly derived from it.

Perhaps the simplest model that captures some of the essential qualitative features of the propagation of an impulse along an axon is the *FitzHugh–Nagumo model*. This model was first suggested by Richard FitzHugh in 1961, and it was investigated by means of electric circuits in the following year by Jin-Ichi Nagumo. It is a simplified version of the Hodgkin–Huxley model, which describes in a more detailed manner the activation and deactivation dynamics of a spiking neuron, and the concomitant propagation of an action potential along the axon.

Before describing the FitzHugh–Nagumo model itself, we offer some loose heuristics for the form of the equations. For now, we write $u(t)$ for the action potential at a particular site on the axon, that is, the difference in voltage between the inside and outside of the tube. We will introduce spatial dependence at a later point.

An important model for a quantity which varies more or less periodically (which the action potential is observed to do in certain cases) but which may decay as time passes is the *damped harmonic oscillator*

$$(5.1) \qquad \ddot{u} + \zeta\dot{u} + cu = 0,$$

where ζ is a constant that determines the degree to which oscillations are damped and tend to die out. Although this model is far too simplistic to describe the behaviour of a neuron, it gives us a place to start. As with any higher-order ordinary differential equation, we can write (5.1) as a system of first-order ODEs; if we introduce an auxiliary variable v, defined by $\dot{v} = cu$, then (5.1) implies that

$$\dot{v} = -\ddot{u} - \zeta\dot{u},$$

and hence $v = -\dot{u} - \zeta u$. Thus (5.1) is equivalent to the first-order system

$$\dot{u} = -\zeta u - v,$$
$$\dot{v} = cu.$$

Allowing the damping coefficient to vary as a function of u leads to various non-linearly damped oscillators; one of the first of these to be studied was the *van der Pol oscillator*, introduced by the Dutch physicist Balthasar van der Pol in 1920 to model certain electrical circuits:

$$(5.2) \qquad \ddot{u} + \zeta(u^2 - 1)\dot{u} + cu = 0.$$

Using the same auxiliary variable as before, we integrate

$$\dot{v} = cu = -\ddot{u} + \zeta(\dot{u} - u^2\dot{u})$$

to obtain $v = -\dot{u} + \zeta(u - u^3/3)$, and thus we can write (5.2) as

$$\dot{u} = \zeta\left(u - \frac{u^3}{3}\right) - v,$$
$$(5.3)$$
$$\dot{v} = cu.$$

Based on the system (5.3) for the van der Pol oscillator, Richard FitzHugh devised the following *Bonhoeffer–van der Pol model* for the behaviour of an action potential $u(t)$ in a neuron, taking into account the presence of an external stimulus $I(t)$:

$$\begin{aligned}
\dot{u} &= u - \frac{u^3}{3} - v + I, \\
\dot{v} &= cu - \gamma - dv.
\end{aligned}$$
(5.4)

Here γ, c, and d are fixed parameters which must be determined empirically.

It is worth noting the particular structure of the system (5.4). The derivative $\dot{v}$ depends linearly on u and v, while the derivative $\dot{u}$ depends linearly on v and the external term I, and depends on itself via a cubic polynomial. This qualitative structure is the key to the behaviour of solutions of (5.4), and so the following abstract form is often of interest:

$$\begin{aligned}
\dot{u} &= ag(u) - bv + I, \\
\dot{v} &= cu - dv.
\end{aligned}$$

Here g is a cubic polynomial, I is the external signal, and the positive parameters a, b, c, and d depend on physical properties of the system.

Finally, we can add a spatial dimension to the system. Model the axon as lying on the positive x-axis, and let $u(x,t)$ denote the action potential at a distance x from the soma and a time t. As before, $v(x,t)$ is an auxiliary recovery variable (whose physical interpretation has to do with the concentration of various ions). Adding a diffusion term to allow the signal to propagate and choosing as our cubic polynomial $g(u) = -u(u - \theta)(u - 1)$, where $0 < \theta < 1$, we obtain the FitzHugh–Nagumo model:

$$\begin{aligned}
\frac{\partial u}{\partial t} &= -au(u - \theta)(u - 1) - bv + I + \kappa \frac{\partial^2 u}{\partial x^2}, \\
\frac{\partial v}{\partial t} &= cu - dv.
\end{aligned}$$
(5.5)

Since the effect of diffusion is small relative to the other processes at work, κ is of higher order than the other parameters.

The system of equations (5.5) is used to simulate the propagation of signals (in the form of traveling waves) in various *excitable media*, such as heart tissue or nerve fibre. The success of this system in the biological sciences is partly due to the fact that some properties of the model are analytically tractable, so that computer simulations need not be used in all cases. However, a full reconstruction of the global behaviour of solutions of (5.5) is too complicated to be done analytically, and so we must turn to numerical computations in order to fully understand the system.

b. Numerical investigations: From continuous to discrete. Having introduced the FitzHugh–Nagumo model (5.5) for the time evolution of an action potential along an axon, we would now like to understand how the solutions of (5.5) behave. Ideally, they ought to reflect the empirically observed behaviour of real neurons.

Because the system of partial differential equations in (5.5) does not admit a closed-form analytic solution (or at least, no such solution has been found), we must resort to numerical analysis to study it. Thus we turn the problem over to a computer; but what will the computer do? It will approximate the solution by discretising the problem: by replacing the various derivatives in (5.5) with their discrete analogues, one obtains a map which can be iterated. So let's follow along with the general scheme the computer will use. ...

Fixing some small value of $\Delta > 0$, the derivative $\partial u/\partial t$ is approximately equal to

$$\frac{u(x, t + \Delta) - u(x, t)}{\Delta},$$

and similarly for $\partial v/\partial t$. The second derivative $\partial^2 u/\partial x^2$ is approximately equal to

$$\frac{u(x + \Delta, t) - 2u(x, t) + u(x - \Delta, t)}{\Delta^2},$$

and so the system (5.5) is approximated by

$$\frac{u(x, t + \Delta) - u(x, t)}{\Delta} = -au(x, t)(u(x, t) - \theta)(u(x, t) - 1) - bv(x, t)$$
$$+ \kappa \frac{u(x + \Delta, t) - 2u(x, t) + u(x - \Delta, t)}{\Delta^2},$$
$$\frac{v(x, t + \Delta) - v(x, t)}{\Delta} = -dv(x, t) + cu(x, t).$$

If we write $u_k(n) = u(k\Delta, n\Delta)$ for the value of u at position $k\Delta$ and time $n\Delta$, and similarly for v, then this becomes

$$u_k(n+1) = u_k(n) - a\Delta u_k(n)(u_k(n) - \theta)(u_k(n) - 1) - b\Delta v_k(n)$$
$$+ \kappa\frac{u_{k+1}(n) - 2u_k(n) + u_{k-1}(n)}{\Delta},$$
$$v_k(n+1) = v_k(n) - d\Delta v_k(n) + c\Delta u_k(n).$$

Writing $g(\mathbf{u}(n))$ for the diffusion term, we have

$$u_k(n+1) = u_k(n) - Au_k(n)(u_k(n) - \theta)(u_k(n) - 1)$$

(5.6)
$$- \alpha v_k(n) + \kappa g(\mathbf{u}(n)),$$
$$v_k(n+1) = \beta u_k(n) + \gamma v_k(n),$$

where $\alpha = b\Delta$, $\beta = c\Delta$, $\gamma = 1 - d\Delta$, and $A = a\Delta$. For our purposes we shall assume that α and β are small, θ is near $1/2$, and γ is near 1. These parameters control the linear terms in the system; the non-linear part of the system, which is responsible for most of the interesting behaviour, is controlled by the parameter A, which is referred to as the *leading parameter*. Different values of A reflect different physical properties of the system being studied, and may lead to different qualititative behaviours of the model; thus a question of interest to us is how the behaviour of solutions of (5.6) changes as A varies.

The discrete system (5.6) is called a *coupled map lattice* (CML). The name comes about as follows:

(1) We only deal with the values of u and v on a discrete set within the original (continuous) domain; in the original FitzHugh–Nagumo model, both u and v take two *real* arguments, and so their domain is $\mathbb{R} \times \mathbb{R}$ (one dimension for space, one for time), while in (5.6), we are only interested in their values on a *lattice* $\Delta\mathbb{Z} \times \Delta\mathbb{Z} \subset \mathbb{R} \times \mathbb{R}$.

(2) At any given time step, the value of u and v at each site k is primarily determined by the value of u and v at that same site at the previous time step, via the *local map* $f\colon \mathbb{R}^2 \to \mathbb{R}^2$ given by $f(u,v) = (f_1(u,v), f_2(u,v))$, where

(5.7)
$$f_1(u,v) = u - Au(u - \theta)(u - 1) - \alpha v,$$
$$f_2(u,v) = \beta u + \gamma v.$$

(3) The local maps at each site are *coupled* together by interactions such as $\kappa g(\mathbf{u})$, which allow the value of u at a site k to be affected by the value at nearby sites (we could also consider diffusion in v, but do not in this particular model). This interaction is typically small compared to the magnitude of the local map, but cannot be neglected when studying the global dynamics as different sites can have large effects on each other over time.

Coupled map lattices were first introduced by the Japanese scientist Kunihiko Kaneko and are in many ways easier to study than the PDEs from which they are often derived. The process of iterating the system (5.6) is quite straightforward, and by so doing, we obtain values $u_k(n)$ and $v_k(n)$ which "ought" to lie near the values $u(k\Delta, n\Delta)$ and $v(k\Delta, n\Delta)$ in the solution of the system of PDEs (5.5).

There will, of course, be some small error term due to the approximations we have made. We would be quite happy if this error term remained small and negligible, but the fact of the matter is that for many of the parameter values in which we are interested, this error term increases relatively quickly, so that the asymptotic behaviour of solutions of the two systems may be quite different. In particular, the behaviour of the discrete system (which the computer calculates) is not necessarily a reliable guide to the behaviour of the continuous system.

One response to this difficulty is to use a more sophisticated numerical method. The approach described above to derive a CML from the corresponding PDE is known as *Euler's method*, and is very simplistic; other, more advanced methods are possible, which yield better approximations. However, these are still approximations, and over time, some accumulation of the error term is unavoidable; indeed, different approximation techniques may lead to solutions whose qualitative properties are quite different. Thus it is not clear just how one is to examine the asymptotic behaviour of the FitzHugh–Nagumo model using numerical methods.

At this point, we recall that what we are really interested in studying is the physical phenomenon itself; thus the important question is not how well the discrete model approximates the continuous model, but how well it approximates the physical propagation of an impulse through a neuron. Rather than comparing the solutions of the CML (5.6) to the solutions of the PDE (5.5), we can compare them directly to the observed data. If they turn out to approximate the real system as well as or better than the solutions of the PDE do, then we may simply consider the discrete model, rather than the continuous one.

The biological literature is divided over whether or not this is the case; both the discrete and the continuous FitzHugh–Nagumo models have their advocates, and in fact, which of the two models is a better approximation to the real system depends on which values of the parameters we are interested in. For our part, we will examine the coupled map lattice (5.6), as it is more tractable and will be seen to be of interest from a purely mathematical point of view as well.

Despite the ease with which the CML can be used in computer simulations (being given as a system of recursive equations), it can actually be very difficult to describe its qualitative global behaviour, due to the interaction between the various sites. An analysis of this problem (which is well beyond the scope of this book) reveals that the global behaviour is to a great degree determined by the local map (5.7), particularly if the local map is chaotic, and so it is to this map that we will now turn our attention.

Lecture 22

a. Studying the local map. We now begin to examine the dynamics of the local map (5.7) for the coupled map lattice associated to the FitzHugh–Nagumo model. We will fix the parameters α, β, γ, and θ, and allow the leading parameter A to vary; thus we are actually studying a *family* of maps, one for each value of A.

This is a fairly common situation; real-world systems do depend on parameters and thus are modeled not by a single map f but a family of maps f_A. In general, the analysis proceeds in two steps.

Step one. For a particular value of the parameter A, give a qualitative description of the global geometric picture of trajectories of f_A.

To do so, we should determine the asymptotic behaviour of a "randomly" chosen orbit, by determining which of the following $f_A^n(x)$ does as $n \to \infty$:

(a) Converge to a fixed point.

(b) Converge to a periodic orbit.

(c) Fluctuate chaotically.

It is also of interest to understand in what manner this asymptotic behaviour is reached; do typical trajectories immediately begin to converge to a fixed point? Do they fluctuate periodically or chaotically for some finite amount of time before converging? Finally, it is possible that different sorts of behaviour coexist; maybe half the trajectories converge to a fixed point, and the other half fluctuate chaotically.

Step two. Describe how this global picture changes when the parameter A varies. Usually when A increases slightly, the qualitative behaviour of trajectories stays more or less the same, with dramatic changes occurring only at some particular (usually isolated) values of the parameter A, called *bifurcation values* (we will give a more formal definition in Chapter 6).

In the course of understanding the global behaviour of trajectories and how it changes with the leading parameter for the FitzHugh–Nagumo model, we will develop a number of tools which allow us to gain some insight as to which of the three types of behaviour dominates for an arbitrary map. Thus certain parts of our discussion will be relevant far beyond the particular context of this single model.

Finding the fixed points of a map $f \colon \mathbb{R}^d \to \mathbb{R}^d$ amounts to solving the equation $f(\mathbf{x}) = \mathbf{x}$. For the map (5.7), this leads to the following equations for the two coordinates:

$$u = u - Au(u - \theta)(u - 1) - \alpha v,$$
$$v = \beta u + \gamma v.$$

From the second of these, we obtain

$$(5.8) \qquad\qquad v = \frac{\beta u}{1 - \gamma},$$

and then the first becomes

$$Au(u - \theta)(u - 1) + \frac{\alpha\beta}{1 - \gamma}u = 0.$$

Thus we have a fixed point at the origin, where $u = v = 0$, and any other fixed point must satisfy

$$A(u - \theta)(u - 1) + \frac{\alpha\beta}{1 - \gamma} = 0.$$

Solving this quadratic equation yields

$$(5.9) \qquad u = \frac{1}{2}\left(\theta + 1 \pm \sqrt{(\theta - 1)^2 - \frac{4\alpha\beta}{A(1 - \gamma)}}\right).$$

The discriminant is non-negative if and only if

$$A \geq A_0 = \frac{4\alpha\beta}{(1 - \gamma)(1 - \theta)^2},$$

and so we see that for $0 < A < A_0$, the origin is the only fixed point of f. For $A = A_0$, there is exactly one more fixed point, and for $A > A_0$, there are two more, given by (5.9) and (5.8); we will denote these by $\mathbf{p}_1$ and $\mathbf{p}_2$.

Thus A_0 marks the boundary between two qualitatively different sorts of behaviour. Imagine a tuning knob which controls the parameter A; if we begin with the knob turned so that $0 < A < A_0$, then the system has only one fixed point, and this general structure persists for a little while as we turn the knob and increase A. However, when we turn the knob far enough that A reaches A_0, the system undergoes a bifurcation, and two new fixed points appear.

b. Stability of fixed points for general maps. Having found the fixed points of a map f, the next step is to determine their stability— whether they attract or repel nearby trajectories, and how that behaviour depends on the value of the parameter A. The tool that we use for this purpose is the Jacobian derivative, which at each point $\mathbf{x} \in \mathbb{R}^2$ is a linear map $Df(\mathbf{x})\colon \mathbb{R}^2 \to \mathbb{R}^2$ given by the matrix

$$(5.10) \qquad Df(\mathbf{x}) = \begin{pmatrix} \frac{\partial f_1}{\partial x_1}(\mathbf{x}) & \frac{\partial f_1}{\partial x_2}(\mathbf{x}) \\ \frac{\partial f_2}{\partial x_1}(\mathbf{x}) & \frac{\partial f_2}{\partial x_2}(\mathbf{x}) \end{pmatrix}.$$

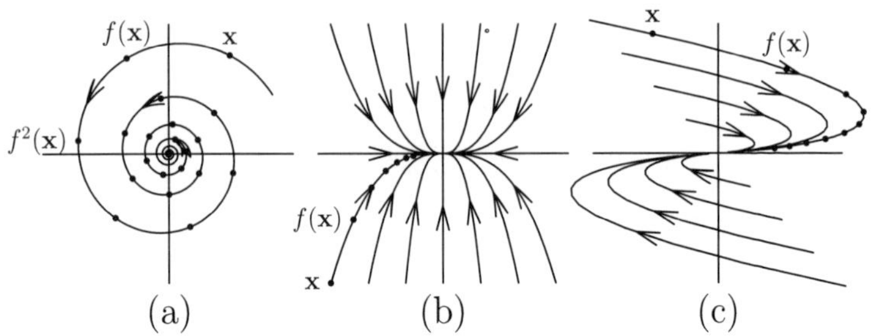

Figure 5.2. Trajectories near an attracting fixed point.

This linear map describes the behaviour of f in a neighbourhood of a fixed point. Specifically, we recall the following result from the calculus of several variables:[1]

Proposition 5.1. *Let $f\colon \mathbb{R}^2 \to \mathbb{R}^2$ be continuously differentiable in a neighbourhood of $\mathbf{p}_0$. Then we have*

$$(5.11) \qquad f(\mathbf{x}) = f(\mathbf{p}_0) + Df(\mathbf{p}_0)(\mathbf{x} - \mathbf{p}_0) + \mathbf{r}(\mathbf{x} - \mathbf{p}_0),$$

where $\mathbf{r}$ is an error term such that

$$(5.12) \qquad\qquad \lim_{\mathbf{y} \to \mathbf{0}} \frac{\|\mathbf{r}(\mathbf{y})\|}{\|\mathbf{y}\|} = 0.$$

Heuristically, this says that in a small neighbourhood of $\mathbf{p}_0$, f behaves like a linear map, with an error term that is small relative to $\mathbf{x} - \mathbf{p}_0$. (The condition (5.12) is often written as $\mathbf{r}(\mathbf{y}) = o(\mathbf{y})$.) In particular, if $\mathbf{p}_0$ is a fixed point, then the action of f near $\mathbf{p}_0$ is close to the action of the linear map given by the matrix $Df(\mathbf{p}_0)$.

Thus we briefly recall the possible qualitative behaviours for a linear map $A\colon \mathbb{R}^2 \to \mathbb{R}^2$. Let λ and μ be the eigenvalues of A, with $|\lambda| \le |\mu|$. Assuming neither λ or μ lies on the unit circle, there are three possibilities for the behaviour of trajectories vis-à-vis the fixed point $\mathbf{0}$.

(1) $|\lambda| \le |\mu| < 1$: All trajectories of A converge to $\mathbf{0}$, making it an attracting fixed point f, also called a *node*. The manner in which trajectories converge depends on λ and μ. If λ and μ

---

[1]For simplicity of exposition, we state all results in the two-dimensional case; higher-dimensional analogues are available.

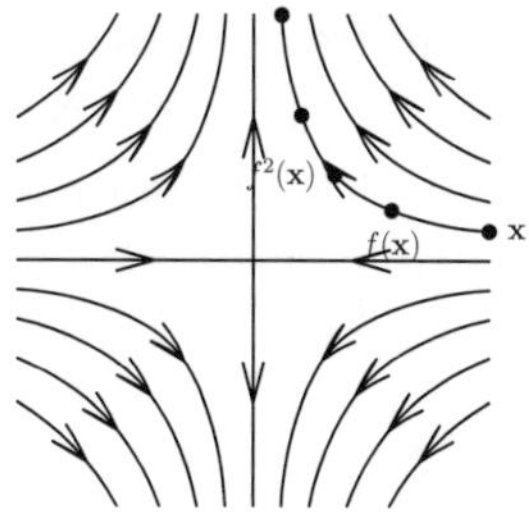

Figure 5.3. Trajectories near a hyperbolic fixed point.

are complex, then trajectories move along a logarithmic spiral, as shown in Figure 5.2(a); if λ and μ are real, then trajectories move along the curves shown in Figure 5.2(b) (if A is diagonalisable) or in Figure 5.2(c) (if A is similar to $\left(\begin{smallmatrix} \lambda & 1 \\ 0 & \lambda \end{smallmatrix}\right)$).

(2) $|\lambda| < 1 < |\mu|$: $\mathbf{0}$ is a *hyperbolic* fixed point, also called a *saddle*, shown in Figure 5.3. From one direction (the stable direction, horizontal in the figure), which corresponds to the eigenline for λ, trajectories approach $\mathbf{0}$ as $n \to +\infty$, while from another direction (the unstable direction, vertical in the figure), corresponding to the eigenline for μ, the *backwards* trajectories approach $\mathbf{0}$ as $n \to -\infty$. All other trajectories follow hyperbola-like paths, at first moving closer to $\mathbf{0}$, and then moving away.

(3) $1 < |\lambda| \le |\mu|$: All trajectories of A move away from $\mathbf{0}$, making it a repelling fixed point. Trajectories move along one of the curves in Figure 5.2, but in the opposite direction.

Remark. We stress that the trajectories of A are discrete collections of points which lie on the curves in Figures 5.2 and 5.3, rather than the curves themselves. Thus these curves are not themselves trajectories, but rather paths along which trajectories "jump" discretely.

Exercise 5.1. Let f be the linear map on $\mathbb{R}^2$ given by $f(x_1, x_2) = (2x_1 + x_2, x_1 + x_2)$. Determine the stability of the fixed point $\mathbf{0}$, and draw the different types of curves along which trajectories of f move.

We do not consider linear maps with eigenvalues lying on the unit circle, because in this case the error term in Proposition 5.1 can affect the asymptotic behaviour of trajectories of f near $\mathbf{p}_0$, and thus the stability of the fixed point may not be entirely determined by $Df(\mathbf{p}_0)$.

However, the condition that $Df(\mathbf{p}_0)$ have an eigenvalue λ with $|\lambda| = 1$ is a very restrictive one, and it is "typically" the case that $Df(\mathbf{p}_0)$ has no eigenvalues lying on the unit circle. In this case, our somewhat vague statement that the behaviour of f near $\mathbf{p}_0$ is determined by the matrix $Df(\mathbf{p}_0)$ may be made precise, as follows.

Theorem 5.2 (Hartman–Grobman theorem). *Let $f \colon \mathbb{R}^2 \to \mathbb{R}^2$ be continuously differentiable in a neighbourhood of a fixed point $\mathbf{p}_0$, and suppose that $|\lambda| \neq 1$ for all eigenvalues λ of $Df(\mathbf{p}_0)$. Then there exist open sets $U \ni \mathbf{p}_0$ and $V \ni \mathbf{0}$ and a homeomorphism $h \colon f(U) \to V$ such that $h(\mathbf{p}_0) = \mathbf{0}$ and for every $\mathbf{x} \in U$ we have*

$$f(\mathbf{x}) = h^{-1} \circ Df(\mathbf{p}_0) \circ h(\mathbf{x}).$$

Proof. See [**HK03**]. $\square$

The following two corollaries of Theorem 5.2 give us tools for determining when a fixed point $\mathbf{p}_0$ is stable or unstable.

Corollary 5.3. *Suppose that $f \colon \mathbb{R}^2 \to \mathbb{R}^2$ is continuously differentiable in a neighbourhood of some fixed point $\mathbf{p}_0$ and that $|\lambda| < 1$ for both eigenvalues λ of $Df(\mathbf{p}_0)$. Then $\mathbf{p}_0$ is stable: there exists $\varepsilon > 0$ such that for all $\mathbf{x} \in \mathbb{R}^2$ with $d(\mathbf{x}, \mathbf{p}_0) < \varepsilon$, we have $\lim_{k \to \infty} f^k(\mathbf{x}) = \mathbf{p}_0$; that is, the orbit of $\mathbf{x}$ is attracted to the fixed point $\mathbf{p}_0$.*

Corollary 5.4. *Suppose that $f \colon \mathbb{R}^2 \to \mathbb{R}^2$ is continuously differentiable in a neighbourhood of some fixed point $\mathbf{p}_0$ and that $|\lambda| > 1$ for both eigenvalues λ of $Df(\mathbf{p}_0)$. Then $\mathbf{p}_0$ is unstable: there exists $\varepsilon > 0$ such that for all $\mathbf{x} \in \mathbb{R}^2$ with $d(\mathbf{x}, \mathbf{p}_0) < \varepsilon$, we have $d(f^k(\mathbf{x}), \mathbf{p}_0) > \varepsilon$ for some $k > 0$; that is, the orbit of $\mathbf{x}$ is repelled from the fixed point $\mathbf{p}_0$.*

Thus in order to determine if a fixed point is stable or unstable, we merely check whether or not the eigenvalues of $Df(\mathbf{p}_0)$ are either all inside or all outside the unit circle.[2] If this is the case—if all the

---

[2] It is possible for $\mathbf{p}_0$ to be stable or unstable even if some or all of the eigenvalues lie *on* the unit circle, but this case is harder to analyse and depends on the higher derivatives of f.

eigenvalues lie on the same side of the unit circle—then the trajectories of f near $\mathbf{p}_0$ lie along the curves shown in Figure 5.2, with axes aligned in the eigendirections of $Df(\mathbf{p}_0)$.

The third alternative is that $Df(\mathbf{p}_0)$ has one eigenvalue inside the unit circle and one outside, in which case we use the corresponding terminology from the linear case and say that $\mathbf{p}_0$ is a *hyperbolic fixed point*, or *saddle*. We write λ^s and λ^u for the eigenvalues of $Df(\mathbf{p}_0)$, with $|\lambda^s| < 1 < |\lambda^u|$, and denote the corresponding eigenvectors by $\mathbf{v}^s$ and $\mathbf{v}^u$.

Now there is one direction in which $Df(\mathbf{p}_0)$ is contracting (the stable direction, parallel to $\mathbf{v}^s$), and one direction in which $Df(\mathbf{p}_0)^{-1}$ is contracting (the unstable direction, parallel to $\mathbf{v}^u$).[3] The lines in these directions are invariant under the action of $Df(\mathbf{p}_0)$, and if we consider their preimages under the conjugating homeomorphism h from Theorem 5.2, we obtain differentiable[4] curves $\gamma^s, \gamma^u \colon (-\varepsilon, \varepsilon) \to \mathbb{R}^2$ such that:

(1) $\gamma^s(0) = \gamma^u(0) = \mathbf{p}$.

(2) The tangent vectors $\dot{\gamma}^s(0)$ and $\dot{\gamma}^u(0)$ are parallel to $\mathbf{v}^s$ and $\mathbf{v}^u$, respectively.

(3) Given $t \in (-\varepsilon, \varepsilon)$ and points $\mathbf{x}^s = \gamma^s(t)$, $\mathbf{x}^u = \gamma^u(t)$, we have $f^k(\mathbf{x}^s) \to \mathbf{p}_0$ and $f^{-k}(\mathbf{x}^u) \to \mathbf{p}_0$ as $k \to \infty$.

The curves $W_\varepsilon^s = \{\gamma^s(t) \mid -\varepsilon < t < \varepsilon\}$ and $W_\varepsilon^u = \{\gamma^u(t) \mid -\varepsilon < t < \varepsilon\}$ are called the *local stable and unstable curves*, respectively. The properties just stated show that W_ε^s and W_ε^u pass through $\mathbf{p}_0$ in directions parallel to the stable and unstable eigenvectors $\mathbf{v}^s$ and $\mathbf{v}^u$, and that trajectories on W_ε^s are attracted to $\mathbf{p}_0$, while trajectories on W_ε^u are repelled (more precisely, attracted in backward time).

Because $Df(\mathbf{p}_0)^{-1}$ is contracting along $\mathbf{v}^u$, we have $f^{-1}(W_\varepsilon^u) \subset W_\varepsilon^u$; consequently, $W_\varepsilon^u \subset f(W_\varepsilon^u)$. Indeed, we may iterate f any number of times and obtain longer and longer curves $f^k(W_\varepsilon^u)$ such that

$$W_\varepsilon^s \subset f(W_\varepsilon^s) \subset \cdots \subset f^k(W_\varepsilon^s) \subset \cdots .$$

---

[3] Note that the condition that $Df(\mathbf{p}_0)^{-1}$ be contracting is a more stringent one than the condition that $Df(\mathbf{p}_0)$ be expanding; see Appendix.

[4] Despite the fact that the conjugacy h is only continuous in general, those curves are indeed differentiable.

Exercise 5.2. Let $f\colon \mathbb{R}^2 \to \mathbb{R}^2$ be invertible. For each $k \in \mathbb{N}$, show that $f^k(W_\varepsilon^u)$ is a curve through $\mathbf{p}_0$, tangent to $\mathbf{v}^u$, such that $f^{-k}(\mathbf{x}) \to \mathbf{p}_0$ for every $\mathbf{x} \in f^k(W_\varepsilon^u)$.

Each of the curves $f^k(W_\varepsilon^u)$ is a local unstable curve in its own right, with similar properties to W_ε^u by Exercise 5.2. Taking their union, we obtain the *global unstable curve* $W^u = \bigcup_{k \geq 0} f^k(W_\varepsilon^u)$, which has the following purely topological characterisation:

$$(5.13) \qquad W^u = \left\{ \mathbf{x} \in \mathbb{R}^2 \;\middle|\; \lim_{k \to \infty} f^{-k}(\mathbf{x}) = \mathbf{p}_0 \right\}.$$

The global stable curve can be constructed in the same way and characterised as the set of points whose forward trajectories approach $\mathbf{p}_0$.

Remark. The existence of stable and unstable curves through a hyperbolic fixed point is actually a special case of the *Hadamard–Perron theorem*, which holds in a more general setting. This result is one of the key tools in the proof of the Hartman–Grobman theorem (despite the fact that here we have given the interpretation of the stable and unstable curves in terms of that theorem).

Exercise 5.3. Find all the fixed points of each of the following maps of the plane, and determine the type of their stability (stable, unstable, or saddle). For each saddle point $\mathbf{p}$, determine the direction of the local stable and unstable curves by finding their tangent vectors at $\mathbf{p}$.

(a) $f(x, y) = (2x + y^2, 2x + 3y)$,

(b) $f(x, y) = (x^2 - 5x + y, x^2)$,

(c) $f(x, y) = \left(\sin\left(\frac{\pi}{3}x \right), \frac{y}{2} \right)$.

Exercise 5.4. Consider the map f of the plane given by

$$f(x, y) = (x^2 + y^2 - 2a^2, x + y).$$

Find all the fixed points of the map and determine the type of their stability depending on the value of the parameter a. Give a geometric description of the system; that is, describe the asymptotic behaviour of trajectories beginning in various regions of $\mathbb{R}^2$.

Lecture 23

a. Stability of fixed points for the FitzHugh–Nagumo model.
We now return our attention to the local map (5.7) for the FitzHugh–
Nagumo coupled map lattice and use the techniques of the previous
lecture to examine the stability of the fixed point $\mathbf{0}$ and of the two
new fixed points $\mathbf{p}_1$ and $\mathbf{p}_2$. For this particular map f, the Jacobian
derivative is

$$(5.14) \qquad Df(u,v) = \begin{pmatrix} 1 - A\theta + 2A(1+\theta)u - 3Au^2 & -\alpha \\ \beta & \gamma \end{pmatrix};$$

the eigenvalues of this matrix at the fixed points will tell us the stability of those fixed points (as long as none of them lie on the unit
circle). However, finding an explicit expression for these eigenvalues
involves a rather complicated computation, which we would prefer to
avoid.

Thus we use a slightly less direct approach; recall that the eigenvalues of a matrix are the roots of its characteristic polynomial, and
so they depend continuously on the coefficients of that polynomial.
Those coefficients in turn depend continuously on the entries of the
matrix (being sums of products of those entries), and so the eigenvalues depend continuously on the entries of Df.

In particular, given $\varepsilon > 0$, there exists δ such that for $\alpha, \beta < \delta$,
the eigenvalues of Df are within ε of the eigenvalues of the diagonal
matrix

$$(5.15) \qquad T = T(u,v) = \begin{pmatrix} 1 - A\theta + 2A(1+\theta)u - 3Au^2 & 0 \\ 0 & \gamma \end{pmatrix}.$$

Since T is diagonal, we can read off its eigenvalues directly. One
eigenvalue is γ, which is between 0 and 1 for the parameters we consider; thus there is always at least one stable (contracting) direction.
The other eigenvalue depends on parameters and the fixed point, and
so we consider these points separately.

For all values of A, the map f has a fixed point at the origin,
and at this point, the second eigenvalue of T is $1 - A\theta$. We have
$|1 - A\theta| < 1$ if and only if $0 < A < 2/\theta$, and since λ and μ are close
to $1 - A\theta$ and γ, there exists A_1 near $2/\theta$ such that the behaviour of
f near the origin is as follows:

(1) For $0 < A < A_1$, both eigenvalues of $Df(\mathbf{0})$ lie inside the unit circle; hence the origin is an attracting fixed point.

(2) For $A > A_1$, one eigenvalue of $Df(\mathbf{0})$ lies inside the unit circle (near γ), and the other lies outside the unit circle (near $1 - A\theta$); hence the origin is a hyperbolic fixed point (a saddle).

One can show that $A_1 > A_0$, and hence by the time A reaches A_1, there are two more fixed points to keep track of, $\mathbf{p}_1$ and $\mathbf{p}_2$.

To determine the behaviour of f near the fixed points $\mathbf{p}_1$ and $\mathbf{p}_2$, we first need to determine their location. Notice that the fixed points of f all lie on the line with equation (5.8); indeed, this line consists of precisely those points (u, v) for which $f_2(u, v) = v$. Similarly, the fixed points all lie on the cubic polynomial with equation

$$v = -\frac{A}{\alpha} u(u - \theta)(u - 1);$$

this curve contains precisely those points for which $f_1(u, v) = u$. Thus the fixed points of f are the points where this curve intersects the line (5.8), as shown in Figure 5.4.

Because β is small, this line is nearly horizontal, and so these points of intersection are very near the points where the graph of the cubic polynomial intersects the x-axis, which occurs at $u = 0$, $u = \theta$, and $u = 1$. Thus $\mathbf{p}_1$ and $\mathbf{p}_2$ are approximately $(\theta, 0)$ and

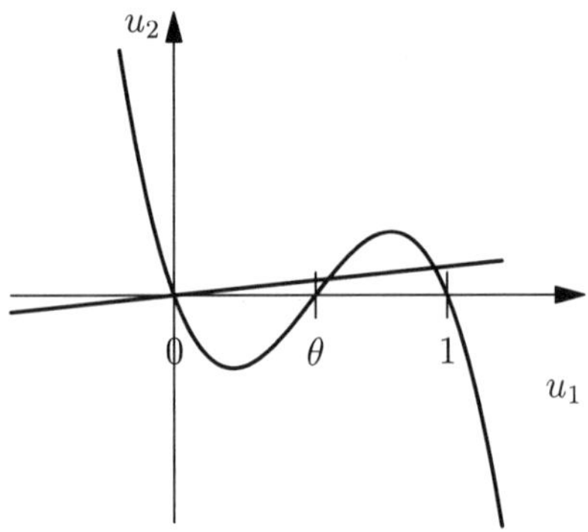

Figure 5.4. Finding the fixed points of f.

$(1, 0)$, respectively; we could also have seen this by looking at the form (5.9) takes when α and β vanish.

Now we can use the approximation (5.15) for Df to estimate the eigenvalues of $Df(\mathbf{p}_1)$ and $Df(\mathbf{p}_2)$. At $(u, v) = (0, 0)$, the matrix T in (5.15) has $\lambda_1 = \gamma$ as one eigenvalue (as always), and the other eigenvalue is

$$\lambda_2 = 1 - A\theta + 2A(1 + \theta)\theta - 3A\theta^2 = 1 + A\theta - A\theta^2.$$

Because $0 < \theta < 1$, we have $\lambda_2 = 1 + A\theta(1 - \theta) > 1$, and since $\lambda_1 = \gamma < 1$ and $Df(\mathbf{p}_1) \approx T(\mathbf{p}_1)$, we see that $\mathbf{p}_1$ is a hyperbolic fixed point for every value of the parameter A, with one stable and one unstable direction.

The situation is different at $\mathbf{p}_2$, where the second eigenvalue of T is

$$\lambda_2 = 1 - A\theta + 2A(1 + \theta) - 3A = 1 + A\theta - A,$$

with $|\lambda_2| < 1$ if and only if $0 < A < 2/(1 - \theta)$. In particular, there is a critical value $A_1' \approx 2/(1 - \theta)$ such that for $A_0 < A < A_1'$, the fixed point $\mathbf{p}_2$ is attracting, while for $A > A_1'$, it is a saddle.

b. Periodic points. The ultimate goal of all the analysis in which we are presently embroiled is to understand the dynamics of f by classifying the possible trajectories, and to describe how the dynamics change as A varies. By knowing the fixed points of f and their stability, we gain information about the *local* behaviour of the system; that is, how trajectories behave near the fixed points.

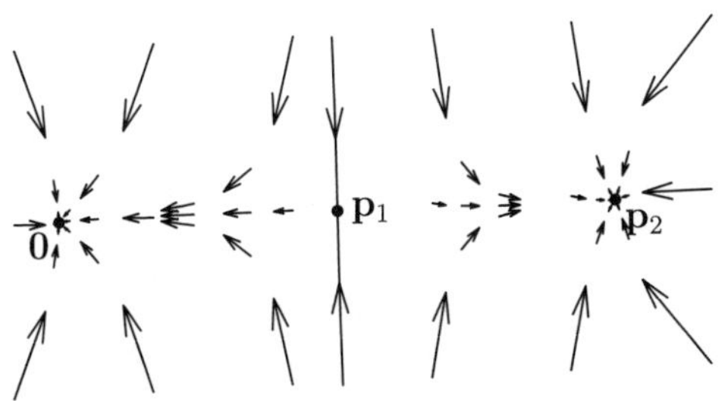

Figure 5.5. Orbits of f for $A_0 < A < \min\{A_1, A_1'\}$.

In general, it may be rather difficult to use this information to piece together a global picture which describes how trajectories behave *everywhere*. For the time being, we omit the details of this particular jigsaw puzzle, and simply assert that for the map f in the parameter ranges we consider, there are two sorts of trajectories: unbounded trajectories, which diverge to ∞, and bounded trajectories, which converge to a fixed point (or possibly, as we will soon see, to a periodic orbit).

Thus for parameter values $A_0 < A < \min\{A_1, A_1'\}$, we have three fixed points: $\mathbf{p}_1$ is a saddle, while $\mathbf{0}$ and $\mathbf{p}_2$ are stable. The bounded trajectories of f are as shown in Figure 5.5, and fall into one of three classes, based on which of the three fixed points they approach. The set of points whose trajectory approaches $\mathbf{p}_1$ is the stable curve through $\mathbf{p}_1$ and separates trajectories that approach $\mathbf{0}$ from trajectories that approach $\mathbf{p}_2$; for this reason, it is also called a *separatrix*. If a bounded trajectory begins to the left of the separatrix, it approaches $\mathbf{0}$; if it begins to the right, it approaches $\mathbf{p}_2$.

As A increases, the behaviour of the trajectories changes. For some value $A = \tilde{A}_1 < A_1$, the eigenvalue $\lambda_2 \approx 1 - A\theta$ is equal to 0. This eigenvalue governs the behaviour of points which lie just to the left or right of $\mathbf{0}$; when it is positive, for $A < \tilde{A}_1$, points to the left remain on the left, and points to the right remain on the right. For $A > \tilde{A}_1$, on the other hand, we have $\lambda_2 < 0$, and so points lying just to the left of $\mathbf{0}$ are mapped to points lying just to the right, and vice versa; in some sense, the orientation of the fixed point reverses at $\tilde{A}_1$. A similar reversal happens at $\mathbf{p}_2$ at the parameter value $\tilde{A}_1' \approx 1/(1 - \theta)$.

Increasing A still further, we eventually reach either A_1 or A_1'. Note that whether $A_1 < A_1'$ or $A_1' < A_1$ dependes on the values of the parameters θ, α, β, and γ; for the sake of concreteness, we will assume that $A_1 < A_1'$, so that $\mathbf{0}$ changes behaviour before $\mathbf{p}_2$ does.

As A approaches A_1 from below, the eigenvalue λ_2 approaches -1 from above. As long as the eigenvalue is greater than -1, the image $f(\mathbf{x})$ is closer to $\mathbf{0}$ than $\mathbf{x}$ itself is, and so $f^n(\mathbf{x}) \to \mathbf{0}$; however, the rate of convergence becomes slower and slower as λ_2 approaches -1. Finally, when $\lambda_2 < -1$, nearby points (in the horizontal direction) are mapped further away by f, and the fixed point 0 is now a saddle, with one stable and one unstable direction.

This change in behaviour means that the current picture is incomplete. For values of A slightly larger than A_1, some trajectories leave $\mathbf{0}$ slowly in a more or less horizontal direction, alternating between the left and right sides of the fixed point. Where do they go? They cannot immediately approach the next fixed point, $\mathbf{p}_1$, since it is also a saddle which is repelling in the horizontal direction.

What is missing? We have found all the fixed points, which represent the simplest possible orbits. The next simplest type of orbit is a periodic orbit, for which the trajectory returns to the initial point after some finite number of iterations.

In the present case, such an orbit appears around the fixed point $\mathbf{0}$ when the stability changes at A_1; for $A > A_1$, there exist two points $\mathbf{q}_1$ and $\mathbf{q}_2$ (which depend on A) such that $f(\mathbf{q}_1) = \mathbf{q}_2$ and $f(\mathbf{q}_2) = \mathbf{q}_1$. This orbit is the missing piece of the puzzle; trajectories which are repelled from $\mathbf{0}$ are attracted to the nearby periodic orbit of period 2.

Are all nearby orbits so attracted? How do we determine the stability of a periodic orbit? We use the fact that both $\mathbf{q}_1$ and $\mathbf{q}_2$ are mapped to themselves by f^2, which is a manifestation of the general fact that periodic points of f with period n correspond to fixed points of f^n. The stability of the periodic orbit for f is given by the stability of the fixed point for f^n, which in this case is determined by the eigenvalues of $Df^2(\mathbf{q}_i)$. One may easily see that it does not matter whether we compute the Jacobian at $\mathbf{q}_1$ or at $\mathbf{q}_2$, since by the chain rule,

$$
\begin{aligned}
Df^2(\mathbf{q}_1) &= Df(f(\mathbf{q}_1))Df(\mathbf{q}_1) \\
&= Df(\mathbf{q}_2)Df(\mathbf{q}_1) \\
&= (Df(\mathbf{q}_1))^{-1}Df(\mathbf{q}_1)Df(\mathbf{q}_2)Df(\mathbf{q}_1) \\
&= Df(\mathbf{q}_1)^{-1}Df^2(\mathbf{q}_2)Df(\mathbf{q}_1);
\end{aligned}
$$

it follows that $Df^2(\mathbf{q}_1)$ and $Df^2(\mathbf{q}_2)$ are similar matrices, and hence have the same eigenvalues.

Exercise 5.5. Find all periodic points of period 2 of the map

$$f(x, y) = \left(x^2 - \frac{3}{8}y^2, 2x\right),$$

and determine the type of their stability (stable, unstable, or saddle).

In order to compute, or even estimate, the eigenvalues of $Df^2(\mathbf{q}_i)$, we would first need to find the points $\mathbf{q}_1$ and $\mathbf{q}_2$, which involves solving the equation $f^2(u, v) = (u, v)$. Even with the approximation $\alpha = \beta = 0$, this leads to a polynomial of degree nine, which is algebraically intractable.

Numerical evidence (for the two-dimensional system), along with some geometric reasoning (which we give later for its one-dimensional approximation), indicates that the eigenvalue of $Df^2(\mathbf{q}_i)$ corresponding to the horizontal direction is slightly smaller than 1 when A is just larger than A_1. Thus when it appears, the period-2 orbit is attracting, and every trajectory of the map f approaches one of the three fixed points or the period-2 orbit.

Definition 5.5. A dynamical system $f \colon \mathbb{R}^d \to \mathbb{R}^d$ is called *Morse–Smale* if it has finitely many periodic orbits, each of which is either attracting, repelling, or a saddle, such that given any initial condition $\mathbf{x}$, the trajectory $\{f^n(\mathbf{x})\}$ approaches one of these orbits.

For all the values of A we have examined so far, the local map (5.7) for the FitzHugh–Nagumo CML is a Morse–Smale system. This behaviour continues for some time as we increase A; as A increases further past A_1, the eigenvalue corresponding to the horizontal direction decreases, and the orbit becomes more strongly attracting. When A increases to the point $\tilde{A}_2$ where the eigenvalue passes 0 and becomes negative, the orientation of trajectories near the period-2 orbit changes, just as happened at $\tilde{A}_1$ for the fixed point $\mathbf{0}$.

Beyond $\tilde{A}_2$, the rate at which nearby trajectories converge to the period-2 orbit decreases, until at some value $A = A_2$, one of the eigenvalues of $Df^2(\mathbf{q}_i)$ passes -1; at this point, the orbit ceases to be attracting and becomes hyperbolic. The map f^2, for which $\mathbf{q}_1$ and $\mathbf{q}_2$ are fixed points, behaves just like f did, and it spawns an attracting period-2 orbit near each of the newly unstable fixed points. Together, these two orbits form an attracting period-4 orbit for the original map f, which has two points near each of $\mathbf{q}_1$ and $\mathbf{q}_2$. Thus the system has gained a new periodic orbit but is still Morse–Smale.

This behaviour continues; the period-4 orbit eventually becomes unstable, at which point a period-8 orbit is born, and so on. In fact, there exists a sequence of parameter values $A_1 < A_2 < A_3 < \cdots$ such that for $A_{n-1} < A < A_n$, f has a stable orbit of period 2^n in the region to the left of the stable curve through $\mathbf{p}_1$, and unstable orbits of period 2^k for all $0 \le k \le n$.

Meanwhile, the same story is unfolding to the right of the separatrix through $\mathbf{p}_1$; at A_1' the fixed point $\mathbf{p}_2$ spawns an attracting period-2 orbit, and we have a sequence $A_1' < A_2' < A_3' < \cdots$ such that for $A_{n-1}' < A < A_n'$, f has a stable orbit of period 2^n in the region to the *right* of the stable curve through $\mathbf{p}_1$, and unstable orbits of period 2^k for all $0 \le k \le n$.

Numerical analysis shows that the bifurcation values A_n (and A_n' as well) get closer and closer together as n grows and, in fact, converge to some value A_∞, at which we have orbits of period 2^n for *any* natural number n. In particular, there are infinitely many coexisting periodic orbits, which marks a fundamental change in the behaviour of f, as the system is no longer Morse–Smale.

The fact that the local map (5.7) is Morse–Smale for parameters in the range $0 < A < A_\infty$ means that in that regime, the behaviour of the system is in some sense relatively simple and easy to understand. For $A \ge A_\infty$, the behaviour is much more intricate, as suggested by the above remarks.

Lecture 24

a. Beyond period-doubling: Down the rabbit hole. In the previous lecture, we described the sequence of period-doubling bifurcations through which the local map (5.7) for the FitzHugh–Nagumo CML passes as A increases. In fact, we saw that there are two such sequences, one corresponding to the fixed point $\mathbf{0}$, for which the resulting periodic orbits lie to the left of the stable curve through $\mathbf{p}_1$, and one corresponding to $\mathbf{p}_2$, for which the periodic orbits lie to the right of that curve.

In order to visualise these concurrent period-doubling cascades, and also to get a sense of what lies beyond them, for $A > A_\infty$, we turn to the *bifurcation diagram* of f (sometimes called the *orbit diagram*). This is a numerically computed graphical device, presented in Figure 5.6, which shows the asymptotic behaviour of the map f across a range of values of the leading parameter A, given fixed values of the other parameters.

The horizontal axis of the diagram represents the parameter A, and the vertical axis represents the first coordinate u of the phase space $\mathbb{R}^2$; a vertical cross-section of the diagram is the projection of the attracting part of phase space onto the u-axis. The diagram is generated as follows:

(1) Fix a value of A and an initial condition (u, v), which should lie near the origin so that its orbit is bounded and to the left of the separatrix through $\mathbf{p}_1$; then compute the iterates $f^n(u, v)$.

(2) Ignore the first few iterates to give the transient part of the orbit time to die away;[5] in the case $A < A_\infty$, for example, this gives the orbit time to converge to a periodic orbit.

(3) Writing the subsequent iterations as (u_n, v_n), plot the points (A, u_n) on the diagram.[6]

(4) Choose a new value of A and repeat the whole procedure until enough points have been filled in to give a sense of the structure of the bottom half of the bifurcation diagram.

(5) Repeat the above procedure in its entirety, this time choosing initial conditions to the right of the separatrix through $\mathbf{p}_1$; this generates the top half of Figure 5.6.

We see in Figure 5.6 that for $A < A_1$, the points which are plotted are all quite close to the fixed point; for $A_1 < A < A_2$, they are close

---

[5] "First few" may mean several dozen, several hundred, or several thousand, depending on how refined a picture is desired and on how quickly the system converges (or is believed to converge) to its asymptotic behaviour.

[6] Again, how many of these points are plotted is in some sense a judgment call.

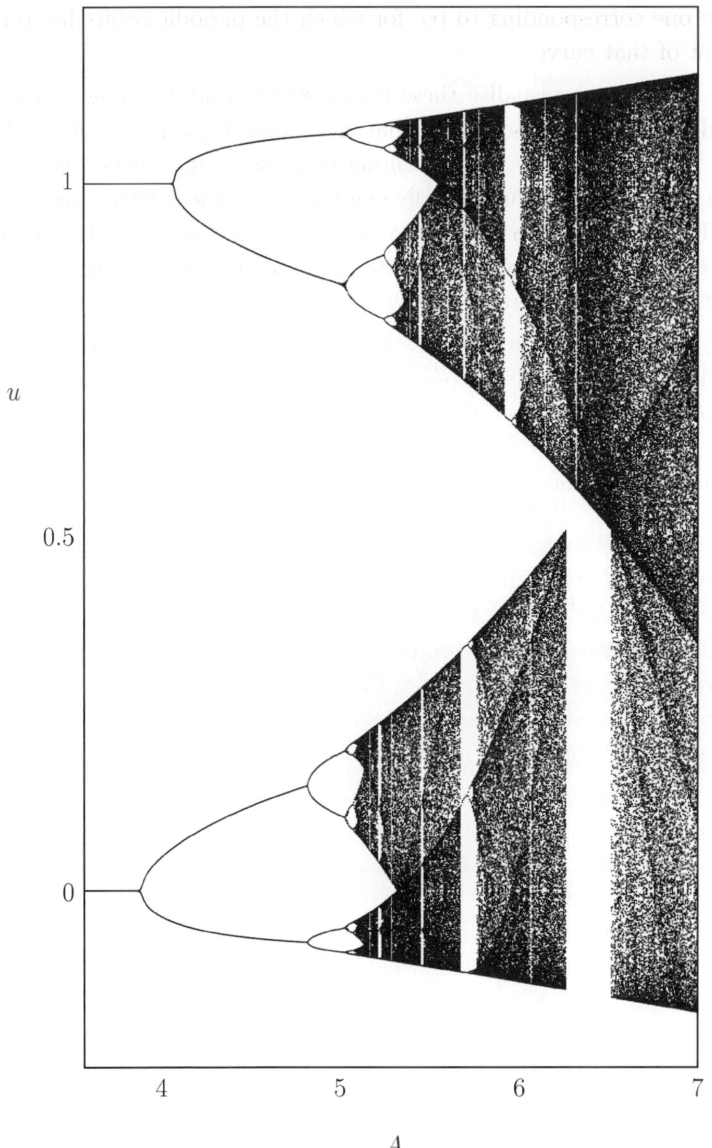

Figure 5.6. The bifurcation diagram for the local map (5.7) of the FitzHugh–Nagumo model with $\theta = .51$, $\alpha = .01$, $\beta = .02$, and $\gamma = .8$, as A varies from 3 to 7.

to the period-2 orbit, and so on. This reflects the fact that the long-term behaviour of any trajectory follows one of the periodic orbits, which is true for any Morse–Smale system.

For $A \geq A_\infty$, the situation is quite different. There are values of A for which the trajectory of a randomly chosen point is dense in a Cantor set (that is, a set homeomorphic to the usual middle-third Cantor set) or even an entire interval, and the map becomes chaotic. However, there are also windows of stability; the most noticeable of these are the two period-3 windows which occur near $A = 5.7$ and $A = 6$ in Figure 5.6. In these parameter ranges, we suddenly return from the chaos which seems to prevail past A_∞ to a more orderly regime; every orbit that we observe is attracted to a cycle of period 3, and the system seems relatively simple once again (although as we will see later, it is not Morse–Smale). There are windows of stability with other orders as well, scattered throughout the bifurcation diagram.

These moments of stability are transitory. Consider a window containing a stable periodic orbit of period n; as A increases, this attracting orbit becomes repelling and spawns an attracting periodic cycle of length $2n$. This too becomes repelling in its turn, shedding an attracting cycle of length $4n$, then $8n$, and so on; the whole period-doubling cascade is repeated, but over a much smaller range of the parameter A than in its original incarnation.

This reappearance of the period-doubling cascade suggests a sort of self-similarity of the bifurcation diagram, and indeed this diagram is self-similar in an asymptotic sense. We mention just one striking manifestation of this self-similarity. Recall that $A_n - A_{n-1}$ gives the length of the parameter interval for which the periodic orbit of length 2^n is attracting. From the bifurcation diagram, we see that this length goes to zero, and in fact it decreases exponentially in n; thus we write

$$\kappa = \lim_{n \to \infty} \frac{1}{n} \log(A_n - A_{n-1}).$$

Writing $\delta = e^\kappa$, we have $A_n - A_{n-1} \approx C\delta^n$ for some constant C. In and of itself, this would not be cause for any special excitement. However, we can do the same calculation for the period-doubling cascade $A_1' < A_2' < A_3' < \cdots$, or for the period-doubling cascade in any of the windows of stability, and the truly striking result is that we get the same value of δ in every case!

Even that remarkable congruence is not the end of the story. There are in fact many one-parameter families of maps which lead to similar bifurcation diagrams, with period-doubling cascades, windows of stability, and so on (we will study one such family, the logistic map, in Chapter 6). Under relatively mild conditions on the family of maps, we find the same old tale in each period-doubling cascade; the rate of decay of the lengths $A_n - A_{n-1}$ exists, and what is more, it is equal to δ.

The number δ is known as *Feigenbaum's constant*, after Mitchell Feigenbaum, who was the first to discover this example of *quantitative universality*.

b. Becoming one-dimensional. Despite the various heuristic justifications which have been offered, our analysis of the FitzHugh–Nagumo model, and in particular the bifurcation diagram in Figure 5.6, is based primarily on numerical evidence. Many of our claims have been based on the fact that f is a small perturbation of the map

$$\tilde{f}(u, v) = (u - Au(u - \theta)(u - 1), \gamma v)$$

and the conviction that since $0 < \gamma < 1$—and hence the v-coordinate goes to 0 under repeated iteration of this map—the essential behaviour is (or ought to be) given by the one-dimensional map $g(x) = x - Ax(x - \theta)(x - 1)$.

Exercise 5.6. Describe the behavior of all trajectories of the map $F\colon \mathbb{R}^2 \to \mathbb{R}^2$ given by $F(x, y) = (x^2 + c, y/2)$ for $c \geq 0$.

Many of the claims we have made so far regarding the structure of the bifurcation diagram of f have been proved for such one-dimensional maps such as the map g on which f is based. However, very little of that theory has been rigorously extended to the two-dimensional case. While most of the results are believed to carry over, that belief is based on numerical evidence and computer experiments rather than rigorous proofs.

Because the one-dimensional case is easier to deal with, we restrict our attention for the next little while to that setting, and consider continuous maps $g\colon \mathbb{R} \to \mathbb{R}$. In this context, a number of rigorous results can be proved regarding the orbit structure of the map g, beginning with the following exercise.

Exercise 5.7. Let $g\colon \mathbb{R} \to \mathbb{R}$ be continuous, and suppose that $a < b$ are such that $g([a, b]) \supset [a, b]$. Show that g has a fixed point in $[a, b]$.

Exercise 5.8. We might attempt to generalise the result of Exercise 5.7 to higher dimensions by claiming that if $X \subset \mathbb{R}^d$ is a closed set and $G\colon X \to \mathbb{R}^d$ is a continuous map such that $G(X) \supset X$, then G has a fixed point in X. Show that this claim is false by giving a counterexample.

Definition 5.6. If I and J are intervals in $\mathbb{R}$ such that $g(I) \supset J$, we say that I *g-covers* J.

Exercise 5.7 shows that any interval which g-covers itself contains a fixed point; as a corollary, we see that any interval which g^n-covers itself contains a periodic point whose period divides n. This gives us a tool for finding periodic orbits, which we use to prove the following result.

Proposition 5.7. *If a continuous map $g\colon \mathbb{R} \to \mathbb{R}$ has a period-3 orbit, then it has periodic points of all orders.*

Proof. Let $g\colon \mathbb{R} \to \mathbb{R}$ be continuous, and let $x_0 = g(x_2)$, $x_1 = g(x_0)$, and $x_2 = g(x_1)$ be the three points in a period-3 orbit. By changing the indices if necessary, we can assume that $x_0 < \min(x_1, x_2)$, and there are now two possibilities:

$$x_0 < x_1 < x_2 \qquad \text{or} \qquad x_0 < x_2 < x_1.$$

We prove the theorem in the case $x_0 < x_1 < x_2$; the proof in the other case is similar.

Let $I_1 = [x_0, x_1]$ and $I_2 = [x_1, x_2]$. It follows immediately from the Intermediate Value Theorem that

$$(5.16) \qquad g(I_1) \supset [g(x_0), g(x_1)] = [x_1, x_2] = I_2,$$

and also that

$$(5.17) \qquad g(I_2) \supset [g(x_2), g(x_1)] = [x_0, x_2] = I_1 \cup I_2.$$

Given $n \geq 1$, we find a periodic orbit of period n as follows. If $n = 1$, then we are after a fixed point of g, whose existence is guaranteed by the result of Exercise 5.7 applied to I_2.

An easy induction argument shows that $g^n(I_1) \supset I_2$ for all $n \geq 1$, and hence $g^n(I_1) \supset I_1$ for all $n \geq 2$. Then the result of Exercise 5.7 applied to I_1 and g^n shows that g^n has a fixed point x in I_1, which is almost enough to complete the proof. However, we want to prove the slightly stronger statement that n is the smallest period of x; that is, that $g^k(x) \neq x$ for $1 \leq k \leq n-1$. For this, we need the following lemma.

Lemma 5.8. *If $h \colon \mathbb{R} \to \mathbb{R}$ is continuous and $[a', b']$ h-covers $[a, b]$, then there exists a subinterval $[a'', b''] \subset [a', b']$ such that $h([a'', b'']) = [a, b]$.*

Proof. Let $E = \{x \in [a, b] \mid h(x) \leq a\}$ and $F = \{x \in [a, b] \mid h(x) \geq b\}$, and consider $a_1 = \sup E$, $a_2 = \sup F$. Without loss of generality, suppose that $a_1 < a_2$ (the proof in the other direction is similar). Then let $a'' = a_1$, $b'' = \inf F \cap [a_1, b']$, and the result follows. $\qquad\square$

From the lemma and (5.16)–(5.17), it follows that there exist intervals $I_{12} \subset I_1$ and $I_{21}, I_{22} \subset I_2$ such that $g(I_{ij}) = I_j$. Continuing, we find nested sequences of basic intervals $I_{w_1 \ldots w_n}$ such that $g(I_{w_1 \ldots w_n}) = I_{w_2 \ldots w_n}$. Because the length of these intervals may not go to zero as $n \to \infty$, we cannot carry out the full Cantor-like construction and obtain a conjugacy with the subshift Σ_A^+, $A = \left(\begin{smallmatrix} 0 & 1 \\ 1 & 1 \end{smallmatrix}\right)$; however, we can observe that for the particular sequence $w_1 = 1$, $w_2 = w_3 = \cdots = w_n = 2$, we have $g^n(I_{w_1 \ldots w_n}) = I_1 \cup I_2 \supset I_{w_1 \ldots w_n}$, and so there exists $x \in I_{w_1 \ldots w_n}$ such that $g^n(x) = x$. Because $g^k(x) \in I_2$ for all $1 \leq k \leq n-1$, n must be the minimal period of x. $\qquad\square$

A similar argument to the one used in the proof of Proposition 5.7 allows us to construct points with arbitrary itineraries, that is, trajectories which enter I_1 and I_2 in any conceivable pattern. This is one of the hallmarks of chaotic behaviour, and indeed, when Proposition 5.7 was first proved by Tien-Yien Li and James Yorke in 1975, the title of their paper was "Period Three Implies Chaos".

Unbeknownst to Li and Yorke, a rather more general result than Proposition 5.7 had been proved eleven years earlier, in 1964, by the Ukranian mathematician Aleksandr Sharkovsky. However, being behind the Iron Curtain, Sharkovsky had little access to the West, and so his result was not widely publicised until Sharkovsky and Yorke met in East Berlin shortly after the publication of Li and Yorke's paper.[7]

Sharkovsky's theorem places surprising restrictions on the combinations of periodic orbits which can exist in a given system, using the following non-standard ordering on the set of positive integers $\mathbb{N}$: given two integers m and n, we say that m precedes n (or equivalently, that n follows m), and write $m \prec n$, if m appears before n in the following list:

$$3, \ 5, \ 7, \ 9, \ldots$$
$$2\cdot 3, \ 2\cdot 5, \ 2\cdot 7, \ 2\cdot 9, \ \ldots$$
$$2^2\cdot 3, \ 2^2\cdot 5, \ 2^2\cdot 7, \ 2^2\cdot 9, \ \ldots$$
$$2^3\cdot 3, \ 2^3\cdot 5, \ 2^3\cdot 7, \ 2^3\cdot 9, \ \ldots$$
$$\ldots$$
$$\ldots, 2^n, \ 2^{n-1}, \ \ldots, 2^2, \ 2, \ 1.$$

Theorem 5.9 (Sharkovsky's theorem). *If a continuous map $g\colon \mathbb{R} \to \mathbb{R}$ has a periodic orbit of period m, then it also has a periodic orbit of period n for every n which follows m in the above ordering; that is, every $m \prec n$.*

Proof. See [**HK03**]. $\qquad\qquad\qquad\qquad\qquad\qquad\qquad\qquad\qquad$ □

The proof of Theorem 5.9 runs along much the same lines as the proof of Proposition 5.7, although the combinatorics are more intricate. The key ingredient in the proof is the Intermediate Value Theorem, which explains why it is crucial that the domain of g be $\mathbb{R}$. Indeed, if $X = \{z \in \mathbb{C} \mid |z| = 1\}$ is the unit circle, and $g\colon X \to X$ is the map $g(z) = e^{2\pi i/n}z$, then every periodic point of g has period n,

---

[7]The historical development of chaos theory, including this episode, is described in [**Gle87**].

in sharp contrast to the situation described in Sharkovsky's theorem; this example also shows that the theorem fails in higher dimensions.

In $\mathbb{R}$, however, the theorem gives a great deal of information about the periodic orbit structure of a continuous map g. For example, if g has a periodic point of period 2^n, then it must have periodic points of period 2^k for every $0 \leq k \leq n$, which is reminiscent of the behaviour we saw in the period-doubling cascade earlier. Similarly, if g has a periodic point whose period is not a power of two, then it must have periodic points of period 2^n for *every* n. In particular, we have the following corollary, which applies to any Morse–Smale system on the line:

Corollary 5.10. *If a continuous map $g \colon \mathbb{R} \to \mathbb{R}$ has only finitely many periodic points, then they all have a period which is a power of two.*

Sharkovsky's theorem leads us to suspect that even within the windows of stability in the bifurcation diagram in Figure 5.6, there are some very complicated dynamics going on. However, because it only applies to one-dimensional maps, we will in the next chapter turn our attention to such maps. In particular, we will study the family of logistic maps, where we discover a similar bifurcation diagram, complete with period-doubling cascades, windows of stability, and in the end, chaos.

Chapter 6

The Bifurcation Diagram for the Logistic Map

Lecture 25

a. Bifurcations of the logistic map. We return now to the logistic map, or rather the family of logistic maps, which was first introduced in Lecture 2; for a given parameter $c \in \mathbb{R}$, the map is

$$f_c(x) = x^2 + c.$$

We have already seen that for $c > 1/4$, all trajectories of f_c go to $+\infty$; we focus on what happens as c decreases. In particular, we make precise the notion of bifurcation, which we have already discussed, and examine the types of bifurcations which occur in the logistic family.

Recall that two continuous maps $f \colon X \to X$ and $g \colon Y \to Y$ are topologically conjugate if there exists a homeomorphism $\phi \colon Y \to X$ such that $f \circ \phi = \phi \circ g$. For example, the maps $f_c \colon x \mapsto x^2 + c$ and $g_\lambda \colon y \mapsto \lambda y(1 - y)$ are topologically conjugated by the homeomorphism $\phi \colon y \mapsto \frac{\lambda}{2}(1 - 2y)$, where c and λ are related by $4c = \lambda(2 - \lambda)$ in the appropriate parameter ranges (see Exercise 1.2).

It is often useful to think of the conjugating homeomorphism ϕ as a change of coordinates, under which f and g display the same dynamics, just as two similar matrices A and B have the same action under a suitable change of basis. Thus we say that two topologically conjugate maps have the same qualitative behaviour.

Exercise 6.1. Show that f_c and $f_{c'}$ are topologically conjugate for any $c, c' > 1/4$.

There are many cases in which changing the value of the parameter slightly does not change the qualitative behaviour of the map; for example, Exercise 6.1 shows that the logistic maps f_c with $c > 1/4$ are all topologically conjugate. Thus although the precise quantitative behaviour varies with the parameter c, the qualitative behaviour is unchanged.

A bifurcation occurs when an arbitrarily small change in the value of the parameter *does* change the qualitative behaviour of the map:

Definition 6.1. A one-parameter family of maps $F_c \colon X \to X$ has a *bifurcation* at c_0 if for all $\varepsilon > 0$, there exists a parameter value $c \in (c_0 - \varepsilon, c_0 + \varepsilon)$ for which F_c and F_{c_0} are not topologically conjugate.

Exercise 6.2. Let $F_c \colon X \to X$ be a one-parameter family of continuous maps on a topological space X, and write $\mathcal{B} \subset \mathbb{R}$ for the set of bifurcation values, that is, the set of values c_0 at which F_c has a bifurcation. Show that $\mathcal{B}$ is closed.

It is easy to see that the periodic orbit structure of a map is an invariant of topological conjugacy; that is, two topologically conjugate maps f and g must have the same numbers of fixed points, points of period 2, period 3, etc. Thus a change in this orbit structure, such as the appearance of any new periodic orbits, immediately heralds a bifurcation in the system. Similarly, because stability is determined by where trajectories converge, a change in the type of stability of a periodic orbit also indicates a bifurcation.

For the logistic family f_c, the first bifurcation occurs at $c = 1/4$; for $c > 1/4$ there are no fixed points, while for $c = 1/4$ there is one, and for $c < 1/4$ there are two, given by

$$p_1 = \frac{1 - \sqrt{1 - 4c}}{2}, \qquad p_2 = \frac{1 + \sqrt{1 - 4c}}{2}.$$

The stability of these fixed points is determined by the absolute value of the derivative, and we see that

$$f'_c(p_1) = 1 - \sqrt{1 - 4c}, \qquad f'_c(p_2) = 1 + \sqrt{1 - 4c}.$$

Thus p_2 is unstable for all values of c, while p_1 is stable for a little while after the bifurcation at $c = 1/4$. As long as this state of affairs persists, any trajectory which begins in the interior of the interval $I = [-p_2, p_2]$ converges to p_1, the two trajectories which begin at the endpoints converge to p_2, and any trajectory which begins outside of I diverges to $+\infty$.

So how long does this state of affairs persist? Observe that $f'_c(p_1)$ decreases as c decreases and also that $f'_c(p_1) = -1$ when $c = -3/4$. Thus for $-3/4 < c < 1/4$, all trajectories in $(-p_2, p_2)$ converge to p_1, but for $c < -3/4$, both fixed points are unstable. This implies that a bifurcation occurs at $-3/4$, since the stability of p_1 changes; to determine the behaviour of trajectories for $c < -3/4$, we look for periodic orbits, since there are no new fixed points.

Recall that a period-2 orbit of f_c corresponds to a fixed point of f_c^2, and so we want to solve the equation

$$f_c^2(x) = (x^2 + c)^2 + c = x;$$

that is, we want to find the roots of

$$x^4 + 2cx^2 - x + c^2 + c = 0.$$

This is made rather easier to solve by the observation that we already know two of the roots; the fixed points p_1 and p_2 of the original map f_c. In fact, the polynomial $f_c(x) - x$ divides $f_c^2(x) - x$, since a root of the former is obviously a root of the latter. Dividing, we obtain

$$\frac{x^4 + 2cx^2 - x + c^2 + c}{x^2 - x + c} = x^2 + x + 1 + c = 0,$$

which has solutions

$$q_1 = -\frac{1}{2} - \sqrt{-\frac{3}{4} - c}, \qquad q_2 = -\frac{1}{2} + \sqrt{-\frac{3}{4} - c}.$$

These are real numbers if and only if $c \leq -3/4$; in other words, f_c has a period-2 orbit on the real line if and only if the fixed point p_1 is unstable! One may easily verify that $f_c(q_1) = q_2$ and $f_c(q_2) = q_1$, and that p_1 lies between q_1 and q_2.

The stability of this period-2 orbit is given by the derivative of the map f_c^2 for which it is a fixed point; using the chain rule, we observe that

$$\begin{aligned}
(f_c^2)'(q_1) &= f_c'(q_2)f_c'(q_1) \\
&= (-1 + \sqrt{-3 - 4c})(-1 - \sqrt{-3 - 4c}) \\
&= 1 - (-3 - 4c) \\
&= 4 + 4c.
\end{aligned}$$

When the period-2 orbit is born, at $c = -3/4$, we have $(f_c^2)'(q_1) = 1$, and this quantity decreases as c decreases, becoming equal to -1 when $c = -5/4$. Thus for $-5/4 < c < -3/4$, the period-2 orbit is stable, and it is possible to show that every trajectory which begins in $(-p_2, p_2)$ asymptotically approaches the period-2 orbit, in the sense that $f^{2n}(x) \to q_i$ for either $i = 1$ or $i = 2$.

This pattern of behaviour continues as c decreases further, with successive periodic orbits of length 2^n becoming unstable and spawning stable orbits of length 2^{n+1}, which become unstable in their turn, and so on *ad infinitum*. However, the algebraic approach we have been following becomes increasingly messy, as we must deal with polynomials of higher and higher degree.

b. Classifying bifurcations. Consider the parameter values $c_0 = 1/4$ and $c_1 = -3/4$. At both of these values, the periodic orbit structure of the logistic map f_c changes, and so a bifurcation occurs; however, the bifurcations are of different sorts. As c decreases through c_0, we go from having no fixed points to having two, one stable and one unstable. At c_1, on the other hand, there is a preexisting fixed point, which persists through the bifurcation; the change is in the stability of that fixed point and in the appearance of an attracting period-2 orbit.

These two types of bifurcations are common enough to merit their own names: the bifurcation at c_0 is an example of a *tangent bifurcation* (sometimes called a *saddle-node* or *fold bifurcation*), and the bifurcation at c_1 is an example of a *period-doubling bifurcation* (sometimes called a *flip bifurcation*). The following definitions make these

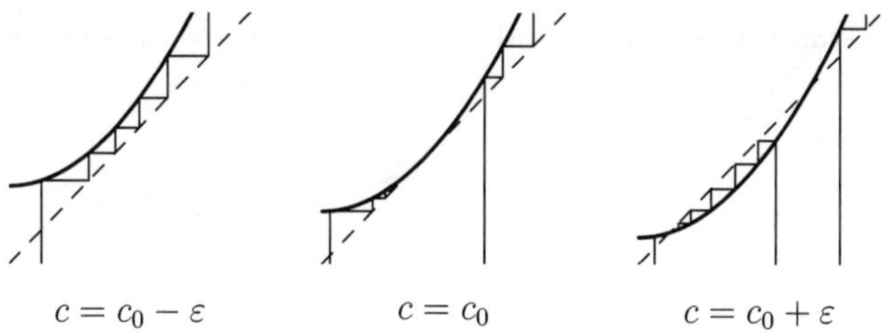

Figure 6.1. A tangent bifurcation.

notions precise; in what follows, $F_c\colon \mathbb{R} \to \mathbb{R}$ is any one-parameter family of continuous maps of the real line.

Definition 6.2. F_c has a *tangent (or saddle-node, or fold) bifurcation* at c_0 if there exists an open interval $I \subset \mathbb{R}$ such that for sufficiently small values of $\varepsilon > 0$,

(1) $F_{c_0-\varepsilon}$ has no fixed points in I;

(2) F_{c_0} has one fixed point in I, which is neutral; and

(3) $F_{c_0+\varepsilon}$ has two fixed points in I, one attracting and one repelling.

The conditions in the above definition imply that the fixed points appear as c increases through c_0. We also say that F_c has a tangent bifurcation at c_0 if the fixed points appear as c decreases, that is, if the above conditions hold with $c_0 - \varepsilon$ and $c_0 + \varepsilon$ interchanged.

A typical picture for a tangent bifurcation is shown in Figure 6.1. Note that at the value $c = c_0$, the graph of F_c is tangent to the bisectrix $y = x$, hence the name.

A tangent bifurcation is a *local* bifurcation, insofar as its definition only involves the behaviour of the map on a small interval I, and centres on the appearance at the critical parameter value c_0 of a fixed point. Indeed, suppose a one-parameter family of maps undergoes a tangent bifurcation at a map $f = F_{c_0}$. Then we have $f'(p) = 1$, as already mentioned; thus the bifurcation occurs precisely at the point where the linearisation of the map around a fixed point has an eigenvalue on the unit circle.

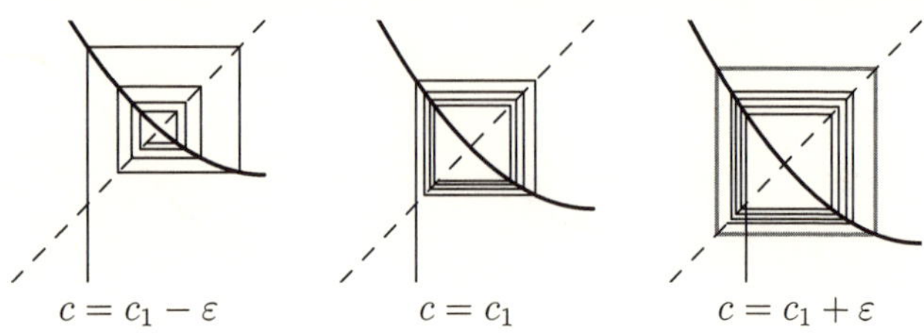

Figure 6.2. A period-doubling bifurcation.

In fact, this condition is necessary in order for a bifurcation involving a fixed point to occur. Consider an arbitrary differentiable map $f \colon \mathbb{R} \to \mathbb{R}$, and suppose that f has a fixed point p with $|f'(p)| \neq 1$. Then a small enough perturbation $\tilde{f}$ of f will have exactly one fixed point $\tilde{p}$ near p, with $\tilde{f}'(\tilde{p}) \approx f'(p)$, and so no bifurcation occurs at f. Thus a local bifurcation can only occur when f has a fixed point p with $f'(p) = \pm 1$ (or in the higher-dimensional setting, when an eigenvalue of $Df(\mathbf{p})$ crosses the unit circle).

The case $f'(p) = 1$ leads to the tangent bifurcation just described; when $f'(p) = -1$, which occurs at c_1 for the logistic map, we have the following sort of bifurcation.

Definition 6.3. F_c has a *period-doubling (or flip) bifurcation* at c_1 if there exists an open interval $I \subset \mathbb{R}$ that contains exactly one fixed point p_c of F_c for values of c near c_1, and which is such that for sufficiently small values of $\varepsilon > 0$,

(1) $p_{c_1 - \varepsilon}$ is attracting, and $F_{c_1 - \varepsilon}$ has no other periodic points in I;

(2) p_{c_1} is neutral, and F_{c_1} has no other periodic points in I; and

(3) $p_{c_1 + \varepsilon}$ is repelling, and $F_{c_1 + \varepsilon}$ has a unique attracting period-2 orbit $\{q_\varepsilon^1, q_\varepsilon^2\}$ such that $\lim_{\varepsilon \to 0} q_\varepsilon^i = p_{c_1}$ for $i = 1, 2$.

The above definition describes the process by which a stable fixed point becomes unstable and sheds a stable orbit of period 2 as c increases through c_1. As before, we also call c_1 a period-doubling bifurcation point if this happens as c decreases through c_1 (which is the case with the logistic map). We also allow the case in which a *repelling* fixed point becomes stable and sheds an unstable period-2 orbit.

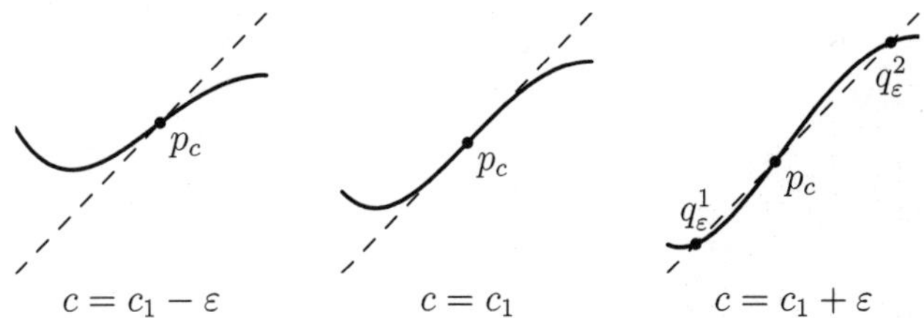

Figure 6.3. A pitchfork bifurcation in the second iterate.

Finally, we will be flexible enough with the terminology to say that F_c has a period-doubling bifurcation at c_1 if its nth iterate F_c^n satisfies the above criteria for some n, that is, if some periodic orbit of F_c with length n changes stability at c_1 and sheds a new periodic orbit of length $2n$ with the original stability properties.

A typical picture for a period-doubling bifurcation is presented in Figure 6.2, which shows the fixed point becoming unstable as $F_c'(p)$ passes through -1. Figure 6.3 shows how the dynamics of F_c^2 change; as c passes through c_1, the fixed point of F_c^2 changes stability and spawns two new attracting *fixed* points. We say that the second iterate F_c^2 has a *pitchfork bifurcation* at c_1.

Exercise 6.3. Describe the type of bifurcation at the given value of the parameter for the following maps:

(a) $f_\lambda(x) = \lambda x + x^2 + 1/4$ for $\lambda = 0$ or $\lambda = 2$;

(b) $g_\mu(x) = \mu x(1 + x^2)$ for $\mu = 1$;

(c) $h_\gamma(x) = \gamma \sin x$ for $\gamma = -1$ or $\gamma = 1$.

Exercise 6.4. Consider the family $F_c(x) = cx - x^3$ of maps. Show that 0 is a fixed point for all F_c, and determine the values of c for which 0 is attracting and for which it is repelling. Show that a period-doubling bifurcation occurs at $c = -1$.

Exercise 6.5. Let $T: \mathbb{R} \to \mathbb{R}$ be the map on the line generated by the function $y = ax^3$, $a > 0$. Find all fixed points and determine the type of their stability. Describe the behaviour of all trajectories of the map.

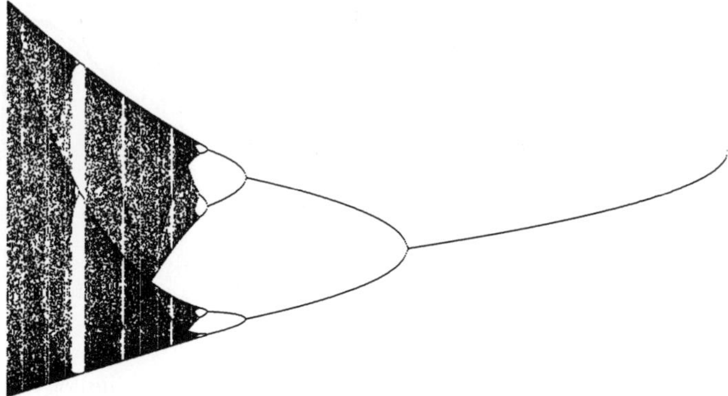

Figure 6.4. The bifurcation diagram for the family of logistic maps.

Lecture 26

a. The period-doubling cascade. Figure 6.4 shows the bifurcation diagram for the family of logistic maps $f_c\colon x \mapsto x^2 + c$; comparing this with the bifurcation diagram for the FitzHugh–Nagumo model in Figure 5.6, we see many of the same qualitative features—period-doubling cascades, windows of stability, etc.

In the previous lecture, we examined two sorts of bifurcations; the tangent bifurcation at $c_0 = 1/4$, where two fixed points are born, one stable and one unstable, and the period-doubling bifurcation at $c_1 = -3/4$, where the stable fixed point becomes unstable and an attracting period-2 orbit appears. Figure 6.4 shows further period-doubling bifurcations at $c_2 > c_3 > c_4 > \cdots$; for $c \in (c_{n+1}, c_n)$, the map f_c has an attracting periodic orbit of length 2^n, and repelling periodic orbits of length 2^k for $0 \le k < n$.

It is apparent from the diagram that the distance between successive bifurcations shrinks as n grows; indeed, it is possible to show that

$$\lim_{n \to \infty} \frac{c_n - c_{n-1}}{c_{n+1} - c_n} = \delta \approx 4.669 \cdots \,;$$

that is, the exponential rate of decay of this distance is Feigenbaum's constant (recall the discussion of the FitzHugh–Nagumo model in Lecture 24).

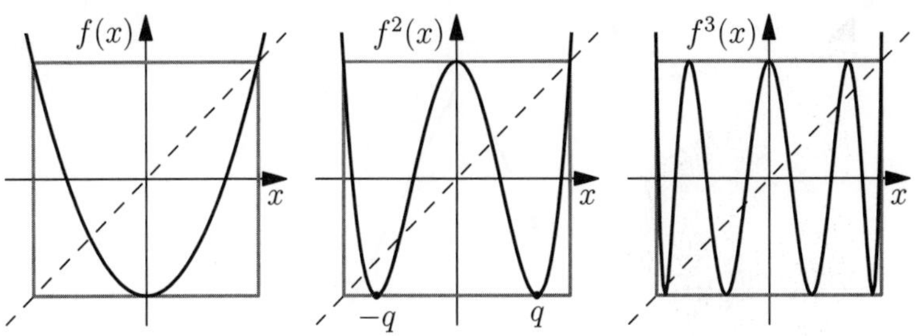

Figure 6.5. The map $f = f_{-2}$ and its iterates.

Because $c_n - c_{n-1}$ decreases exponentially, the sequence $\{c_n\}$ converges to a limit c_∞, the *Feigenbaum parameter*. This parameter lies at the end of the period-doubling cascade, and so f_{c_∞} has periodic points of order 2^n for every natural number n. In particular, it is no longer Morse–Smale, and the same continues to be true of f_c for $c < c_\infty$.

What do the dynamics of these non-Morse–Smale maps look like? How do we describe the structure of the bifurcation diagram in the regime $c \leq c_\infty$? These maps are difficult to analyse, and we begin by jumping ahead a little way to examine what happens for $c \leq -2$, before returning to study the truly intricate part of the picture.

b. Chaos at the end of the bifurcation diagram. Consider the parameter value $c = -2$, which is the smallest value of c shown in Figure 6.4. For simplicity of notation, in this section we write f for the map $f_{-2}\colon x \mapsto x^2 - 2$. Observe that f has fixed points at $p_1 = -1$ and $p_2 = 2$. Furthermore, if $|x| > 2$, then $f^n(x) \to +\infty$ as $n \to \infty$, and so the only interesting trajectories are those which remain within the interval $[-2, 2]$.

One may easily check (either by computing minimum and maximum values, or by looking at Figure 6.5) that the interval $[-2, 2]$ is invariant for the map f. Furthermore, f is monotonic on each of the intervals $[-2, 0]$ and $[0, 2]$; in fact, it maps each of these intervals homeomorphically to the entire interval $[-2, 2]$. Hence f is a one-dimensional full-branched Markov map, whose action may be thought of as a combination of stretching and folding; the interval

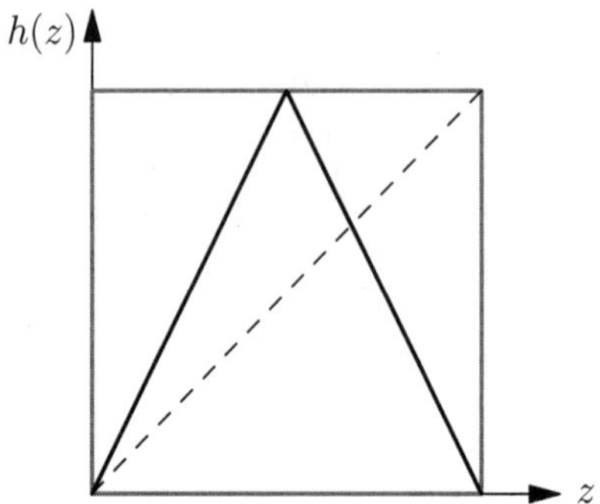

Figure 6.6. The tent map $h\colon [0,1] \to [0,1]$.

$[-2,2]$ is first stretched out, and then folded in half, so that each half of the original interval has been stretched out to cover the whole thing.

Upon iterating the map f, we see that if J is either of the intervals whose image under f covers $[-2,2]$ once (that is, $f\colon J \to [-2,2]$ is a homeomorphism), then the image of J covers $[-2,2]$ twice under the action of f^2. In particular, as shown in Figure 6.5, there are points $-q < 0 < q$ such that each of the intervals $I_1 = [-2,-q]$, $I_2 = [-q,0]$, $I_3 = [0,q]$, and $I_4 = [q,2]$ has $f^2(I_j) = [-2,2]$, and f^2 is a bijection from each I_j to $[-2,2]$. A similar observation holds for f^3, where we have eight intervals, and for higher iterates f^n, where we have 2^n intervals which are mapped homeomorphically onto $[-2,2]$.

It follows from Exercise 5.7 that the map f^n has at least 2^n fixed points, and examination of the graph of f^n shows that this number is exact. Thus for every $n \geq 1$, the map f has 2^n periodic points of period n (of course, some of these are also periodic points of period k for some $k < n$). This is a far cry from the limited number of periodic orbits found in the Morse–Smale case. Indeed, the exponential growth rate of the number of period n orbits is somehow indicative of the chaotic behaviour of the map; this growth rate is in many cases related to the topological entropy of the map f, which we introduced in Lecture 18.

Before leaving the parameter value $c = -2$, we observe that by some felicitious alignment of the stars (really, by virtue of the fact that it is a one-dimensional full-branched Markov map), the map f can be put into a rather simpler form—indeed, a piecewise linear form—via a clever change of coordinates. First, as a special case of the observation at the beginning of Lecture 25, f is conjugated to the map $g\colon y \mapsto 4y(1-y)$ from $[0,1]$ to itself by the change of coordinates

$$\phi\colon [0,1] \to [-2,2],$$

$$y \mapsto 2 - 4y.$$

Then, using the further change of coordinates

$$\psi\colon [0,1] \to [0,1],$$

$$z \mapsto \sin^2\left(\frac{\pi z}{2}\right),$$

the following diagram commutes:

(6.1)

$$
\begin{array}{ccc}
[0,1] & \xrightarrow{\ h\ } & [0,1] \\
\downarrow{\scriptstyle\psi} & & \downarrow{\scriptstyle\psi} \\
[0,1] & \xrightarrow{\ g\ } & [0,1] \\
\downarrow{\scriptstyle\phi} & & \downarrow{\scriptstyle\phi} \\
[-2,2] & \xrightarrow{\ f\ } & [-2,2]
\end{array}
$$

Here $h\colon [0,1] \to [0,1]$ is the *tent map* defined by

(6.2)
$$h(z) = \begin{cases} 2z & 0 \le z \le 1/2, \\ 2(1-z) & 1/2 \le z \le 1, \end{cases}$$

whose graph is shown in Figure 6.6. To see that the top half of (6.1) commutes, observe that

$$g(\psi(z)) = 4\sin^2\left(\frac{\pi z}{2}\right)\left(1 - \sin^2\left(\frac{\pi z}{2}\right)\right) = 4\sin^2\left(\frac{\pi z}{2}\right)\cos^2\left(\frac{\pi z}{2}\right)$$

$$= \left(2\sin\frac{\pi z}{2}\cos\frac{\pi z}{2}\right)^2 = \sin^2\pi z,$$

while for $0 \le z \le 1/2$,

$$\psi(h(z)) = \sin^2\frac{\pi \cdot 2z}{2} = \sin^2\pi z,$$

and for $1/2 \le z \le 1$,

$$\psi(h(z)) = \sin^2(\pi - \pi z) = \sin^2\pi z.$$

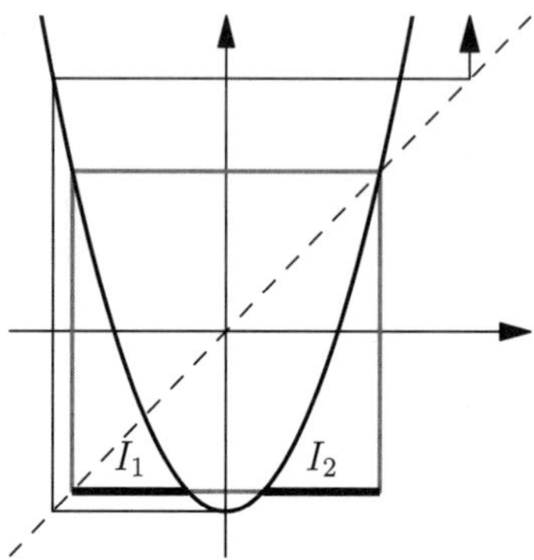

Figure 6.7. Trajectories escaping to infinity for $c < -2$.

This conjugacy allows us to answer certain questions about the non-linear map f by first answering them for the piecewise linear map h, in which context they are often more tractable. In fact, we can go one step further and use the Markov property of h to define a coding map $\varphi \colon \Sigma_2^+ \to [0, 1]$ such that $h \circ \varphi = \varphi \circ \sigma$; the difference between this case and the situation we are used to is that the coding map is no longer a homeomorphism, since it fails to be one-to-one for sequences which terminate in an infinite string of ones or twos. This is because the basic intervals on which h is defined—$[0, 1/2]$ and $[1/2, 1]$—are not disjoint, but share an endpoint, and is the same phenomenon we observed with the Sierpiński gasket. As was the case there, it only affects countably many points, and thus does not affect any questions concerning Hausdorff dimension.

Exercise 6.6. Give a complete analysis of orbits for the transformation $F \colon \mathbb{R} \to \mathbb{R}$ given by $F(x) = x^4 - 4x^2 + 2$.

c. The centre cannot hold: escape to infinity. In the previous section, we were able to give a relatively thorough analysis of the logistic map $f = f_c$ in the case $c = -2$. How much of this goes through for other values of c?

For any value of $c < 1/4$, there are two fixed points $p_1 < p_2$, and only the interval $[-p_2, p_2]$ is of interest, since any trajectory which leaves this interval diverges to infinity, as in Figure 6.7.

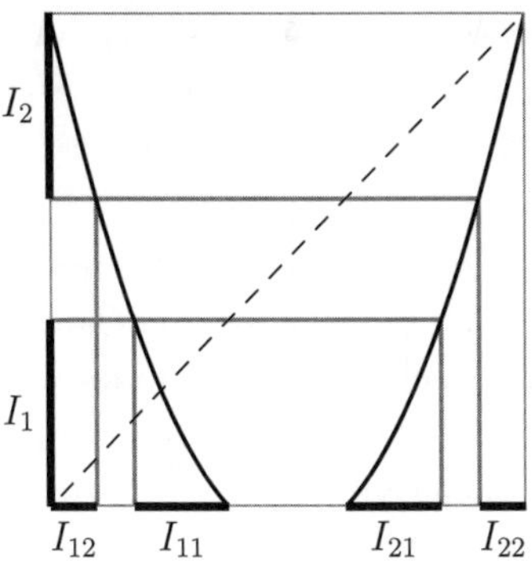

Figure 6.8. Finding an invariant set for a map with escape.

For $-2 \leq c < 1/4$, the interval $[-p_2, p_2]$ is invariant; any trajectory which begins there stays there, and so we only need to exclude trajectories which begin outside the interval of interest. To see if this behaviour continues for $c < -2$, we must examine $f_c(0) = c$, since this is the minimal value assumed by f_c, and indeed, $f_c([-p_2, p_2]) \subset [-p_2, p_2]$ if and only if $f_c(0) = c \geq -p_2$.

We compare these values by observing that c is the constant term in the fixed point equation $f_c(x) - x = x^2 - x + c = 0$, which is a quadratic polynomial, and hence it is the product of the roots of that polynomial, which are p_1 and p_2. Similarly, the sum of the roots is determined by the linear coefficient: $p_1 + p_2 = 1$, whence $c = p_2(1 - p_2)$. It follows that $[-p_2, p_2]$ is invariant if and only if $p_2(1 - p_2) \geq -p_2$, that is, if and only if $p_2(2 - p_2) \geq 0$.

The fixed point p_2 is always positive, and so the interval is invariant for $p_2 \leq 2$, while for $p_2 > 2$, some points in $[-p_2, p_2]$ have images outside that interval, and thus have a trajectory which escapes to infinity. This happens precisely when $c < -2$; in this parameter range, we have the picture shown in Figure 6.7, where any point not in the intervals I_1 or I_2 is mapped outside of $[-p_2, p_2]$ by f_c.

This should start to sound familiar by now; we have a map from an interval to itself, but we can only start at points whose images remain in the interval. The intervals I_1 and I_2 contain those points x for which $f_c(x) \in [-p_2, p_2]$; Figure 6.8 shows the construction of four intervals I_{ij} containing all points x for which $f_c^2(x) \in [-p_2, p_2]$. Indeed, f_c is a one-dimensional full-branched Markov map, and the set of points with a bounded trajectory is precisely the repelling Cantor set for f_c.

Even though the map f_c is non-linear, it is not hard to show that the Hausdorff dimension of this repelling Cantor set is less than 1; in particular, the Cantor set has Lebesgue measure zero. This means that Lebesgue-a.e. point $x \in \mathbb{R}$ has a trajectory which diverges to ∞; consequently, the bifurcation diagram is invisible for $c < -2$ because the algorithm described in Lecture 24 only lets us see the asymptotic behaviour of randomly chosen bounded trajectories.

Lecture 27

a. Finding the relevant part of phase space: ω-limit sets. In our investigations of the logistic map f_c for various values of the parameter c, we have found that some parts of the phase space $\mathbb{R}$ are more interesting than others, in the sense that they capture the essential long-term behaviour of f_c. For example, when $c > c_\infty$, almost every trajectory tends to the stable periodic orbit, and so the points in that orbit are the most important part of phase space for f_c. When $c = -2$, the points in the interval $[-2, 2]$ are important, but those outside it are not so interesting because their iterates tend to infinity. Finally, when $c < -2$, there is a Cantor set C which contains all the points whose orbits remain bounded, and hence captures all the interesting behaviour of f_c.

How do we formalise these ideas? How can we define what makes some sets capture interesting aspects of the dynamics, while others are somehow negligible? What properties should these "interesting" sets have?

The first important property is invariance; we want to consider a set $E \subset \mathbb{R}$ which is mapped into itself by f, so that no trajectories escape from E; otherwise E does not contain the long-term behaviour of all the trajectories which begin in E.

Secondly, we want E to be minimal in some sense. For f_c with $c > c_\infty$, the interval $[-p_2, p_2]$ is certainly mapped into itself by f_c, but it is too big: given an initial condition $x \in [-p_2, p_2]$, there are many open sets in $[-p_2, p_2]$ which the trajectory of x never reaches and which are therefore of no importance in describing the long-term behaviour of that trajectory. In particular, we would like the orbit of some well-chosen point x to be dense in E.

These considerations motivate the following definition.

Definition 6.4. Let X be a metric space and $f\colon X \to X$ a continuous map. Given $x \in X$, the ω-*limit set* of x is

$$\omega(x) = \bigcap_{N=0}^{\infty} \overline{\bigcup_{n=N}^{\infty} \{f^n(x)\}}.$$

To put this in words rather than symbols, a point $y \in X$ is in the ω-limit set of x if and only if y is in the closure of the forward trajectory[1] of every iterate of x, that is, if and only if there exists a sequence of natural numbers $n_k \to \infty$ such that $\lim_{k \to \infty} f^{n_k}(x) = y$.

Example 6.5. If x is a periodic point with $f^p(x) = x$, then $\omega(x) = \{x, f(x), \ldots, f^{p-1}(x)\}$. Similarly, if x approaches a periodic orbit— that is, if there exists $y = f^p(y)$ such that $\lim_{n \to \infty} f^{np}(x) = y$—then $\omega(x) = \{y, f(y), \ldots, f^{p-1}(y)\}$.

Exercise 6.7. Consider the map $f\colon \mathbb{C} \to \mathbb{C}$ given by $f(z) = e^{2\pi i \alpha} z$, where α is an irrational real number; f is the map which rotates the complex plane by $2\pi\alpha$ around the origin. Show that for every $z_0 \in \mathbb{C}$,

$$\omega(z_0) = \{z \in \mathbb{C} \mid |z| = |z_0|\}.$$

Exercise 6.8. Consider the shift $\sigma\colon \Sigma_2^+ \to \Sigma_2^+$, and show that there exists a sequence $x = (x_1, x_2, \ldots) \in \Sigma_2^+$ such that $\omega(x) = \Sigma_2^+$; in particular, x has a dense orbit.

Recall the construction of the bifurcation diagram in Lecture 24. The procedure described there is similar to the definition just given of ω-limit set; indeed, what is plotted by the bifurcation diagram is an approximation to the ω-limit set of a random point for each parameter value c.

---

[1] The forward trajectory of x, also called the positive semi-trajectory, is the set of all iterates $f^n(x)$ for which $n \geq 0$.

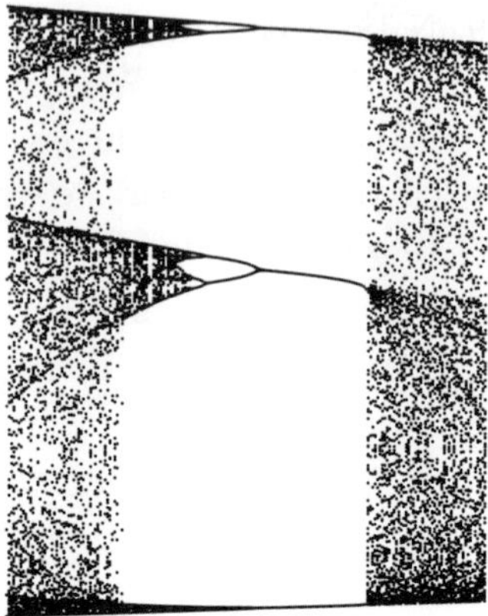

Figure 6.9. The bifurcation diagram in the period-3 window.

Does it matter which point we choose? Of course, if we choose an unstable periodic point x, then $\omega(x)$ will not capture the same information as $\omega(y)$, where y is not on the periodic orbit. In general, then, $\omega(x)$ and $\omega(y)$ may differ for $x \neq y$.

It turns out, though, that in many important situations, *almost every* point x (with respect to Lebesgue measure) has the same ω-limit set; that is, there exists a set $E \subset \mathbb{R}$ of Lebesgue measure zero such that $\omega(x) = \omega(y)$ for all $x, y \notin E$. For example, in the family of logistic maps, almost every initial condition x has ω-limit set equal to the unique stable periodic orbit when $c > c_\infty$, and for $c = -2$, it is possible to show that almost every initial condition x has $\omega(x) = [-2, 2]$.

b. Windows of stability in the bifurcation diagram. Let us return to the bifurcation diagram shown in Figure 6.4 and finally turn our attention to the truly interesting part of the picture, the parameter values $-2 \leq c \leq c_\infty$. Within this range, we find a number of *windows of stability*, intervals for c within which the map f_c suddenly has a stable periodic orbit once again, to which almost every point is attracted.

The largest and most conspicuous of these windows of stability is the period-3 window between $c \approx -1.791$ and $c \approx -1.748$, which is shown in Figure 6.9. Notice that while the period-3 orbit is stable at the beginning of this window (on the right), it eventually becomes unstable and gives birth to a stable period-6 orbit, and the whole period-doubling cascade occurs here just as it did for $c > c_\infty$, leading to another occurrence of Feigenbaum's constant δ.

Because the map has a period-3 orbit in this range of parameters, we can apply Proposition 5.7. This result shows that despite the long-term stability of orbits in the period-3 window, the system is not Morse–Smale, because there are infinitely many periodic orbits.

In fact, we observe what is known as *transient chaos* in this parameter regime; the trajectory of a randomly chosen point x may follow an unstable periodic orbit for quite some time before being repelled, at which time it may follow some other unstable orbit for a spell, and so on and so forth, perhaps taking a very long time to actually settle down to the stable periodic orbit. Thus the trajectory we observe may initially appear chaotic by spending a long while wandering through the intricately intertwined tangle of periodic orbits before it becomes regular.

c. Chaos outside the windows of stability. We see a similar picture to the one described above when we look at the windows of stability corresponding to periodic orbits of other lengths. Firstly, Sharkovsky's theorem implies the existence of infinitely many periodic orbits and prescribes which lengths must appear. Secondly, the stable periodic orbit undergoes a period-doubling cascade as c decreases and eventually moves out of the window of stability.

Let S be the set of parameter values c for which the map f_c has a stable periodic orbit; S is the union of all the windows of stability. It is natural to ask how big S is; do typical parameter values c lie inside a window of stability or somewhere else?

If we perturb f in such a way that both f and f' vary continuously, then fixed points and periodic points also vary continuously, as does the value of f' at such points. Windows of stability are characterised by the existence of stable periodic orbits, which are characterised by the conditions $f^n(p) = p$ and $|(f^n)'(p)| < 1$. Since p and

$(f^n)'(p)$ vary continuously with the parameter c, a small perturbation of c will not destroy these conditions, and thus each window of stability is open.

This shows that S is open; it can also be shown that S is dense in $[-2, 1/4]$ (although the proof is rather hard). So in some sense, the set S is quite large; indeed, topologically speaking, it is as large as can be.

However, there are other sorts of behaviour possible for the logistic map f_c. It turns out that these are epitomised by the two cases $c = -2$ and $c = c_\infty$. In the first case, the ω-limit set $\omega(x)$ contains an interval (and in particular, has Hausdorff dimension equal to 1) for almost every x. In the second case, we find that $\omega(x)$ is a Cantor set for almost every x (and hence has Hausdorff dimension less than 1).

Denote by A the set of all parameter values for which the former behaviour is observed—for which $\omega(x)$ contains an interval for a.e. x. How big is the set A? We have already said that S is open and dense, and obviously $A \cap S = \emptyset$; thus A is a nowhere dense set. However, a celebrated result due to Michael Jakobson shows that A has positive Lebesgue measure, and so there is a non-negligible set of parameter values for which the trajectory of a randomly chosen point fills out an interval, and the map f_c is chaotic (see Figure 1.11).

Chapter 7

Chaotic Attractors and Persistent Chaos

Lecture 28

a. Trapping regions. The gap between what has been proved regarding the family of logistic maps and what is believed to be true
based on numerical results remains substantial; it becomes even wider
when we consider the local map (5.7) for the FitzHugh–Nagumo
model.

Despite the lack of rigorous results, the empirical evidence over-
whelmingly suggests that what was proved in very restricted circumstances (one-dimensional quadratic maps in a limited parameter
range) holds much more generally, as is suggested, for example, by
the bifurcation diagram in Figure 5.6. We see a period-doubling cascade leading to the onset of chaos at A_∞, beyond which there are
windows of stability surrounded by maps with chaotic behaviour (although as mentioned at the end of the previous chapter, even within
these windows of stability the map exhibits transient chaos).

Throughout all this, the map $f \colon \mathbb{R}^2 \to \mathbb{R}^2$ has three fixed points,
$\mathbf{0}$, $\mathbf{p}_1$, and $\mathbf{p}_2$, two of which began life as stable fixed points, and
then lost their stability at the bifurcation values A_1 and A_1'. During
the period-doubling cascade, all trajectories wind up approaching a

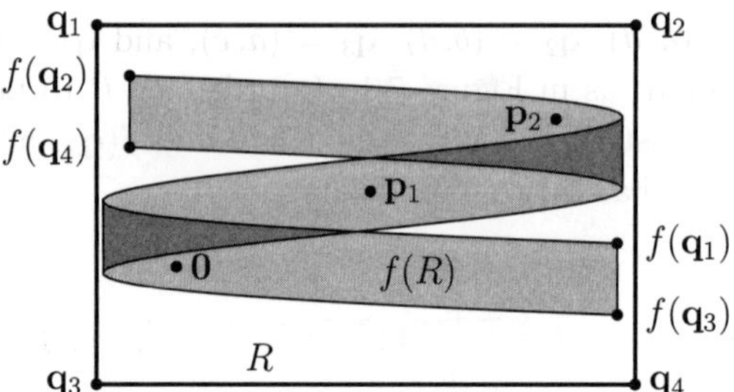

Figure 7.1. A trapping region for the FitzHugh–Nagumo map.

stable periodic orbit of length 2^n; after the onset of chaos, however, there are no stable periodic orbits to approach, except in the windows of stability. So where do the orbits go?

Even if trajectories of f are not asymptotically stable or periodic, it may happen that they are confined within some bounded region. This is certainly the case if there is some bounded region $R \subset \mathbb{R}^2$ for which $f(R) \subset R$, since then the forward trajectory of any point in R remains in R. The following definition codifies this idea, adding a small topological requirement.

Definition 7.1. An open set R is a *trapping region* for the map f if $\overline{R}$ is compact and $\overline{f(R)} \subset R$.

We will explore the dynamical consequences of the existence of a trapping region after first proving that such a region exists for the FitzHugh–Nagumo map with certain values of A.

Proposition 7.2 (Orendovici–Pesin). *Fix parameters $0 < \gamma < 1$ and $\beta > 0$. There exists an open rectangle $R = (a, b) \times (c, d) \subset \mathbb{R}^2$ and $A' > 0$ such that for $\alpha > 0$ sufficiently small, θ sufficiently near $1/2$, and $0 < A < A'$, the rectangle R is a trapping region for the local map (5.7) for the FitzHugh–Nagumo model. Furthermore, R contains the three fixed points $\mathbf{0}$, $\mathbf{p}_1$, and $\mathbf{p}_2$.*

Proof. *(See [OP00] and [PY04].)* We will guarantee containment of the fixed points by taking $a < 0 < 1 < b$ and appropriate values of c and d.

Let $\mathbf{q}_1 = (a, d)$, $\mathbf{q}_2 = (b, d)$, $\mathbf{q}_3 = (a, c)$, and $\mathbf{q}_4 = (b, c)$ be the four corners of R, as in Figure 7.1. In order for R to be a trapping region, $f(\mathbf{q}_2)$ must lie below the line $v = d$ and $f(\mathbf{q}_3)$ must lie above the line $v = c$. That is, we require that

$$\beta b + \gamma d < d, \qquad \beta a + \gamma c > c,$$

or equivalently,

$$d > \frac{\beta b}{1 - \gamma}, \qquad c < \frac{\beta a}{1 - \gamma}.$$

Thus $\mathbf{q}_2$ must lie above the line $v = \beta u/(1 - \gamma)$ and $\mathbf{q}_3$ must lie below it; observe that this is the line which passes through the three fixed points 0, $\mathbf{p}_1$, and $\mathbf{p}_2$, and hence for $a < 0 < 1 < b$ all three fixed points lie in R.

In order to guarantee that R is a trapping region, it remains to show that the images of both the top and bottom edges of R are themselves in R. If we write $k(t) = t - At(t - \theta)(t - 1)$, then these images are as follows:

$$\{(k(u) - \alpha d, \beta u + \gamma d) \mid a \le u \le b\},$$
$$\{(k(u) - \alpha c, \beta u + \gamma c) \mid a \le u \le b\}.$$

That the v-coordinate will lie between c and d follows from the linearity of the v-component of the map and the fact that the corners $\mathbf{q}_i$ are mapped into R. Thus we require only the following inequalities for every $u \in [a, b]$:

$$(7.1) \qquad\qquad a < k(u) - \alpha d < k(u) - \alpha c < b.$$

The idea now is to prove these in the unperturbed case where $\alpha = 0$ and $\theta = 1/2$, and then to use a continuity argument to extend the result to small values of α and values of θ near $1/2$. In the unperturbed case, (7.1) amounts to choosing a and b such that

$$(7.2) \qquad\qquad a < k(t_1) < k(t_2) < b,$$
$$(7.3) \qquad\qquad a < k(b),$$
$$(7.4) \qquad\qquad k(a) < b,$$

where t_1 and t_2 are the unique local minimum and local maximum, respectively, of the cubic polynomial k. We see immediately from the form of $k(t)$ that $k(t_1)$ decreases as A increases and $k(t_2)$ increases as A increases; a little computation shows that for $A = 8$ we have $k(t_1) \approx -.207$ and $k(t_2) \approx 1.207$. Thus taking $a = -.21$ and $b = 1.21$, we see that (7.2) is satisfied for $0 < A < 8$.

Now the inequality $k(a) < b$ may be written as

$$a - Aa(a - \theta)(a - 1) < b,$$

or equivalently,

$$A < \frac{a - b}{a(a - \theta)(a - 1)} \approx 7.98,$$

where the computation is for the particular case $\theta = 1/2$. Similarly, $k(b) < a$ is equivalent to

$$A < \frac{b - a}{b(b - \theta)(b - 1)} \approx 7.98.$$

Thus taking $A' = 7.5$ to give ourselves a bit of room to play with, we see that (7.2)–(7.4) all hold for every $0 < A < A'$, with $\theta = 1/2$ and $\beta = 0$, and hence R is a trapping region for these particular parameter values.

Finally, note that all the inequalities which guarantee that R is a trapping region involve only continuous functions of the parameters; in particular, they all hold for values of θ sufficiently close to $1/2$ and values of $\beta > 0$ sufficiently small. $\qquad\square$

b. Attractors. What are the dynamical implications of the existence of a trapping region? First, we observe that a trajectory which enters a trapping region will never leave it—hence the name. Furthermore, the property $\overline{f(R)} \subset R$ is inherited by the images of R, thanks to the following exercise.[1]

Exercise 7.1. Given a continuous map f and an arbitrary domain R such that $\overline{R}$ is compact, show that $\overline{f(R)} = f(\overline{R})$.

Using the result of the exercise, we see that

$$\overline{f^2(R)} = \overline{f(f(R))} = f(\overline{f(R)}) \subset f(R).$$

Continuing in this way, we obtain a nested sequence of compact sets

$$\overline{R} \supset \overline{f(R)} \supset \overline{f^2(R)} \supset \cdots \supset \overline{f^n(R)} \supset \overline{f^{n+1}(R)} \supset \cdots .$$

---

[1]Observe that $f(R)$ may not be open (see Figure 7.1) and hence may not be a trapping region in its own right.

We may "take the limit" of this sequence by taking the intersection of all these sets, and we obtain

$$\tag{7.5} \Lambda = \bigcap_{n \geq 0} f^n(R).$$

The intersection Λ is called an *attractor* for the map f.

Theorem 7.3. *Let R be a trapping region for a continuous map f, and define an attractor Λ by (7.5). Then Λ has the following properties:*

(1) Λ is compact and non-empty.

(2) Λ is f-invariant: $f(\Lambda) = \Lambda$.

(3) Λ is the largest f-invariant subset of R; that is, $Z \subset \Lambda$ for every f-invariant set $Z \subset R$.

(4) Λ attracts every orbit of f which enters R: $\omega(x) \subset \Lambda$ for every $x \in R$.

(5) Λ contains all the fixed points and periodic points of f in R.

Proof.

(1) Λ is the intersection of nested compact sets, and hence is compact and non-empty.

(2) Since $f(\overline{R}) \subset R$, we have

$$f(\Lambda) = f\left(\bigcap_{n \geq 0} f^n(\overline{R})\right) = \bigcap_{n \geq 0} f(f^n(\overline{R})) = \bigcap_{n \geq 1} f^n(\overline{R}) = \Lambda.$$

(3) If $Z \subset R$ is f-invariant ($f(Z) = Z$), then for every $n \in \mathbb{N}$ we have $Z = f^n(Z) \subset f^n(R)$, and hence $Z \subset \Lambda$.

(4) The set $\omega(x)$ is f-invariant, so by Property (3), $\omega(x) \subset \Lambda$.

(5) If $p \in R$ is a fixed point, then $\{p\}$ is f-invariant, so Property (3) applies. Similarly, if $f^k(p) = p$, then $\{p, f(p), \ldots, f^{k-1}(p)\}$ is f-invariant. $\qquad \square$

Exercise 7.2. Let R be an open domain such that $\overline{f(R)} \subset \overline{R}$. Show that Theorem 7.3 remains true in this case.

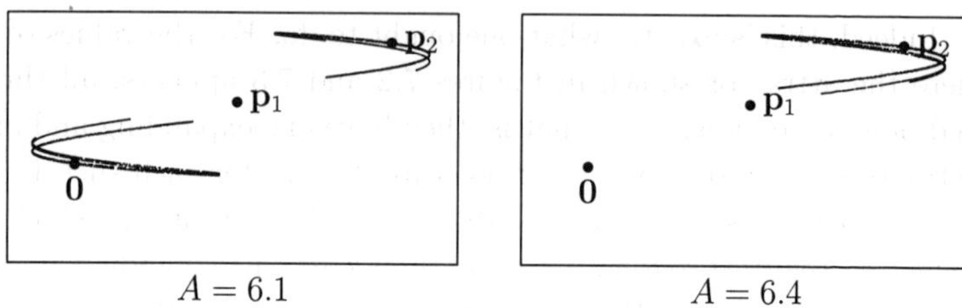

Figure 7.2. The attractor for the FitzHugh–Nagumo map with $\alpha = .01, \beta = .02, \theta = .51, \gamma = .2$, and varying A.

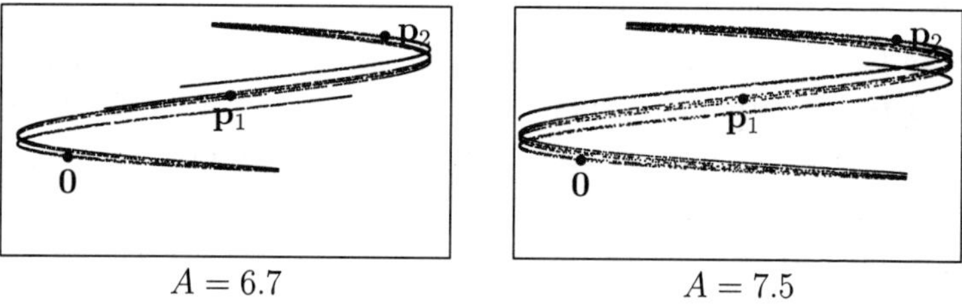

Figure 7.3. Changes in the attractor as A increases.

We return now to the specific example of the FitzHugh–Nagumo system, for which Proposition 7.2 guarantees the existence of a trapping region R and, hence, an attractor Λ. Figure 7.2 shows a numerically computed approximation to the attractor for two different parameters of A, drawn by plotting long orbit segments to approximate $\omega(x)$. Notice that for certain values of A, corresponding to the gap in the bottom half of Figure 5.6, the attractor lies entirely in the top right quadrant of the trapping region.

As A continues to increase, the attractor "grows", as shown in Figure 7.3, to occupy more and more[2] of the trapping region R; here one also sees the "grainy" structure which is inevitably associated with the orbit-plotting method of producing such images. The human eye, on viewing Figure 7.3, immediately wants to connect the dots and view Λ as a union of curves, rather than simply a collection of points.

---

[2]One may legitimately ask, though, in just what sense it becomes larger, beyond the obvious statement that its diameter increases. Throughout all of this, the Lebesgue measure of Λ is zero; thus a useful quantification of the attractor's size ought to involve some dimensional quantity.

Indeed, this is exactly what one ought to do. For the values of A where the attractor shown in Figures 7.2 and 7.3 appears, all three fixed points are hyperbolic; that is, they have one expanding and one contracting direction. We saw in Lecture 22 that for each such fixed point $\mathbf{p}$ there exists a local unstable curve W_ε^u through $\mathbf{p}$ which is tangent to the unstable eigenvector. This curve is expanding in the sense that the image $f(W_\varepsilon^u)$ is a curve which contains W_ε^u.

Taking the union of the curves $f^n(W_\varepsilon^u)$, we obtain, as we did before, the global unstable curve W^u, which is f-invariant and contained in R; by Theorem 7.3, this implies that $W \subset \Lambda$.

In fact, it is conjectured (and widely believed) that for an appropriate range of the parameters, W^u is dense in Λ for the FitzHugh–Nagumo system; however, this remains an open problem.

Lecture 29

a. The Smale–Williams solenoid. From the discussion in the previous lecture, it is apparent that the FitzHugh–Nagumo model is very rich in intricate and interesting behaviour, but is also quite difficult to analyse. We thus turn our attention to simpler model examples, which exhibit a similar richness of behaviour but are rather more tractable.

Our first such example is a map from the solid torus to itself. Abstractly, the solid torus is

$$P = D^2 \times S^1,$$

the direct product of a disc and a circle. Writing the disc as

$$D^2 = \{(x, y) \in \mathbb{R}^2 \mid x^2 + y^2 \leq 1\},$$

and the circle as $S^1 = \mathbb{R}/2\pi\mathbb{Z}$, we may use coordinates (x, y, θ) on P; x and y give the coordinates on the disc, and θ is the angular coordinate on the circle.

We may visualise P via its embedding in $\mathbb{R}^3$ as the standard torus of revolution together with the region it encloses:

$$\rho(P) = \left\{ (x, y, z) \in \mathbb{R}^3 \ \middle|\ \left(\sqrt{x^2 + y^2} - 2\right)^2 + z^2 \leq 1 \right\};$$

Figure 7.4. A map from the solid torus to itself.

here $\rho\colon D^2 \times S^1 \to \mathbb{R}^3$ is the map given by

$$\rho(x, y, \theta) = ((2 + x)\cos\theta, (2 + x)\sin\theta, y).$$

Fixing parameters $r \in (0, 1)$ and $\alpha, \beta \in (0, \min\{r, 1 - r\})$, we define a map $f\colon P \to P$ by

$$(7.6) \qquad f(x, y, \theta) = (\alpha x + r\cos\theta, \beta y + r\sin\theta, 2\theta).$$

The images of P under the first two iterates f and f^2 are shown in Figure 7.4. The action of f on P may be described as follows:

(1) Take the torus and slice it along a disc so that it becomes a tube.

(2) Squeeze this tube so that its cross-sections are no longer circles of radius 1, but ellipses with axes of length α and β.

(3) Stretch the tube along its axis until it is twice its original length.

(4) Wrap the resulting longer, skinnier tube twice around the z-axis within the original solid torus.

(5) Glue the ends of the tube together.

We see that $f(P) \subset \operatorname{int} P$, and so we may repeat the procedure in the previous section, obtaining an attractor by taking the intersection of all images of P:

$$(7.7) \qquad\qquad \Lambda = \bigcap_{n \geq 0} f^n(P).$$

The attractor Λ is known as the *Smale–Williams solenoid*; in order to investigate the structure of Λ, we look at a vertical cross-section of the solid torus $P = D^2 \times S^1$ by fixing the angular coordinate θ and considering the disc $D^2 \times \{\theta\}$.

From Figure 7.4, it is clear that the image $f(P)$, which is a long skinny tube wrapped twice around the z-axis, intersects this disc in

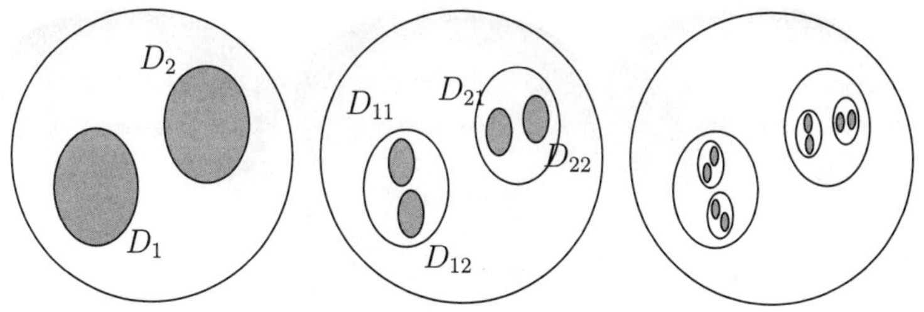

Figure 7.5. A cross-section of the Smale–Williams solenoid.

two ellipses D_1 and D_2, whose axes have length α and β (see Figure 7.5). The second image $f^2(P)$ is an even longer and skinnier tube which is wrapped *four* times around the z-axis, and intersects the disc in four ellipses D_{11}, D_{12}, D_{21}, and D_{22}, whose axes have lengths α^2 and β^2.

Continuing in this manner, we see that $f^n(P) \cap (D^2 \times \{\theta\})$ is the union of 2^n ellipses $D_{w_1 \ldots w_n}$ whose axes have lengths α^n and β^n. By now the reader should not be too shocked to discover that this is yet another example of a Cantor-like construction;[3] the basic sets at each step are the ellipses just mentioned, and the cross-section $C = \Lambda \cap (D^2 \times \{\theta\})$ is a Cantor set obtained as the intersection of the basic sets at all levels.

Exercise 7.3. Consider the cross-section $\theta = 0$ of the solid torus P, and describe the location of the centres of the ellipses $D_{w_1 \ldots w_n}$.

Each basic set is the intersection of a tube with the disc $D^2 \times \{\theta\}$; as n increases, the diameters of the tubes decrease exponentially, and so upon passing to the limit set C, we see that each point in C is contained in precisely one curve which meets $D^2 \times \{\theta\}$ transversely (indeed, orthogonally). Thus in a neighbourhood of each cross-section (a slice out of the torus), the attractor is the direct product $C \times (-\varepsilon, \varepsilon)$. However, this product structure is only local; if we follow one of these curves all the way around the torus, we will in general return to a different point of C than the one we left (see Figure 7.6).

---

[3]Indeed, one could obtain the exact construction shown in Figure 1.20 by modifying f so that $f(P)$ wraps around the z-axis *three* times, and allowing $\alpha = \beta$ to depend on θ.

Figure 7.6. Following an unstable curve once around the torus.

The local product structure of the attractor Λ has more than just a geometric significance; it also helps us describe the dynamics of the map f. Through each point $\mathbf{p} = (x, y, \theta) \in \Lambda$, we have a disc $W^s = D^2 \times \{\theta\}$ and a curve $W^u_\varepsilon = \{(x, y)\} \times (-\varepsilon, \varepsilon)$, as shown in Figure 7.6. The former is contracting while the latter is repelling, as follows: given $\mathbf{q} \in W^s$, we have

$$d(f(\mathbf{p}), f(\mathbf{q})) \leq \max\{\alpha, \beta\} d(\mathbf{p}, \mathbf{q}),$$

while for $\mathbf{q}' \in W^u_\varepsilon$, the orbits are driven further apart:

$$d(f(\mathbf{p}), f(\mathbf{q}')) = 2d(\mathbf{p}, \mathbf{q}').$$

Thus every point looks like a saddle; it has two stable directions (forming the disc) and one unstable direction (the curve). Notice, however, that since $\mathbf{p}$ may not be fixed, the reference to which or from which the orbit $\{f^n(\mathbf{q})\}$ is attracted or repelled is not the point $\mathbf{p}$ itself, but the trajectory of $\mathbf{p}$.

b. Uniform hyperbolicity. The behaviour exhibited by the Smale–Williams solenoid Λ, wherein hyperbolic behaviour exists at *every* point, not just fixed points, is an important enough and widespread enough phenomenon to warrant the following general definition.

Definition 7.4. Let $U \subset \mathbb{R}^d$ be open, and let $f \colon U \to f(U)$ be $\mathcal{C}^1$ with $\mathcal{C}^1$ inverse (see Appendix). A compact f-invariant set $\Lambda \subset U$ is called *hyperbolic* if for every $\mathbf{x} \in \Lambda$, there exists a direct sum decomposition $\mathbb{R}^d = E^s(\mathbf{x}) \oplus E^u(\mathbf{x})$ such that the subspaces $E^s(\mathbf{x})$ and $E^u(\mathbf{x})$ have the following properties:

(1) *Uniform contraction/expansion:* There exist $\lambda \in (0,1)$ and $C > 0$, independent of $\mathbf{x}$, such that for every $n \geq 0$, $\mathbf{v}^s \in E^s(\mathbf{x})$, and $\mathbf{v}^u \in E^u(\mathbf{x})$, we have

$$\|Df^n(\mathbf{x})\mathbf{v}^s\| \leq C\lambda^n\|\mathbf{v}^s\|,$$
$$\|Df^{-n}(\mathbf{x})\mathbf{v}^u\| \leq C\lambda^n\|\mathbf{v}^u\|.$$

(2) *Invariance of stable and unstable subspaces:* For every $\mathbf{x} \in \Lambda$, we have $Df(\mathbf{x})E^s(\mathbf{x}) = E^s(f(\mathbf{x}))$ and $Df(\mathbf{x})E^u(\mathbf{x}) = E^u(f(\mathbf{x}))$.

(3) There exists an open neighborhood U of Λ such that

$$\Lambda = \bigcap_{n\in\mathbb{Z}} f^n(U).$$

Roughly speaking, this definition says that the hyperbolicity we previously observed at hyperbolic fixed points can be found at every point of Λ. (Indeed, if $\Lambda = \{\mathbf{p}\}$ is a single fixed point, then the definition reduces to the definition of a hyperbolic fixed point.)

The first condition states that the map is contracting in the direction of E^s and expanding in the direction of E^u (since contraction along backward orbits corresponds to expansion along forward orbits). The second condition accounts for the fact that although most points $\mathbf{x} \in \Lambda$ are not fixed, the directions in which expansion and contraction occur should still be consistent along an orbit.

The third condition means that Λ is *locally maximal*—that is, if $Z \subset U$ is an invariant set, then $Z \subset \Lambda$. Traditionally, the definition of hyperbolic sets includes only the first two conditions, but many principal results (for example, Theorem 7.7 below) require the third condition as well, and so we will only consider locally maximal hyperbolic sets.

The hyperbolicity just described represents a fundamentally new type of behaviour compared with the Morse–Smale systems we found for the logistic map in the period-doubling cascade, where we observed hyperbolic behaviour only at a finite number of fixed points. Here, by contrast, the hyperbolicity is ubiquitous.

Definition 7.5. A hyperbolic set Λ is a *hyperbolic attractor* if there exists an open set $U \supset \Lambda$ such that $\Lambda = \bigcap_{n\geq0} f^n(U)$. The open set $\bigcup_{n\geq0} f^{-n}(U)$ is the *basin of attraction* for Λ.

Figure 7.7. The north-south map.

The Smale–Williams solenoid is an important example of a hyperbolic attractor; the basin of attraction in this case is the entire solid torus.

Exercise 7.4. The "north-south" map f of the unit sphere is defined so that trajectories move in the directions shown in Figure 7.7, from the north pole (which is a repelling fixed point) to the south pole (which is an attracting fixed point). Find the largest hyperbolic set and the attractor for f. Describe the basin of attraction.

What does the pervasive hyperbolicity just described mean for the dynamics of f? If $\mathbf{p}$ and $\mathbf{q}$ are two points in Λ which do not lie on the same stable disc W^s or unstable curve W^u_ε, then repeated iteration by f will decrease the distance between $f^n(\mathbf{p})$ and $f^n(\mathbf{q})$ in the stable direction (corresponding to the coordinates x and y) but will increase it in the unstable direction (corresponding to θ). In particular, the trajectory of $\mathbf{q}$ is repelled from the trajectory of $\mathbf{p}$.

So almost every pair of trajectories moves apart under the action of f; however, Λ is bounded, so they cannot move too far apart. Indeed, the definition of a hyperbolic set is given in terms of local properties (expanding and contracting subspaces for the linear map $Df(\mathbf{x})$), and so it ceases to give any information about the relationship of the two trajectories once they are no longer close. After they separate, a similar line of reasoning shows that the trajectory of $\mathbf{q}$ is constantly being repelled from whatever trajectories it finds itself near at any given time, and eventually is repelled back towards the

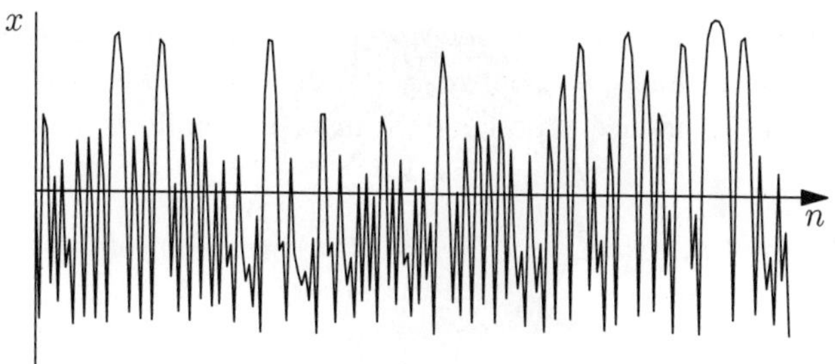

Figure 7.8. The x-coordinates of a trajectory as a chaotic signal.

trajectory of $\mathbf{p}$; at this point it is once again repelled from the trajectory of $\mathbf{p}$, and the whole cycle repeats itself.

This behaviour, this unending dispersal and return, is characteristic of hyperbolic dynamics. If we plot the x-coordinate of the trajectory of a point $\mathbf{p} \in \Lambda$ as a function of n, we see a chaotic signal, without periodicity or pattern, as shown in Figure 7.8. Various quantitative properties of this signal and of the dispersal and return of orbits of f are related to the dimensional quantities we have studied. For example, the rate at which nearby trajectories are repelled is given by the Lyapunov exponent of the map, which relates the entropy and the dimension. Furthermore, the statistical properties of recurrence (the times at which the trajectory beginning at $\mathbf{p}$ returns to a neighbourhood of $\mathbf{p}$), and of the correlations between measurements of a chaotic signal at different times, turn out to be related to various dimensional quantities of the hyperbolic set, and also to the multifractal analysis mentioned at the end of Chapter 4.

c. Symbolic dynamics. Another fundamental manifestation of the chaotic nature of the map $f \colon \Lambda \to \Lambda$ is its connection to symbolic dynamics. In Lecture 15 we saw that one-dimensional expanding Markov maps can be modeled by one-sided subshifts of finite type; this correspondence put at our disposal all the machinery of symbolic dynamics, allowing us to investigate the original map in terms of entropy, invariant measures, and so on.

These expanding maps had the characteristics of uniformly hyperbolic behaviour but were non-invertible; hence the one-sided shift spaces were good models for their dynamics. It turns out that if we use a *two-sided* shift space, on which the shift map is invertible, then a similar correspondence is possible for invertible uniformly hyperbolic maps.

Definition 7.6. Given $k \in \mathbb{N}$, the *two-sided symbolic space on k symbols* is the space

$$\Sigma_k = \{1, \ldots, k\}^{\mathbb{Z}}$$
$$= \{w = (w_j)_{j=-\infty}^{\infty} \mid w_j \in \{1, \ldots, k\} \text{ for every } j \in \mathbb{Z}\},$$

where it is convenient to write members of Σ_k in the form

$$w = (\ldots, w_{-2}, w_{-1} | w_0 | w_1, w_2, \ldots)$$

in order to highlight which entry of the sequence is the "centre" (this was not necessary in the one-sided case because the sequence had a beginning). The *full two-sided shift on k symbols* is Σ_k together with the shift map

$$\sigma \colon \Sigma_k \to \Sigma_k,$$
$$(\ldots, w_{-2}, w_{-1} | w_0 | w_1, w_2, \ldots) \mapsto (\ldots, w_{-1}, w_0 | w_1 | w_2, w_3, \ldots).$$

As in the one-sided case, we may fix $a > 1$ and define a metric on Σ_k by

$$d_a(w, w') = \sum_{j \in \mathbb{Z}} \frac{|w_j - w_j'|}{a^{|j|}}.$$

This metric once again induces a topology in which the open sets are unions of cylinders;[4] the latter take the form

$$C_{w_{-n} \ldots w_{-1} | w_0 | w_1 \ldots w_n} = \{w' \in \Sigma_k \mid w_j' = w_j \text{ for all } -n \leq j \leq n\}.$$

More generally, we may consider cylinders of the form

$$C_{w_a \ldots w_b} = \{w' \in \Sigma_k \mid w_j' = w_j \text{ for all } a \leq j \leq b\},$$

where $a \leq b \in \mathbb{Z}$ are arbitrary.

---

[4]As in the one-sided case, cylinders may not actually be open balls for small values of a. In this case, we need to take $a > 3$ to guarantee that $B(w, a^{-n}) = C_{w_{-n} \ldots w_{-1} | w_0 | w_1 \ldots w_n}$.

For the one-sided shift, every sequence has k preimages, because the shift "forgets" the first element of the sequence. The fundamental novelty of the two-sided shift is that nothing is forgotten; all the elements of the sequence remain, with a shifted reference point, and so the map is invertible. This makes the two-sided shift well suited for modeling invertible maps, such as the Smale–Williams solenoid, while the one-sided shift is well suited for modeling non-invertible maps, such as one-dimensional Markov maps.

The idea, then, is to partition Λ into disjoint sets $\Lambda_1, \ldots, \Lambda_k$, and to code trajectories of f by recording which partition element the iterate $f^n(x)$ lands in. (This is exactly what was done for one-dimensional Markov maps, using the forward trajectory with $n \geq 0$.) Thus to a trajectory $\{f^n(x)\}$ we associate the sequence

$$w = (\ldots, w_{-2}, w_{-1} | w_0 | w_1, w_2, \ldots) \in \Sigma_k,$$

where w_j is such that $f^j(x) \in \Lambda_{w_j}$ for each $j \in \mathbb{Z}$. Conversely, we may begin with a sequence $w \in \Sigma_k$ and look for a point whose trajectory is given by w; the set of all such points is

$$(7.8) \qquad \bigcap_{j \in \mathbb{Z}} f^{-j}(\Lambda_{w_j}).$$

What sequences in Σ_k do we obtain as codings of trajectories in Λ? The answer depends on which partition we choose; for example, if $i, j \in \{1, \ldots, k\}$ are such that $f(\Lambda_i) \cap \Lambda_j = \emptyset$, then no sequence w which contains the symbol i followed by the symbol j can correspond to a trajectory in Λ; these sequences are not admissible.

In general, it is not possible to find a partition such that all sequences are admissible, just as we found for one-dimensional Markov maps, where not every map could be modeled by the full shift. However, there is the following remarkable result: for a uniformly hyperbolic map, it *is* possible to find sets $\{\Lambda_1, \ldots, \Lambda_k\}$ (with disjoint interiors) such that for the $k \times k$ transition matrix A given as in (3.15) by

$$(7.9) \qquad a_{ij} = \begin{cases} 0 & f(\Lambda_i) \cap \Lambda_j = \emptyset, \\ 1 & f(\Lambda_i) \cap \Lambda_j \neq \emptyset, \end{cases}$$

the admissible sequences are precisely those which lie in

$$\Sigma_A = \{w \in \Sigma_k \mid a_{w_j w_{j+1}} = 1 \text{ for every } j \in \mathbb{Z}\}.$$

This is made precise by the following theorem, which relates the map $f: \Lambda \to \Lambda$ to the two-sided subshift $\sigma: \Sigma_A \to \Sigma_A$, and which is the fundamental vehicle for most of what is known about dynamics on uniformly hyperbolic sets.

Theorem 7.7. *Let Λ be a hyperbolic set for f. Then there exists a cover of Λ by sets $\Lambda_1, \ldots, \Lambda_k \subset \Lambda$ with disjoint interiors such that for the $k \times k$ transition matrix A given by (7.9) and the coding map $h: \Sigma_A \to \Lambda$ given by (7.8), the following hold:*

(1) h is continuous, onto, and one-to-one on a residual set.[5]

(2) The following diagram commutes:

$$
\text{(7.10)} \qquad
\begin{array}{ccc}
\Sigma_A & \xrightarrow{\ \sigma\ } & \Sigma_A \\
\big\downarrow{\scriptstyle h} & & \big\downarrow{\scriptstyle h} \\
\Lambda & \xrightarrow{\ f\ } & \Lambda
\end{array}
$$

Proof. See [**KH95**] or [**BS02**]. $\qquad\qquad\qquad\qquad\qquad\qquad\square$

This correspondence between hyperbolic sets and subshifts of finite type provides a bridge via which key symbolic results can be transported to the hyperbolic regime. In this manner, symbolic dynamics can be used to establish many properties of hyperbolic maps which are characteristic of chaos (see Lecture 31(c)).

Lecture 30

a. Dimension of direct products. Having discussed some of the qualitative properties of hyperbolic attractors, such as the Smale–Williams solenoid Λ, and the implications of these properties for the dynamics of f, we turn our attention to quantitative questions. In particular, since Λ has a fractal structure, we ask the natural question: what is the Hausdorff dimension of Λ?

---

[5]Recall that a set is *residual* if it is a countable intersection of open dense sets, and hence comprises "almost everything" in a topological sense. In this case, the residual set in question is the union $\bigcup_{n \in \mathbb{Z}} f^n(B)$, where B is the union of the boundaries of the sets Λ_i.

Locally, Λ is a direct product of a Cantor set and an interval; since we know the Hausdorff dimension for both of these sets, we would like to have a general expression for $\dim_H(A \times B)$ in terms of $\dim_H A$ and $\dim_H B$.

Intuitively, we expect dimension to be additive with respect to direct products; after all, the direct product of $\mathbb{R}^d$ and $\mathbb{R}^p$ is $\mathbb{R}^{d+p}$, and so it seems natural to conjecture that in general,

$$(7.11) \qquad \dim_H(A \times B) = \dim_H A + \dim_H B.$$

Remark. When A and B are subsets of the Euclidean spaces $\mathbb{R}^d$ and $\mathbb{R}^p$, their direct product $A \times B$ lies in $\mathbb{R}^{d+p}$, and hence inherits the Euclidean metric. From an abstract point of view, when A and B are arbitrary metric spaces, there are a number of natural metrics that one might use. Chief among these are the following: here $x, x' \in A$ and $y, y' \in B$, and d_A and d_B are the metrics on A and B, respectively.

$$d_1((x, y), (x', y')) = d_A(x, x') + d_B(y, y'),$$
$$d_2((x, y), (x', y')) = \sqrt{d_A(x, x')^2 + d_B(y, y')^2},$$
$$d_\infty((x, y), (x', y')) = \max\{d_A(x, x'), d_B(y, y')\}.$$

The standard metric that we obtain in the Euclidean case is d_2; however, if we endow Euclidean space with the alternate metric (2.11), we obtain d_1, and similarly (2.12) leads to d_∞. As in Exercise 2.5, the three metrics d_1, d_2, and d_∞ are strongly equivalent, and hence any of the three may be used for computations of Hausdorff dimension.

Exercise 7.5. Using the product measures $m_H(\cdot, \alpha) \times m_H(\cdot, \beta)$, show that for any two sets A and B,

$$(7.12) \qquad \dim_H(A \times B) \geq \dim_H A + \dim_H B.$$

Exercise 7.5 establishes one half of (7.11). However, the reverse inequality is not true in general; a counterexample to this effect was first produced by Besicovitch.[6]

Example 7.8. We produce two sets $A, B \subset [0, 1]$ which both have Hausdorff dimension equal to 0, but are large enough that

$$A + B = \{x + y \mid x \in A, y \in B\} \supset [0, 1].$$

---

[6]The example we give here is slightly less general than the one given in [**BM45**], but it follows the same idea.

Since the map $f\colon \mathbb{R}^2 \to \mathbb{R}$ given by $f(x,y) = x+y$ is Lipschitz, this will imply that

$$\dim_H(A \times B) \geq \dim_H f(A \times B) = \dim_H(A+B) \geq \dim_H[0,1] = 1.$$

To construct A and B, we first fix an increasing sequence of positive integers n_k, which is to satisfy a certain growth condition to be defined below. We write $x = 0.x_1 x_2 x_3 \cdots$ for the binary expansion of $x \in [0,1]$; if x is a dyadic rational, so that its binary expansion terminates in an infinite string of zeros or ones, we choose the expansion which ends in zeros. Now define sets $Z_k \subset [0,1]$ as follows:

$$Z_k = \{x \in [0,1] \mid x_i = 0 \text{ for all } n_k < i \leq n_{k+1}\}.$$

Finally, define A and B by

$$A = Z_1 \cap Z_3 \cap Z_5 \cap \cdots,$$
$$B = Z_2 \cap Z_4 \cap Z_6 \cap \cdots.$$

That is, if we think of the sequence $\{n_k\}$ as partitioning $\mathbb{N}$ into a sequence of intervals $(n_k, n_{k+1}]$, then A is the set of numbers whose binary digits x_i are zero whenever i lies in an odd interval; similarly, numbers in B have binary expansions which vanish on the even intervals.

It is immediately apparent that any $w \in [0,1]$ may be written as $w = x+y$ where $x \in A$ and $y \in B$; simply take x and y to be the numbers whose binary expansions are given by

$$x_i = \begin{cases} 0 & i \in (n_{2k-1}, n_{2k}], \\ 1 & i \in (n_{2k}, n_{2k+1}], \end{cases} \qquad y_i = \begin{cases} 1 & i \in (n_{2k-1}, n_{2k}], \\ 0 & i \in (n_{2k}, n_{2k+1}]. \end{cases}$$

Thus we need only choose n_k such that $\dim_H A = \dim_H B = 0$. Observe that for odd values of k, the set $Z_1 \cap Z_3 \cap \cdots \cap Z_k$ is a union of 2^{m_k} intervals of length 2^{-n_k}, where

$$m_k = (n_2 - n_1) + (n_4 - n_3) + \cdots + (n_{k-1} - n_{k-2}).$$

Since $A \subset Z_1 \cap Z_3 \cap \cdots \cap Z_k$ for every odd k, we thus have a family of covers of A by intervals of length 2^{-n_k}, whence

$$\dim_H A \leq \underline{\dim}_B A \leq \varliminf_{k\to\infty} \frac{\log 2^{m_k}}{-\log 2^{-n_k}} = \varliminf_{k\to\infty} \frac{m_k}{n_k}.$$

Since m_k only depends on $n_1, \ldots, n_{k-1}$ and not on n_k itself, we may choose n_k to be increasing rapidly enough that this last quantity tends to 0, and hence $\dim_H A = 0$. Similar considerations give us $\dim_H B = 0$, and hence we have strict inequality in (7.12).

Besicovitch also proved that (7.11) *does* hold under an additional assumption on the sets involved.

Theorem 7.9. *If A is such that $\dim_H A = \underline{\dim}_B A = \overline{\dim}_B A$, then equality holds in (7.11) for any B.*

Proof. This is a consequence of Exercise 7.5 and the following general inequality, which holds for arbitrary A and B:

$$(7.13) \qquad \dim_H(A \times B) \le \overline{\dim}_B A + \dim_H B.$$

To prove (7.13), we fix $s > \overline{\dim}_B A$ and $t > \dim_H B$; thus

$$\varlimsup_{\varepsilon \to 0} \frac{\log N(A, \varepsilon)}{-\log \varepsilon} < s,$$

$$\lim_{\varepsilon \to 0} m_H(B, t, \varepsilon) = 0.$$

In particular, there exists $\varepsilon_0 > 0$ such that for all $0 < \varepsilon < \varepsilon_0$,

$$(7.14) \qquad N(A, \varepsilon) < \varepsilon^{-s},$$

$$(7.15) \qquad m_H(B, t, \varepsilon) < 1.$$

Now by (7.15) there exists an ε-cover $\{U_i \mid i \in \mathbb{N}\}$ of B such that $\sum_i (\operatorname{diam} U_i)^t < 1$. For each i, (7.14) guarantees the existence of a cover $\{U_{i,j} \mid 1 \le j \le N(A, \operatorname{diam} U_i)\}$ of A such that $\operatorname{diam} U_{i,j} = \operatorname{diam} U_i$ for all j. Using the metric d_∞ on the direct product $A \times B$, we see that

$$\operatorname{diam}(U_{i,j} \times U_i) = \max\{\operatorname{diam} U_{i,j}, \operatorname{diam} U_i\} = \operatorname{diam} U_i < \varepsilon,$$

and hence

$$m_H(A \times B, s+t, \varepsilon) \leq \sum_{i,j} \operatorname{diam}(U_{i,j} \times U_i)^{s+t} = \sum_{i,j} (\operatorname{diam} U_i)^{s+t}$$

$$= \sum_i N(A, \operatorname{diam} U_i)(\operatorname{diam} U_i)^{s+t}$$

$$< \sum_i (\operatorname{diam} U_i)^{-s}(\operatorname{diam} U_i)^{s+t}$$

$$= \sum_i (\operatorname{diam} U_i)^t < 1.$$

Taking the limit as $\varepsilon \to 0$, we see that $m_H(A \times B, s+t) < \infty$, and hence $\dim_H(A \times B) \leq s+t$. Since $s > \overline{\dim}_B A$ and $t > \dim_H B$ were arbitrary, this establishes (7.13). $\qquad \square$

Exercise 7.6. Let C_1 and C_2 be the limit sets of two Moran constructions on the line. Show that $\dim_H(C_1 \times C_2) = \dim_H C_1 + \dim_H C_2$.

b. Quantifying the attractor. Returning to the Smale–Williams solenoid Λ, we write

$$\Lambda(\theta, \varepsilon) = \Lambda \cap (D^2 \times (\theta - \varepsilon, \theta + \varepsilon))$$

for the ε-wedge of the attractor around angle θ. Since Λ can be written as a finite union of such wedges, we can compute $\dim_H \Lambda$ by computing $\dim_H \Lambda(\theta, \varepsilon)$.

Writing $C = \Lambda \cap (D^2 \times \{\theta\})$ for a cross-section of the attractor, we recall that $\Lambda(\theta, \varepsilon)$ is homeomorphic to $C \times (-\varepsilon, \varepsilon)$. In fact, the homeomorphism can be chosen to be bi-Lipschitz, and so

$$\dim_H \Lambda(\theta, \varepsilon) = \dim_H(C \times (-\varepsilon, \varepsilon)).$$

Theorem 7.9 only requires coincidence of the Hausdorff and box dimensions for *one* of the two sets A and B. Since these quantities coincide for the interval $(-\varepsilon, \varepsilon)$, we have

$$\dim_H \Lambda(\theta, \varepsilon) = (\dim_H C) + 1,$$

and since $\dim_H C$ does not depend on θ or ε,

$$(7.16) \qquad\qquad \dim_H \Lambda = (\dim_H C) + 1.$$

Of course, we still need to compute $\dim_H C$. In the simplest case where $\alpha = \beta$, the construction of C is exactly of the sort dealt with by Moran's theorem, and we have

$$\dim_H C = \frac{\log 2}{-\log \alpha}.$$

Thus we get

$$(7.17) \qquad \dim_H \Lambda = 1 + \frac{\log 2}{-\log \alpha} = \log 2 \left(\frac{1}{\log 2} + \frac{1}{-\log \alpha} \right);$$

this is reminiscent of (4.17), which related Hausdorff dimension and topological entropy for a one-dimensional Markov map with constant slope. Indeed, one may show that the topological entropy of the Smale–Williams solenoid is $h_{\text{top}}(\Lambda, f) = \log 2$, and that the measure of maximal entropy is the product of the $(1/2, 1/2)$-Bernoulli measure on W^s and Lebesgue measure on W_ε^u.

In (4.17), the scaling factor relating the Hausdorff dimension and the topological entropy was the reciprocal of the Lyapunov exponent. In (7.17), this factor is the somewhat odd-looking expression $(1/\log 2) - (1/\log \alpha)$. What are we to make of this?

c. Lyapunov exponents in multiple dimensions. When f is a one-dimensional map, the definition of the Lyapunov exponent is relatively simple: at a given point x, the Lyapunov exponent $\lambda_f(x)$ is the asymptotic rate of expansion along the orbit of x (provided the limit exists). For maps in more than one dimension, the situation is somewhat more complicated, as f may have different rates of expansion in different directions. Indeed, from the definition of a hyperbolic set Λ we see that along the orbit of any point $\mathbf{x} \in \Lambda$, there are some directions in which f is expanding (corresponding to a positive Lyapunov exponent) and some in which f is contracting (corresponding to a negative Lyapunov exponent).

Thus in general, the Lyapunov exponent depends not only on the point $\mathbf{x}$ but also on the direction $\mathbf{v}$ in which expansion is measured.

Definition 7.10. Let $U \subset \mathbb{R}^d$ be the domain of an invertible differentiable map $f\colon U \to U$. Given a point $\mathbf{x} \in U$ and a vector $\mathbf{v} \in \mathbb{R}^d$, the *forward Lyapunov exponent* of f at the point $\mathbf{x}$ in the direction of $\mathbf{v}$ is

$$\lambda_f^+(\mathbf{x}, \mathbf{v}) = \lim_{k \to \infty} \frac{1}{k} \log \|Df^k(\mathbf{x})\mathbf{v}\|,$$

and the *backward Lyapunov exponent* is

$$\lambda_f^-(\mathbf{x}, \mathbf{v}) = \lim_{k \to \infty} \frac{1}{k} \log \|Df^{-k}(\mathbf{x})\mathbf{v}\|,$$

provided the limits exist. If $\lambda_f^+(\mathbf{x}, \mathbf{v}) = -\lambda_f^-(\mathbf{x}, \mathbf{v})$, then we call this value the *Lyapunov exponent* of f at the point $\mathbf{x}$ in the direction of $\mathbf{v}$, and denote it by $\lambda_f(\mathbf{x}, \mathbf{v})$.

Example 7.11. Let $f: \mathbb{R}^2 \to \mathbb{R}^2$ be the linear map defined by the matrix $\left(\begin{smallmatrix} \alpha & 0 \\ 0 & \beta \end{smallmatrix}\right)$, where $0 < \alpha < 1 < \beta$. Then the trajectories of f are as shown in Figure 5.3, and there are three possibilities for $\lambda(\mathbf{0}, \mathbf{v})$:

(1) $\mathbf{v}$ lies along the x-axis. In this case, $\lambda_f^+(\mathbf{0}, \mathbf{v}) = -\lambda_f^-(\mathbf{0}, \mathbf{v}) = \log \alpha < 0$, so the Lyapunov exponent exists and is negative.

(2) $\mathbf{v}$ lies along the y-axis. In this case, $\lambda_f^+(\mathbf{0}, \mathbf{v}) = -\lambda_f^-(\mathbf{0}, \mathbf{v}) = \log \beta > 0$, so the Lyapunov exponent exists and is positive.

(3) $\mathbf{v}$ does not lie along either axis. In this case, $\lambda_f^+(\mathbf{0}, \mathbf{v}) = \log \beta$, as the forward trajectory is repelled vertically, and $\lambda_f^-(\mathbf{0}, \mathbf{v}) = -\log \alpha$, as the backward trajectory is repelled horizontally. Thus the Lyapunov exponent does not exist.

As Example 7.11 shows, the directions in which the Lyapunov exponents of a linear map exist are the eigenspaces of the map; in such a direction, the Lyapunov exponent is the logarithm of the corresponding eigenvalue. By finding a basis for $\mathbb{R}^d$ consisting of generalised eigenvectors, we can decompose $\mathbb{R}^d$ as the direct sum of subspaces along which the Lyapunov exponents exist. Such subspaces are called *Lyapunov subspaces*, and are the generalisation of eigenspaces to non-linear maps.

Exercise 7.7. Compute the Lyapunov exponent of the linear map from $\mathbb{R}^2$ to itself given by the rotation matrix

$$R_\theta = \begin{pmatrix} \cos \theta & -\sin \theta \\ \sin \theta & \cos \theta \end{pmatrix}.$$

Exercise 7.8. Compute the Lyapunov exponents of the north-south map of the unit sphere (see Exercise 7.4), where the rate of expansion around the north pole is $\alpha > 1$, and the rate of contraction around the south pole is $\beta < 1$.

Returning to the Smale–Williams solenoid $f\colon \Lambda \to \Lambda$, we see that f has two Lyapunov subspaces at any given point. The first subspace is tangent to the local unstable curve W_ε^u; any vector in this direction is expanded by a factor of 2 under the action of f, and so the corresponding Lyapunov exponent is $\log 2 > 0$. The second subspace is orthogonal to the first, and contains the stable disc W^s; any vector in this subspace is contracted by a factor of α under the action of f, and so the corresponding Lyapunov exponent is $\log \alpha < 0$.

Now we may interpret the ratio in (7.17) in terms of Lyapunov exponents; it is the sum of the reciprocals of the absolute values of the Lyapunov exponents in the expanding and contracting directions.

d. The non-conformal case. So far, we have studied the quantitative properties of the map f given in (7.6) only in the case $\alpha = \beta$. In this case, the map is *conformal*: every stable direction has the same rate of contraction, and every unstable direction has the same rate of expansion.

For $\alpha \neq \beta$, we are in the non-conformal case, which is much more difficult. Because the basic sets in Figure 7.5 are no longer similar to the basic sets at previous steps, we cannot use Moran's theorem. This case was studied by the German mathematician Hans Bothe, who considered a more general class of maps f, in which the functions $\cos \theta$ and $\sin \theta$ in (7.6) are replaced by arbitrary periodic functions $z_1(\theta)$ and $z_2(\theta)$, which changes the geometry of how the image $f(P)$ wraps around the z-axis. Bothe obtained a general formula for the Hausdorff dimension of the attractor for "typical" functions z_1 and z_2, but it was not until 1997 that the Hungarian mathematician Károly Simon proved that sin and cos belong to this "typical" class [**Sim97**]. He established that for the Smale–Williams solenoid Λ with $\beta < \alpha < 1/8$, we have[7]

$$\dim_H \Lambda = 1 + \frac{\log 2}{-\log \alpha} = \log 2 \left(\frac{1}{\log 2} + \frac{1}{-\log \alpha} \right).$$

---

[7]This was later extended to include all $\alpha < 1/2$ by Jörg Schmeling.

Somewhat surpisingly, the smaller value β, which corresponds to a direction of faster contraction, does not affect the Hausdorff dimension of the attractor! Thus only one of the two negative Lyapunov exponents plays a role in this particular situation.

e. The attractor for the FitzHugh–Nagumo map. Returning to the local map (5.7) for the FitzHugh–Nagumo model, we recall that for a particular range of values of A, the map f has a trapping region R, as shown in Figure 7.1. This ensures the existence of an attractor $\Lambda \subset R$ as in (7.5), and it is natural to ask what features Λ shares with the Smale–Williams solenoid, since both are attractors.

We saw that the Smale–Williams solenoid is hyperbolic, and hence Theorem 7.7 applies, allowing us to use all the tools of symbolic dynamics to study the solenoid and obtain many properties which are characteristic of hyperbolicity and chaotic behaviour, such as density of periodic points.

The attractor Λ for the FitzHugh–Nagumo system is more difficult to study, because it is not a hyperbolic set; there are no uniformly contracting and expanding subspaces which satisfy the definition of hyperbolicity. However, this does not preclude the possibility that there may exist two Lyapunov subspaces at "typical" points $\mathbf{x} \in \Lambda$, which are given by vectors $\mathbf{v}^s$ and $\mathbf{v}^u$ such that the map f is asymptotically contracting in the direction given by $\mathbf{v}^s$ and asymptotically expanding in the direction given by $\mathbf{v}^u$; that is, $\lambda(\mathbf{x}, \mathbf{v}^s) < 0 < \lambda(\mathbf{x}, \mathbf{v}^u)$.

Computer simulations strongly suggest that this is in fact the case, but no rigorous proofs are available.

Chapter 8

Horseshoes and Intermittent Chaos

Lecture 31

a. The Smale horseshoe: A trapping region that isn't. Let R be the trapping region for the FitzHugh–Nagumo map (5.7) which is given in Proposition 7.2. Figure 8.1 shows how the set $\overline{f(R)}$ changes as A increases. The "arms" of the image become longer and longer, until when A is large enough, $\overline{f(R)}$ is no longer contained in R, as in the third picture. Thus R is no longer a trapping region; since we have no information on the behaviour of trajectories outside the region R, we can no longer completely describe the dynamics of the map in terms of what happens within R. However, we can still describe the trajectories which remain in R, and so we may consider the set of all points $\mathbf{x}$ such that $f^k(\mathbf{x}) \in \mathbb{R}$ for all k.

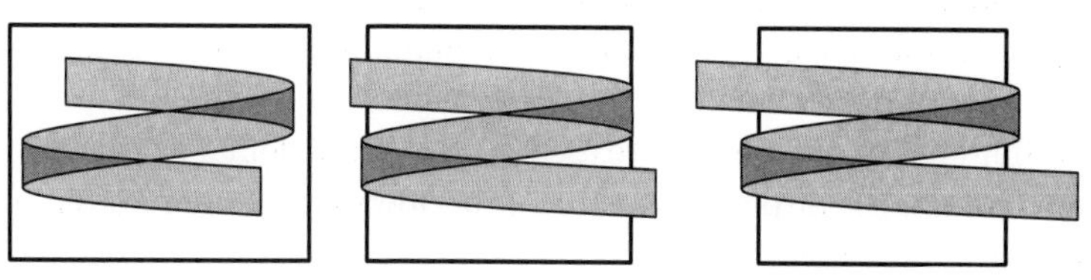

Figure 8.1. Escape from the trapping region.

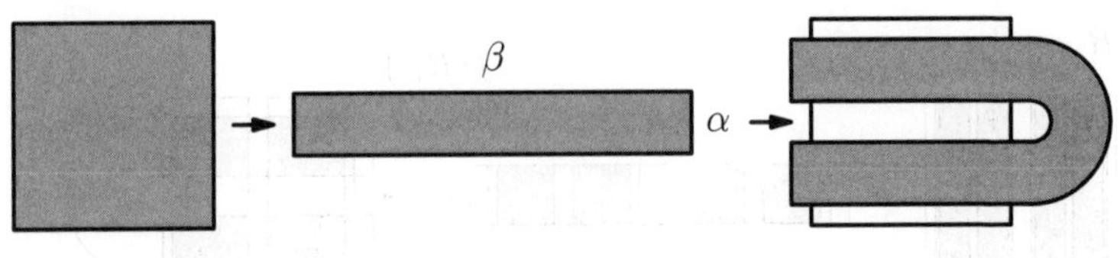

Figure 8.2. A horseshoe.

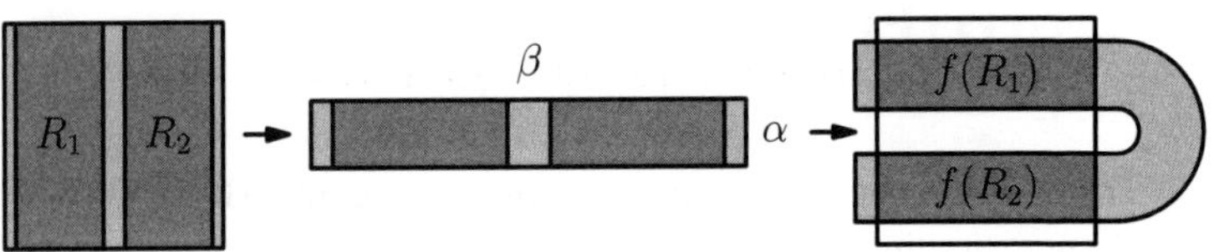

Figure 8.3. Points with two iterates.

Rather than try to describe this set for the FitzHugh–Nagumo map, which leads to rather complicated geometric considerations, we begin by considering a model case in which the picture is much cleaner. Consider a map f in $\mathbb{R}^2$ which acts on the square $R = [0, 1] \times [0, 1]$ as shown in Figure 8.2; first the square is squeezed in the vertical direction by a factor of $\alpha < 1/2$ and stretched in the horizontal direction by a factor of $\beta > 2$, then it is bent and positioned so that $f(R) \cap R$ consists of two rectangles, each of width 1 and height α.

Notice that significant portions of the square R are mapped to the area outside R; this is reminiscent of the one-dimensional Markov maps we have studied, including the logistic map for $c < -2$, under which some points could only be iterated a few times because their trajectories were carried outside the domain of definition of the map. For these maps, we found that Lebesgue-a.e. point can only be iterated finitely many times before escaping from the domain of definition and that the set of points which can be iterated infinitely often is a repelling Cantor set.

This motivates us to ask the same question for the horseshoe map f shown in Figure 8.2; which points in R can be iterated infinitely often? That is, what is the largest set of points whose trajectories remain in R for all time?

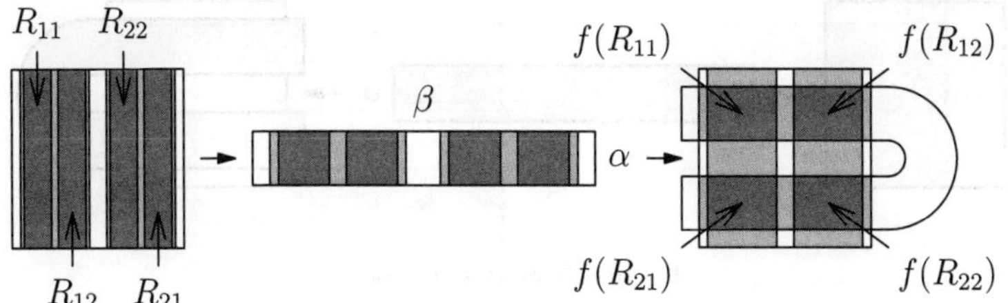

Figure 8.4. Points with three iterates.

Figure 8.3 duplicates Figure 8.2 but highlights the parts of R whose image lies in R; in particular, we see that the set of points for which f can be iterated at least twice is the union of the two rectangles R_1 and R_2, each of which has height 1 and width β^{-1}.

So for which points can f be iterated three times? In order to have $f^2(\mathbf{x}) \in R$, we must have $f(\mathbf{x}) \in R_1 \cup R_2$; Figure 8.4 shows the set of points whose image lands in R_1 or R_2. Observe that

$$f^{-1}(R_1) = R_{11} \cup R_{21},$$
$$f^{-1}(R_2) = R_{12} \cup R_{22},$$

and that $R_{w_1 w_2} \subset R_{w_1}$ for all $w_1, w_2 \in \{1, 2\}$. Continuing this process, we see that the set of points for which f can be iterated n times is the union of 2^n rectangles $R_{w_1 \dots w_n}$, each of width β^{-n} and height 1, which are characterised by

$$(8.1) \qquad R_{w_1 \dots w_n} = R_{w_1} \cap f^{-1}(R_{w_2}) \cap \cdots \cap f^{-(n-1)}(R_{w_n}).$$

Letting n go to infinity, we see that the non-escaping set is

$$(8.2) \qquad \Gamma^+ = \{\mathbf{x} \in \mathbb{R}^2 \mid f^n(\mathbf{x}) \in R \text{ for all } n \geq 0\} = C_{\beta^{-1}} \times [0, 1],$$

where $C_{\beta^{-1}} \subset [0, 1]$ is a Cantor set with both ratio coefficients equal to β^{-1}; in particular, this implies that

$$\dim_H \Gamma^+ = \left(\dim_H C_{\beta^{-1}}\right) + 1 = \frac{\log 2}{\log \beta} + 1.$$

The story so far has largely been a retelling of a familiar tale which we have already seen play out for one-dimensional Markov maps. There is a twist in the plot, however: unlike those maps, the horseshoe map f is one-to-one and, hence, invertible on its image. Consequently,

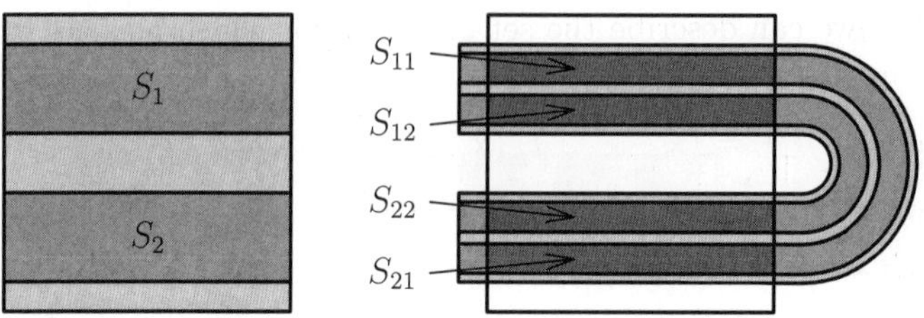

Figure 8.5. Points with two backwards iterates.

we are interested in points for which both the forward *and backward* trajectories remain in R. That is, we are also interested in

$$\Gamma^- = \{\mathbf{x} \in \mathbb{R}^2 \mid f^n(\mathbf{x}) \in R \ \forall n \le 0\}.$$

Since $f(R)$ does not cover the entire square R, we see that the only points in R with any preimages at all are those which lie in $f(R_1)$ or $f(R_2)$ (see Figure 8.3). Write $S_i = f(R_i)$; then the set of points in R with one backwards iterate is the union of the two rectangles S_1 and S_2, each of which has width 1 and height α.

The set of points with two backwards iterates is

$$f(f(R) \cap R) \cap R = f(S_1 \cup S_2) \cap R = S_{11} \cup S_{12} \cup S_{22} \cup S_{21},$$

as shown in Figure 8.5, where $S_{w_1 w_2} = S_{w_1} \cap f(S_{w_2}) = f^2(R_{w_2 w_1})$. Continuing, we obtain rectangles $S_{w_1 \ldots w_n}$ of width 1 and height α^n characterised by

$$(8.3) \quad S_{w_1 \ldots w_n} = S_{w_1} \cap f(S_{w_2}) \cap \cdots \cap f^{(n-1)}(S_{w_n}) = f^n(R_{w_n \ldots w_1}).$$

We see that

$$\Gamma^- = [0, 1] \times C_\alpha,$$

where $C_\alpha \subset [0, 1]$ is a Cantor set with both ratio coefficients equal to α; hence

$$\dim_H \Gamma^- = 1 + \frac{\log 2}{-\log \alpha}.$$

Now we can describe the set of points for which all forward and backward iterates remain in R:

$$\Gamma = \{\mathbf{x} \in \mathbb{R}^2 \mid f^n(\mathbf{x}) \in R \text{ for all } n \in \mathbb{Z}\}$$
$$= \Gamma^+ \cap \Gamma^-$$
$$= (C_{\beta^{-1}} \times [0,1]) \cap ([0,1] \times C_\alpha)$$
$$= C_{\beta^{-1}} \times C_\alpha.$$

The set Γ is known as the *Smale horseshoe*, and has many important dynamical properties; chief among these is the fact that it is a hyperbolic set, with all the attendant consequences of that fact.

Remark. Some of the motivation for the Smale horseshoe came from the van der Pol equation mentioned in Chapter 5, which was studied by Norman Levinson in the 1940s. He later brought the dynamical behaviour of that system to the attention of Stephen Smale, who in 1961 extracted the salient geometric details to produce the horseshoe map f, which captures many of the essential features of a large family of "chaotic" maps, as we will later see.

b. Hausdorff dimension of the horseshoe. Using Moran's theorem and Theorem 7.9, we can compute the Hausdorff dimension of the Smale horseshoe:

$$\dim_H \Gamma = \dim_H C_{\beta^{-1}} + \dim_H C_\alpha$$

(8.4)
$$= \log 2 \left(\frac{1}{\log \beta} + \frac{1}{-\log \alpha} \right).$$

Exercise 8.1. Show that the Smale horseshoe Γ is the limit set of a geometric construction in the square R. Describe the basic sets of this construction, and explain why Moran's formula (2.14) cannot be applied in general to compute the Hausdorff dimension of Γ directly. Find the particular case when it can be applied, and show that in that case the answer given by (2.14) agrees with (8.4).

In light of our previous experience, we expect the terms in (8.4) to include the topological entropy and the Lyapunov exponents of f.

Indeed, just as we found for the Smale–Williams solenoid, we have $h_{\mathrm{top}}(\Gamma, f) = \log 2$, and the measure of maximal entropy is the product of the $(1/2, 1/2)$-Bernoulli measures on each of $C_{\beta^{-1}}$ and C_α.

For the Lyapunov exponents, we observe that at every point $\mathbf{x} \in \Gamma$, we have $Df(\mathbf{x}) = \pm \left(\begin{smallmatrix} \beta & 0 \\ 0 & \alpha \end{smallmatrix} \right)$, and so the Lyapunov exponents are $\log \alpha$ in the vertical (contracting) direction, and $\log \beta$ in the horizontal (expanding) direction.

Thus we see that (8.4) has the same form as (7.17); this suggests that some underlying process is at work here, relating the various quantities of dimension, entropy, and Lyapunov exponent. We give an informal explanation of this relationship. The first step is to observe that both these equations relate the global quantities and both have local versions; for example, (8.4) is completely analogous to the local equation

$$(8.5) \qquad d_\mu(\mathbf{x}) = h_{\mu,f}(\mathbf{x}) \left(\frac{1}{\lambda_f^u(\mathbf{x})} + \frac{1}{-\lambda_f^s(\mathbf{x})} \right),$$

where $\lambda_f^u(\mathbf{x})$ and $\lambda_f^s(\mathbf{x})$ are the positive and negative Lyapunov exponents, respectively, of the map f at the point $\mathbf{x}$, and μ is any f-invariant measure.

We saw an equation almost exactly like this earlier, in (4.28). The difference is that now the map f is invertible; consequently, if we use the definition of Bowen balls in (4.13), the Bowen balls $B_f(\mathbf{x}, n, \delta)$ will not converge to the point $\mathbf{x}$ as $n \to \infty$, but rather to the local stable curve through $\mathbf{x}$. In order to have convergence to a single point, we must consider two-sided Bowen balls,

$$B_f(\mathbf{x}, m, n, \delta) = \{ \mathbf{y} \in \Gamma \mid d(f^j(\mathbf{x}), f^j(\mathbf{y})) < \delta \text{ for all } m \leq j \leq n \},$$

where typically we will take $m < 0 < n$. Then if we choose $m < 0 < n$, $\delta > 0$, and $\varepsilon > 0$ such that

$$\varepsilon \approx \delta e^{-n\lambda_f^u(\mathbf{x})} \approx \delta e^{-m\lambda_f^s(\mathbf{x})},$$

we find that

$$B(\mathbf{x}, \varepsilon) \approx B_f(\mathbf{x}, m, n, \delta),$$

and (8.5) follows upon taking logarithms and limits, since for an invertible map f, the local entropy is

$$h_{\mu,f}(\mathbf{x}) = \lim_{n-m \to \infty} \frac{1}{n-m} \log \mu(B_f(\mathbf{x}, m, n, \delta)).$$

What we have given is only a bare-bones sketch of the argument, and the reader is strongly encouraged to work through the details of the computation.

c. Symbolic dynamics on the Smale horseshoe. Because Γ is a hyperbolic set, Theorem 7.7 guarantees the utility of symbolic dynamics to describe the map $f\colon \Gamma \to \Gamma$. In fact, we can make the correspondence explicit.

Recall that the set of points in R whose image is also in R is $R_1 \cup R_2$, where R_1 and R_2 are the rectangles shown in Figure 8.3. It follows that $\Gamma \subset R_1 \cup R_2$, and so given a point $\mathbf{x} \in \Gamma$, we have $f^n(\mathbf{x}) \in R_1 \cup R_2$ for all n. Define a sequence $w^+ \in \Sigma_2^+$ such that $f^n(\mathbf{x}) \in R_{w_n}$ for all $n \geq 0$; in this manner we can associate to each point $\mathbf{x} \in \Gamma$ a sequence in the symbolic space Σ_2^+.

Observing further that $f^{(n-1)}(f(\mathbf{x})) = f^n(\mathbf{x}) \in R_{w_{n+1}}$, we see that the point $f(\mathbf{x})$ is coded by the sequence

$$\sigma(\omega^+) = (w_1, w_2, w_3, \dots),$$

and it looks like we are well on our way to establishing a topological conjugacy between the map $f\colon \Gamma \to \Gamma$ and the shift $\sigma\colon \Sigma_2^+ \to \Sigma_2^+$.

Once again, however, there is a twist in the plot. If we follow the recipe from our previous encounters with symbolic dynamics, then the sequence w^+ should determine the point x uniquely as the only point in the infinite intersection

$$W^s(w^+) = R_{w_0} \cap R_{w_0 w_1} \cap \cdots \cap R_{w_0 \dots w_n} \cap \cdots .$$

However, each rectangle $R_{w_0 \dots w_n}$ has width $\beta^{-(n+1)}$ and height 1 (see Figures 8.3 and 8.4); this means that the intersection $W^s(w^+)$ is a vertical line, rather than a single point! Indeed, any point on the vertical line $W^s(\mathbf{x}) = W^s(w^+)$ passing through $\mathbf{x}$ has a forward trajectory which is coded by the same sequence w^+ (and so this vertical line is the local stable curve through $\mathbf{x}$).

We see, then, that the coding of the forward trajectory is not enough to determine $\mathbf{x}$ uniquely. We also need to code the backward trajectory $\{f^{-n}(\mathbf{x})\}_{n=0}^{\infty}$; thus we define a sequence

$$w^- = (w_{-1}, w_{-2}, \dots)$$

such that $f^n(x) \in R_{w_n}$ for all $n \le -1$. Once again, w^- does not determine $\mathbf{x}$ uniquely; we have

$$W^u(w^-) = S_{w_{-1}} \cap S_{w_{-1}w_{-2}} \cap \cdots \cap S_{w_{-1}\cdots w_{-n}} \cap \cdots,$$

where each rectangle $S_{w_{-1}\cdots w_{-n}}$ has width 1 and height α^n, and so the intersection $W^u(\mathbf{x}) = W^u(w^-)$ is the horizontal line through $\mathbf{x}$ (the local unstable curve).

Although neither the forward itinerary w^+ nor the backward itinerary w^- of $\mathbf{x}$ is by itself enough to determine $\mathbf{x}$ uniquely, the combination of the two does suffice. Indeed, the vertical line $W^s(\mathbf{x})$ and the horizontal line $W^u(\mathbf{x})$ meet in a single point, $\mathbf{x}$ itself. Thus we can define a coding map

$$h\colon \Sigma_2 \to \Gamma,$$
$$w = (w^-, w^+) \mapsto W^u(w^-) \cap W^s(w^+),$$

where (w^-, w^+) denotes the concatenation

$$(w^-, w^+) = (\ldots, w_{-2}, w_{-1}|w_0|w_1, w_2, \ldots).$$

Thus $\mathbf{x} = h(w)$ is the unique point in Γ for which $f^n(\mathbf{x}) \in R_{w_n}$ for all $n \in \mathbb{Z}$.

Proposition 8.1. *The coding map $h\colon \Sigma_2 \to \Gamma$ is a homeomorphism.*

Proof. h is a bijection, and $\mathbf{x} = h(w)$ is defined by the fact that for every n, we have the inclusion $\mathbf{x} \in R_{w_0 w_1 \ldots w_n} \cap S_{w_{-1} \ldots w_{-n}}$. It follows that $h(C_{w_{-n} \ldots w_{-1}|w_0|w_1 \ldots w_n}) = R_{w_0 w_1 \ldots w_n} \cap S_{w_{-1} \ldots w_{-n}}$. This implies that h and h^{-1} both map closed sets to closed sets, and hence h is a homeomorphism. $\qquad\square$

Applying the shift map $\sigma\colon \Sigma_2 \to \Sigma_2$ to a sequence w shifts the "centre" to the right by one:

$$\sigma(\ldots, w_{-2}, w_{-1}|w_0|w_1, w_2, \ldots) = (\ldots, w_{-2}, w_{-1}|w_0|w_1, w_2, \ldots).$$

Thus it is transparent from the construction of the coding map h that the following diagram commutes:

$$
\begin{array}{ccc}
\Sigma_2 & \xrightarrow{\ \sigma\ } & \Sigma_2 \\
\downarrow{\scriptstyle h} & & \downarrow{\scriptstyle h} \\
\Gamma & \xrightarrow{\ f\ } & \Gamma
\end{array}
$$

(8.6)

As mentioned above, the horizontal lines $W^u(\mathbf{x})$ and the vertical lines $W^s(\mathbf{x})$ have a dynamical meaning. f contracts each vertical line uniformly by a factor of α, and so given $\mathbf{y} \in W^s(\mathbf{x})$, we have

$$d(f^n(\mathbf{x}), f^n(\mathbf{y})) = \alpha^n d(\mathbf{x}, \mathbf{y}) \xrightarrow{n \to +\infty} 0.$$

Thus $W^s(\mathbf{x})$ is the local stable curve through $\mathbf{x}$ for the map f. Similarly, observe that f^{-1} contracts horizontal lines uniformly by a factor of β, and so for $\mathbf{y} \in W^u(\mathbf{x})$, we have

$$d(f^{-n}(\mathbf{x}), f^{-n}(\mathbf{y})) = \beta^{-n} d(\mathbf{x}, \mathbf{y}) \xrightarrow{n \to +\infty} 0.$$

Thus $W^u(\mathbf{x})$ is the local stable curve through $\mathbf{x}$ for the map f^{-1}, and the local *unstable* curve through $\mathbf{x}$ for the map f.

We see from all this that the dynamics of the symbolic space encode the hyperbolic structure of the horseshoe map; the stable curve for f through a point $\mathbf{x}$ consists of those points whose forward itineraries eventually agree with the forward itinerary of $\mathbf{x}$, as given by the sequence w^+, and the unstable curve is given similarly, as those points whose backwards itineraries eventually agree with w^-.

Exercise 8.2. Using the fact that $f \colon \Gamma \to \Gamma$ is conjugate to $\sigma \colon \Sigma_2 \to \Sigma_2$, show that f has 2^n points of period n (that is, fixed points of f^n) lying in Γ, for all $n \geq 1$.

Exercise 8.3. Show that the Smale horseshoe contains a point whose orbit is everywhere dense in the horseshoe.

Lecture 32

a. Variations on a theme: Other horseshoes.

a.1. *Non-linear horseshoes.* As was the case with the one-dimensional Markov maps we discussed earlier, the symbolic approach does not capture all the quantitative geometric information about the horseshoe Γ. In the first place, neither Σ_2 nor the map σ depends on the value of the parameters α and β, while varying these quantities certainly changes the set Γ, and in particular, changes its Hausdorff dimension. This parallels the situation for one-dimensional Markov

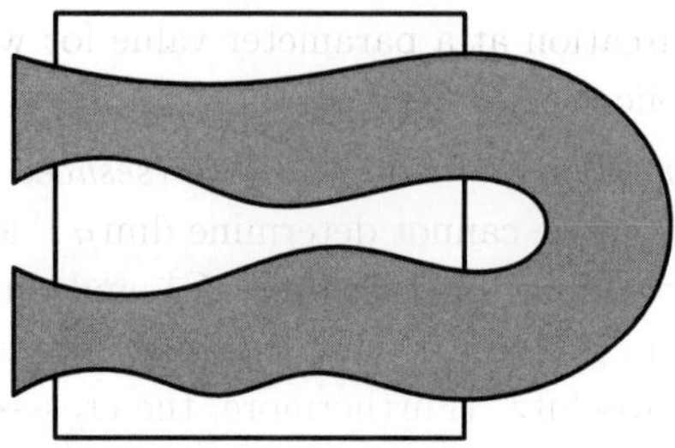

Figure 8.6. A non-linear horseshoe.

maps, where the conjugacy with the one-sided shift carries no information about the ratio coefficients λ_i (in the linear case) or the Lyapunov exponents $\lambda_f(x)$ (in the non-linear case).

Furthermore, we can still carry out the whole procedure if we perturb the map slightly, obtaining a non-linear map f such as the one in Figure 8.6. The set of points which can be iterated once is the union of two "rectangles" with curved sides; the set of points with two forward iterates is the union of four such regions, and so on. Passing to the limit by taking an intersection over all n, the set of points whose forward trajectories remain in the domain of f is a family of more or less vertical curves; each horizontal cross-section of this family is a Cantor set.

Similarly, the set of points with infinitely many backwards iterates is a family of relatively horizontal curves; each vertical cross-section of this family is a Cantor set. Taking the intersection of the two families, the set of points with whose forwards and backwards trajectories remain in the domain of f for all time is a non-linear horseshoe which is homeomorphic to the direct product of two Cantor sets $C^+ \times C^-$ and which is still modeled by Σ_2.

This persistence of qualitative behaviour under a small perturbation of the map f, known as *structural stability*, is characteristic of uniformly hyperbolic maps such as the Smale–Williams solenoid and the Smale horseshoe. We observed a similar phenomenon in Chapter 6, in the fact that windows of stability for the logistic map are open; the essence of structural stability is that a family of systems cannot have a bifurcation at a parameter value for which the map is uniformly hyperbolic.

a.2. *Hausdorff dimension of non-linear horseshoes.* If Γ is a non-linear horseshoe, then we cannot determine $\dim_H \Gamma$ as we did for linear horseshoes, by adding together $\dim_H C^+$ and $\dim_H C^-$, because the homeomorphism between Γ and the direct product $C^+ \times C^-$ is not in general bi-Lipschitz. (Furthermore, the cross-sections C^+ and C^- may vary depending on which cross-section we choose, and we do not have any *a priori* guarantee that the Hausdorff dimension is the same for each cross-section.)

Nevertheless, the fact that the map $f\colon \Gamma \to \Gamma$ is modeled by the two-sided shift $\sigma\colon \Sigma_2 \to \Sigma_2$ still gives us some tools with which to work. For example, the Non-uniform Mass Distribution Principle allows us to get information about $\dim_H \Gamma$ by examining various measures on Γ, while the conjugacy with the symbolic system gives us many measures with which to work, since any shift-invariant measure on Σ_2 determines an f-invariant measure on Γ.

For example, on any horseshoe, whether linear or non-linear, we have Bernoulli measures and Markov measures, defined much as in the one-sided case. A probability vector (p_1, p_2) defines a Bernoulli measure μ on Σ_2 by

$$\mu(C_{w_a \ldots w_b}) = p_{w_a} p_{w_{a+1}} \cdots p_{w_b};$$

a stochastic matrix $P = (p_{ij})$ together with a stationary probability vector π defines a Markov measure m on Σ_2 by

$$m(C_{w_a \ldots w_b}) = p_{w_a} p_{w_a w_{a+1}} p_{w_{a+1} w_{a+2}} \cdots p_{w_{b-1} w_b}.$$

As in the one-sided case, the support of a Markov measure is the subshift of finite type

$$\Sigma_A = \{w \in \Sigma_2 \mid a_{w_j w_{j+1}} = 1 \text{ for all } j \in \mathbb{Z}\},$$

where A is the 2×2 transition matrix given by

$$a_{ij} = \begin{cases} 0 & p_{ij} = 0, \\ 1 & p_{ij} > 0. \end{cases}$$

Σ_A is shift-invariant, and so $\Gamma_A = h(\Sigma_A) \subset \Gamma$ is f-invariant; thus the horseshoe contains all manner of intricate fractals which are preserved by the dynamics of f. Once we have some information about the pointwise dimension of measures supported on these sets, we can investigate their Hausdorff dimension using the Non-uniform Mass Distribution Principle.

Just as in the one-sided case, the local entropy of a Bernoulli measure is given by (4.14), and that of a Markov measure is given by (4.20). Furthermore, if μ is an ergodic measure on Γ (in particular, if μ is a Bernoulli measure or a Markov measure with primitive transition matrix), then the Lyapunov exponents $\lambda_f^u(\mathbf{x})$ and $\lambda_f^s(\mathbf{x})$ exist and are constant μ-a.e.[1] Denoting the constant values by $\lambda^u(\mu, f)$ and $\lambda^s(\mu, f)$, the relationship (8.5) from the previous lecture can be written in terms of the global dimensional quantities for measures:

$$(8.7) \qquad \dim_H \mu = h(\mu, f) \left(\frac{1}{\lambda^u(\mu, f)} + \frac{1}{-\lambda^s(\mu, f)} \right).$$

By the Non-uniform Mass Distribution Principle, we may obtain a (hopefully tight) lower bound for $\dim_H \Gamma_A$ by maximising the quantity on the right-hand side of (8.7) over all Markov measures supported on Γ_A. In the case where f is linear (such as the Smale horseshoe), the Lyapunov exponents $\lambda_f^u(\mathbf{x})$ and $\lambda_f^s(\mathbf{x})$ are constant everywhere, not just almost everywhere, and so the measure of maximal dimension coincides with the measure of maximal entropy.

When f is non-linear, the situation is more complicated, and with the exception of certain atypical cases, no measure of maximal dimension exists (there is a measure which has maximal dimension in the stable direction, and another which has maximal dimension in the unstable direction, but they do not typically coincide).

a.3. *Horseshoes with more branches.* Finally, we can consider the horseshoe-like map $g\colon R \to \mathbb{R}^2$ which acts on the square R as shown in Figure 8.7. Arguing as we did for f, one finds that the set Γ of non-escaping points is homeomorphic to $\Sigma_3 = \{1, 2, 3\}^{\mathbb{Z}}$, and so the restriction of f to Γ is conjugate to the full shift on *three* symbols.

---

[1]This is a consequence of the seminal *Multiplicative Ergodic Theorem*, due to Oseledets. The corresponding statement for one-dimensional maps is a special case of the classical *Pointwise Ergodic Theorem*, due to Birkhoff.

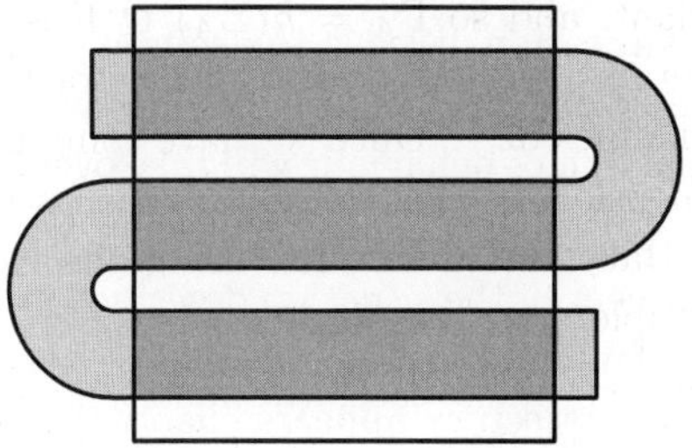

Figure 8.7. A horseshoe with three branches.

If α and β are the contraction and expansion ratios in the vertical and horizontal directions, respectively, then Γ is the direct product of Cantor sets in the interval with ratio coefficients α and β^{-1}, and so

$$\dim_H \Gamma = \frac{\log 3}{-\log \alpha} + \frac{\log 3}{\log \beta} = \log 3 \left(\frac{1}{-\log \alpha} + \frac{1}{\log \beta} \right).$$

This has the same form as (8.4), giving Hausdorff dimension in terms of entropy and Lyapunov exponents; here the topological entropy of the map is $\log 3$. In general, we could consider a horseshoe-like map with k branches, and would find the same formula, with entropy $\log k$. This again reflects the fact that entropy somehow measures the *complexity* of the system. For these maps, the system is entirely linear on the invariant horseshoe; thus all the complexity comes from how many times that set is folded back into itself by the map, and the dimension is proportional to the entropy.

b. Intermittent chaos vs. persistent chaos. If we choose a "typical"[2] point $\mathbf{x} = (x_1, x_2)$ in the Smale horseshoe Γ and observe its trajectory $f^n(\mathbf{x}) = (x_1^{(n)}, x_2^{(n)})$ by plotting either $x_1^{(n)}$ or $x_2^{(n)}$ as a function of n, we see a chaotic signal which persists as $n \to \infty$, without ever settling down into periodic behaviour.

However, even though the Smale horseshoe Γ is the largest f-invariant set in the square R, it has zero Lebesgue measure, and so if we simply choose a point from the square R "at random", it will lie outside of Γ with probability 1. In particular, after some number of

---

[2]Here "typical" means that if μ is any invariant measure on Γ whose support is not a periodic orbit, then the property stated holds for μ-a.e. $\mathbf{x}$.

iterations, the trajectory will leave the square; as long as it remains in the square, it appears chaotic, but once it leaves R, all bets are off, and some other pattern of behaviour will take over from the dynamics of the horseshoe.

This is reminiscent of the situation described in Lecture 27, where we considered the family of logistic maps in a window of stability. In that case, a typical trajectory appears chaotic for some finite period of time, during which it follows various unstable periodic orbits, but eventually settles down to a stable periodic orbit. We referred to this behaviour as *transient chaos*, indicating that although a typical trajectory does behave chaotically, it also eventually leaves the chaotic part of phase space and settles down to a regular pattern of behaviour.

In the present case, we have not defined the horseshoe map f outside of R, and so we cannot say just what a typical trajectory will do once it leaves the rectangle. Indeed, that behaviour is highly contingent upon how f is extended to $\mathbb{R}^2$. For some extensions, a typical trajectory does just what the trajectories for the logistic map do—it leaves the rectangle of the horseshoe and bids farewell to its wandering ways, approaching a stable fixed point. This is once again a manifestation of transient chaos.

For other extensions, a typical trajectory leaves the rectangle of the Smale horseshoe, but does not approach a stable fixed point; rather, it enters the rectangle of some *other* horseshoe of the map f, elsewhere in the plane, and once again displays chaotic behaviour for a while. Eventually, it leaves the rectangle containing that horseshoe, and the whole scenario repeats itself; it may well be that the trajectory stumbles somewhat drunkenly from horseshoe to horseshoe, never persisting in one chaotic regime for very long, but never settling down either. In this case, we have what is called *intermittent chaos*; although the chaotic behaviour never goes away entirely, as the trajectory always finds a neighbourhood of another horseshoe in which to sojourn, there may be long stretches of regular behaviour in between these chaotic intervals.

Thus beyond the basic distinction between regular behaviour and chaotic behaviour, we find that there are various sorts of chaotic behaviour that we may observe in the wild. To begin with, there is *persistent chaos*, where a randomly chosen trajectory lies in the basin of attraction of a chaotic attractor, from which it never escapes. Certain quantitative features of this chaotic behaviour are related to the dimensional properties of the attractor; since the orbit remains near the same attractor, these quantitative features (which we will not discuss) remain the same for all time.

Then there is *intermittent chaos*, in which a typical trajectory moves from horseshoe to horseshoe; the quantitative features mentioned above are related to the dimensional properties of the horseshoe which the trajectory is following at any given time, and so they vary when the orbit changes chaotic regimes, whether it moves to another horseshoe and remains intermittent or to a chaotic attractor and becomes persistent.

Finally, there is *transient chaos*, in which chaotic behaviour eventually gives way to regular behaviour. In practice, our knowledge of a system comes from empirically observed or numerically computed trajectories, and so we can only know a finite segment of the trajectory. Consequently, we may not know the true asymptotic behaviour, which renders the above distinction between transient and intermittent chaos largely a matter of semantics. For this reason, the term "intermittent chaos" is often used to refer to any sort of chaotic system whose trajectories move into and out of various chaotic regimes, whether they go from horseshoes to horseshoes, horseshoes to attractors, or horseshoes to fixed points.

c. Homoclinic orbits and horseshoes. The presence of a horseshoe leads to intermittent chaos; but what leads to the presence of a horseshoe? Can we find some condition on a map f which will guarantee that it acts as a horseshoe map on some subset R of phase space?

To motivate the answer, let us first consider the Smale horseshoe itself, the maximal invariant set Γ for the map $f\colon R \to \mathbb{R}^2$ shown in Figure 8.2. As we saw in (8.6), the map $f\colon \Gamma \to \Gamma$ is topologically conjugate to the full shift $\sigma\colon \Sigma_2 \to \Sigma_2$. This implies that it has two fixed points, corresponding to the sequences $w_1 = (\ldots 1|1|1 \ldots)$ and $w_2 = (\ldots 2|2|2 \ldots)$.

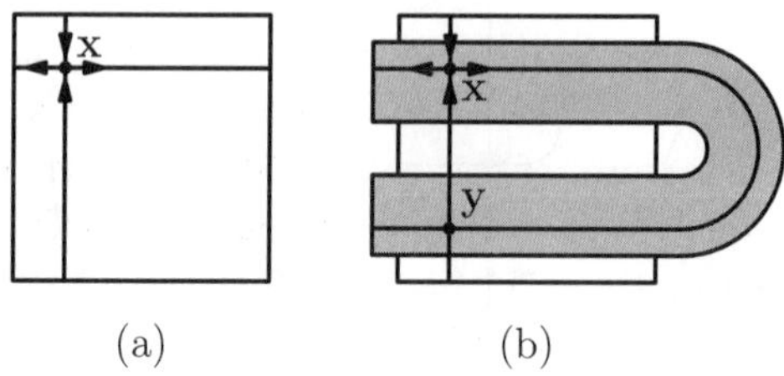

Figure 8.8. Finding a homoclinic orbit in a horseshoe.

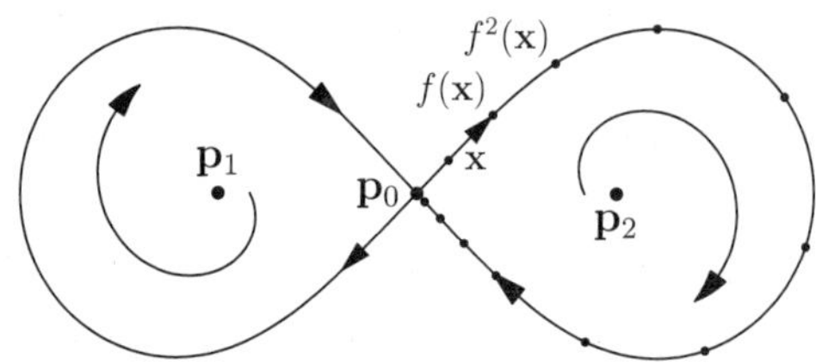

Figure 8.9. A homoclinic orbit with no transversality.

Write $\mathbf{x} = h(w_1)$, and recall that the (local) stable and unstable curves at $\mathbf{x}$ are the vertical and horizontal lines, respectively, passing through $\mathbf{x}$, shown in Figure 8.8(a). Because the global stable and unstable curves are invariant under the action of f (being defined in terms of asymptotic behaviour of trajectories), the unstable curve contains the entire image of the horizontal line through $\mathbf{x}$, which is the sideways "U"-shaped curve shown in Figure 8.8(b).

In particular, the point $\mathbf{y}$ lies on both the stable and unstable curves of $\mathbf{x}$. Thus the trajectory of $\mathbf{y}$ approaches $\mathbf{x}$ along the stable curve as $n \to \infty$, and along the unstable curve as $n \to -\infty$; such an orbit, which is asymptotic to the same fixed point in both the forwards and backwards directions, is known as a *homoclinic orbit*. We see from this argument that if f has a horseshoe, then it also has a homoclinic orbit.[3]

---

[3]This establishes the existence of a homoclinic orbit for both fixed points of f. In fact, every *periodic* point also has intersecting stable and unstable curves, and the points of intersection lie on homoclinic orbits. Since periodic points are dense in the horseshoe, we see that homoclinic orbits are really quite ubiquitous.

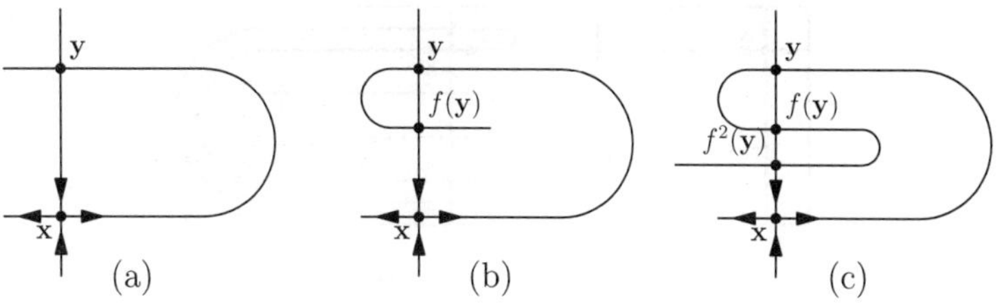

Figure 8.10. Consequences of a transverse homoclinic point.

Remark. An important feature of the homoclinic point $\mathbf{y}$ in Figure 8.8 is that the stable and unstable curves intersect transversally. A different sort of homoclinic orbit arises in the system depicted in Figure 8.9, where $\mathbf{p}_0$ is a hyperbolic fixed point whose stable and unstable curves coincide. We will be concerned primarily with *transverse homoclinic intersections*, where the stable and unstable curves are distinct and cross each other at a non-zero angle.

This relationship between horseshoes and homoclinic orbits might remain a mere curiosity, were it not for the fact that the implication actually runs both ways, as we shall now see. Setting aside the particular form of the map defining the Smale horseshoe, let us consider an arbitrary map f with a transverse homoclinic point $\mathbf{y}$ for a hyperbolic fixed point $\mathbf{x}$, as shown in Figure 8.10(a).

The unstable curve is invariant, and so it also passes through $f(\mathbf{y})$, which lies on the stable curve between $\mathbf{y}$ and $\mathbf{x}$; as Figure 8.10(b) shows, this forces the unstable curve to fold back on itself. A similar argument applies to $f^2(\mathbf{y})$, as Figure 8.10(c) shows, and indeed to any $f^n(\mathbf{y})$; thus the unstable curve folds back on itself infinitely often and is stretched further and further between successive intersections with the unstable curve. Not only that, but the stable curve is left invariant by the action of f^{-1} and must pass through all the points $f^n(\mathbf{y})$ for $n < 0$, so it folds back on itself infinitely often as well. The resulting picture is known as a *homoclinic tangle*; part of a typical homoclinic tangle is shown in Figure 8.11, which captures no more than the very beginning of the complexity of the entire situation.

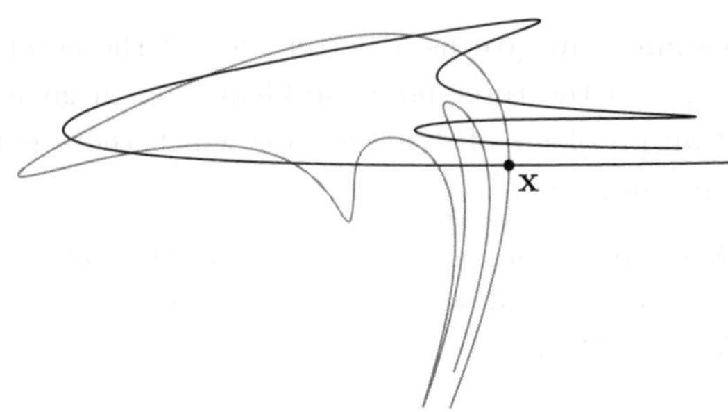

Figure 8.11. A homoclinic tangle.

Observe that the picture in Figure 8.10 is somewhat idealised, and that the intersections between the stable and unstable curves may not be orthogonal initially; however, as $n \to \pm\infty$, the angle of intersection at $f^n(\mathbf{y})$ goes to $\pi/2$. In particular, this implies that the angle of intersection between the stable and unstable curves is uniformly bounded away from 0, so that the transversality we have at $\mathbf{y}$ is in fact uniform across all points of intersection.

One sees immediately from Figure 8.11 that the geometric stucture of the stable and unstable curves, and hence of the orbits of the system, is fantastically complicated; Henri Poincaré, who first discovered this picture in 1889 in conjunction with his work on the three-body problem, remarked,[4]

> When we try to represent the figure formed by these two curves and their infinitely many intersections, each corresponding to a doubly asymptotic solution, these intersections form a type of trellis, tissue, or grid with infinitely fine mesh. Neither of the two curves must ever cut across itself again, but it must bend back upon itself in a very complex manner in order to cut across all of the meshes in the grid an infinite number of times.
>
> One must be struck by the complexity of this shape, which I do not even attempt to illustrate.

---

[4]As quoted in [**KH95**].

Nothing can give us a better idea of the complication of the three-body problem, and in general of all problems of dynamics for which there is no uniform integral.

What Poincaré missed is that the homoclinic tangle actually contains a horseshoe: this fundamental result is due to Smale (for a proof, see [**HK03**] or [**KH95**]).

Theorem 8.2. *Given a transverse homoclinic intersection for a hyperbolic fixed point* $\mathbf{x}$, *one can find a rectangle R containing* $\mathbf{x}$ *and an integer n such that f^n acts on R in a manner similar to the canonical horseshoe map; in particular, there exists a Cantor set $\Gamma \subset R$ which is closed and invariant, which is a hyperbolic set for f^n, and on which the action of f^n is conjugate to a shift.*[5]

Recalling our previous discussion of horseshoes and intermittent chaos, we conclude that the existence of a transversely homoclinic point implies the existence of a horseshoe, which in turn implies the presence of intermittent chaos. On the face of it, this is quite a powerful result given the simplicity of the assumption!

---

[5]In fact, Γ is the closure of the set of intersections of the stable and unstable curves at $\mathbf{x}$.

Chapter 9

Continuous-Time Systems: The Lorenz Model

Lecture 33

a. Continuous-time systems: Basic concepts. With the exception of a brief discussion in Chapter 5 of the differential equations for the FitzHugh–Nagumo model, all the systems we have studied up to this point have been given in terms of a map from some domain to itself. Such systems are known as *discrete-time* systems, since "time" moves in discrete increments, corresponding to how many times the map has been iterated. In this final chapter, we will focus our attention on *continuous-time* systems, which are specified by ordinary differential equations (ODEs) rather than maps. Before examining the connections between the discrete and continuous-time cases, we will point out a striking difference between the two: for discrete-time systems, we were able to produce chaotic behaviour in maps of any dimension, while we will see that continuous-time systems cannot be chaotic in fewer than three dimensions.

We begin by recalling some of the basic notions regarding differential equations. By adding auxiliary variables if necessary, the system of ODEs specifying a continuous-time system may be written as a single first-order ODE for a vector-valued function $\mathbf{x}$. That is, we consider functions $\mathbf{x} \colon (a, b) \to \Omega \subset \mathbb{R}^d$ which solve the equation

$$(9.1) \qquad\qquad \dot{\mathbf{x}}(t) = \mathbf{F}(\mathbf{x}(t)),$$

where $\mathbf{F} \colon \Omega \to \mathbb{R}^d$ is the function defining the ODE on the domain[1] $\Omega \subset \mathbb{R}^d$. For the remainder of this lecture, we consider the case $d = 2$; if we write $\mathbf{x} = (x_1, x_2)$ and $\mathbf{F}(\mathbf{x}) = (F_1(\mathbf{x}), F_2(\mathbf{x}))$, then (9.1) may be written coordinate-wise as the system of differential equations

$$(9.2) \qquad \begin{aligned} \dot{x}_1(t) &= F_1(x_1(t), x_2(t)), \\ \dot{x}_2(t) &= F_2(x_1(t), x_2(t)). \end{aligned}$$

Together with an initial condition $\mathbf{x}(t_0) = \mathbf{x}_0$ at time t_0, the equations (9.2) (or (9.1)) define an *initial value problem* which we are interested in solving.

If the function $\mathbf{F}$ is continuously differentiable—that is, if all the partial derivatives $\frac{\partial F_i}{\partial x_j}$ exist and are continuous on all of Ω—then the Fundamental Theorem on Existence and Uniqueness from the theory of ODEs (see [**Liu03**]) implies that given an initial condition $\mathbf{x}_0 \in \Omega$ and a time t_0, there exists $\varepsilon = \varepsilon(\mathbf{x}) > 0$ such that the system has a unique solution on some interval $t_0 - \varepsilon < t < t_0 + \varepsilon$.

By gluing together the solutions on these small intervals, the solution may be extended to some maximal interval $a < t < b$; the endpoints a and b are either infinite or the points at which $\mathbf{x}(t)$ reaches the boundary of Ω. If Ω is unbounded, then it is possible for a solution $\mathbf{x}(t)$ to reach infinity in finite time; this phenomenon can be avoided by requiring, for example, that

$$(9.3) \qquad\qquad \sup \{\|D\mathbf{F}(\mathbf{x})\| \mid \mathbf{x} \in \Omega\} < \infty.$$

a.1. *ODEs as geometric objects.* The ODE (9.1) can also be given a geometric interpretation. The vector $\mathbf{F}(\mathbf{x})$ specifies a direction and length; placing this vector so that it originates at the point $\mathbf{x}$, we obtain a *vector field* in the plane. Solutions of the ODE (9.1) correspond to *integral curves* of the vector field—that is, curves whose tangent vector at each point is exactly the element of the vector field at that point, as shown in Figure 9.1.

---

[1] To be precise, we take our domain Ω to be homeomorphic to an open ball in $\mathbb{R}^d$.

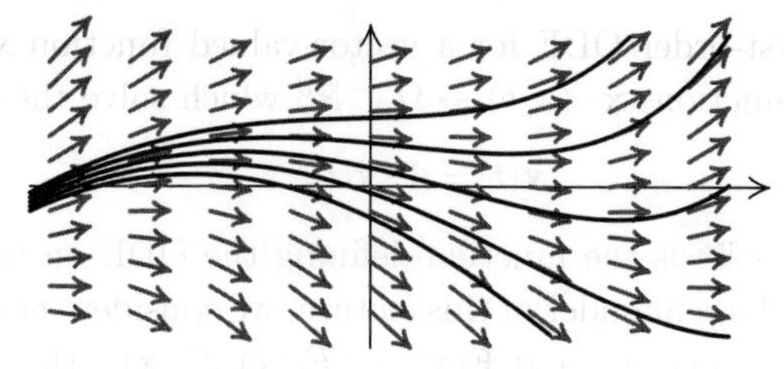

Figure 9.1. Some integral curves for a vector field.

Interpreting an ODE in terms of its associated vector field can be of great utility in answering certain global questions regarding fixed points, as results from index theory, etc. can be brought into play.

a.2. *ODEs as dynamical objects.* The integral curves $\gamma\colon (a,b) \to \mathbb{R}^2$ which represent solutions of the ODE admit various parametrisations, as do all curves; however, the requirement that the tangent vector $\dot\gamma(t)$ have the same length as the vector $\mathbf{F}(\gamma(t))$ fixes a unique natural parametrisation. Writing $\gamma_{\mathbf{x}}$ for the unique curve with $\gamma_{\mathbf{x}}(0) = \mathbf{x}$ and $\dot\gamma(t) = \mathbf{F}(\gamma(t))$, we see that $\gamma_{\mathbf{x}}(t)$ is the point in $\mathbb{R}^2$ which the system with initial condition $\mathbf{x}$ reaches after a time t has elapsed. The map $\varphi_t\colon \mathbf{x} \to \gamma_{\mathbf{x}}(t)$ is called the *time-t map* of the system.

This defines a one-parameter family of maps $\varphi_t\colon \Omega(t) \to \Omega$, where $\Omega(t)$ consists of the points $\mathbf{x} \in \Omega$ such that $\gamma_{\mathbf{x}}(\tau)$ remains in Ω for $0 \leq \tau \leq t$. If $\Omega = \mathbb{R}^2$ and $\|D\mathbf{F}(x)\|$ is bounded as in (9.3), then φ_t is defined on all of $\mathbb{R}^2$ for every t.

Exercise 9.1. Show that the maps φ_t depend smoothly on t. In particular, show that

$$\frac{d\varphi_t(\mathbf{x})}{dt}\bigg|_{t=0} = \mathbf{F}(\mathbf{x})$$

for every $\mathbf{x} \in \Omega$.

Each of these maps corresponds to evolving the system forward (or backward) by the appropriate amount of time, and they are related to each other by the *group property*

$$(9.4) \qquad \varphi_{t+s} = \varphi_t \circ \varphi_s = \varphi_s \circ \varphi_t;$$

that is, $\varphi_{t+s}(\mathbf{x}) = \varphi_t(\varphi_s(\mathbf{x})) = \varphi_s(\varphi_t(\mathbf{x}))$ for all $\mathbf{x} \in \mathbb{R}^2$ and $t, s \in \mathbb{R}$.

Definition 9.1. A *flow* on $\mathbb{R}^d$ is a one-parameter family of one-to-one differentiable maps $\varphi_t \colon \mathbb{R}^d \to \mathbb{R}^d$ such that

(1) $\varphi_0(\mathbf{x}) = \mathbf{x}$ for all $\mathbf{x} \in \mathbb{R}^d$.

(2) The group property (9.4) holds for all $s, t \in \mathbb{R}$.

(3) The function $\varphi(t, \mathbf{x}) = \varphi_t(\mathbf{x})$ is differentiable in t for all $\mathbf{x} \in \Omega$.

Flows provide the third way of looking at continuous-time systems; the three descriptions of such a system in terms of an ODE, a vector field, and a flow are all equivalent, and which one is most suitable depends on the circumstances.

The relationship between discrete-time and continuous-time systems is clearest when we consider the latter in terms of flows. If we restrict our attention to integer values of s and t, then (9.4) reduces to (1.2), which reflects the fact that for a fixed value of t, say $t = 1$, the time-t map of a continuous-time system defines a discrete-time system.

b. Fixed points of continuous-time systems. To describe the trajectories of the solutions to an ODE, we begin (as always), by finding the fixed points. In the dynamical language of the flow φ_t, these are the points $\mathbf{x}$ such that $\varphi_t(\mathbf{x}) = \mathbf{x}$ for all $t \in \mathbb{R}$; in the geometric language of vector fields, these are the points at which the vector field vanishes; and in the language of ODEs, these are the zeros of the function $\mathbf{F}$, the points at which $\mathbf{F}(\mathbf{x}) = \mathbf{0}$, or in terms of the coordinate functions,

$$(9.5) \qquad F_1(x_1, x_2) = F_2(x_1, x_2) = 0.$$

Suppose for the present that (9.5) has only finitely many solutions $\mathbf{x}^{(1)}, \dots, \mathbf{x}^{(n)}$. Then near the fixed point $\mathbf{x}^{(i)}$, Proposition 5.1 gives

$$(9.6) \qquad \mathbf{F}(\mathbf{x}) = \mathbf{F}(\mathbf{x}^{(i)}) + D\mathbf{F}(\mathbf{x}^{(i)})(\mathbf{x} - \mathbf{x}^{(i)}) + \mathbf{r}(\mathbf{x}),$$

where $\mathbf{r}(\mathbf{x})$ is an error term of order $o(\|\mathbf{x} - \mathbf{x}^{(i)}\|)$. Because $\mathbf{x}^{(i)}$ is fixed, the ODE (9.1) becomes

$$\dot{\mathbf{x}}(t) = \mathbf{F}(\mathbf{x}(t)) = D\mathbf{F}(\mathbf{x}^{(i)})(\mathbf{x}(t) - \mathbf{x}^{(i)}) + \mathbf{r}(\mathbf{x}(t)),$$

and using coordinates $\mathbf{v} = \mathbf{x} - \mathbf{x}^{(i)}$, we have

$$\dot{\mathbf{v}}(t) = D\mathbf{F}(\mathbf{x}^{(i)})\mathbf{v}(t) + o(\|\mathbf{v}\|).$$

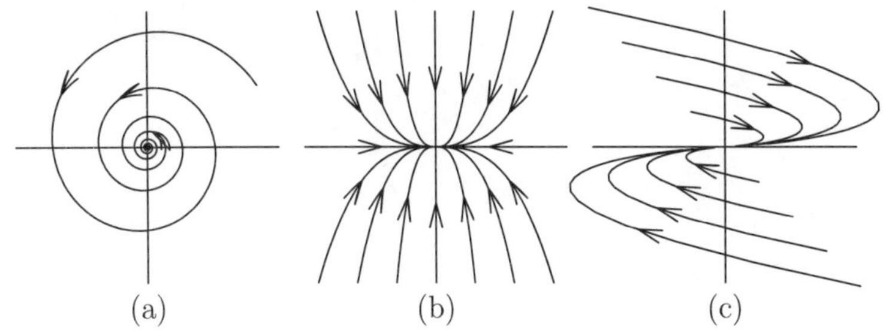

(a)   (b)   (c)

Figure 9.2. Trajectories near an attracting fixed point.

Since the perturbation from the linear map is small in a neighbourhood of a fixed point $\mathbf{x}^{(i)}$, we may hope to describe the solutions
of (9.1) near $\mathbf{x}^{(i)}$—and in particular, determine the stability of $\mathbf{x}^{(i)}$—
by first describing the solutions of the linear system

$$\dot{\mathbf{v}}(t) = D\mathbf{F}(\mathbf{x}^{(i)})\mathbf{v}(t),$$

which depend on the real part of the eigenvalues of $D\mathbf{F}(\mathbf{x}^{(i)})$. In two
dimensions, there are only three non-degenerate possibilities:

(1) Both eigenvalues have negative real part. Then the trajectories
    $\mathbf{v}(t)$ are the curves shown in one of the three phase portraits in
    Figure 5.2. If the eigenvalues have a non-zero imaginary part,
    $\mathbf{x}^{(i)}$ is called an *attracting focus* and the trajectories are those
    shown in Figure 9.2(a). If the eigenvalues lie on the real line, $\mathbf{x}^{(i)}$
    is called an *attracting node* and the trajectories are the curves
    shown in Figure 9.2(b) or (c).

(2) Both eigenvalues have positive real part. Then the trajectories
    are the same curves as in the previous case, but move in the
    opposite direction, away from the fixed point instead of towards
    it. In this case the fixed point is either a *repelling focus* or a
    repelling node.

(3) The eigenvalues are real, and have different signs (one eigenvalue
    is negative and the other positive). In this case the trajectories
    follow the curves shown in Figure 9.3, and the fixed point is called
    a *saddle*. The attracting and repelling directions (horizontal and
    vertical in the picture) correspond to the eigenvectors of $D\mathbf{F}(\mathbf{x}^{(i)})$.

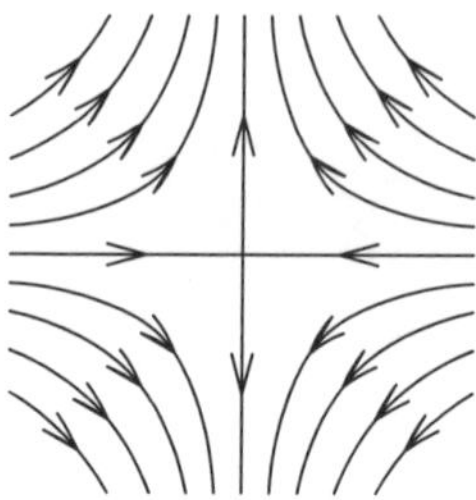

Figure 9.3. Trajectories near a saddle.

Remark. In the discrete-time case, we were at pains to point out that the curves in Figures 9.2 and 9.3 are not themselves trajectories of the system; in the continuous-time case, by contrast, these curves *are* trajectories. This changes the topological nature of things, and will have ramifications later on.

The three possibilities above correspond directly to the three possible behaviours of a fixed point for a discrete-time system which we discussed in Lecture 22. However, the criteria here differ from the criteria in that case, where it was the absolute value of the eigenvalues which determined the stability.

To see why this is, observe that the stability of $\mathbf{x}^{(i)}$ for the continuous-time flow φ_t coincides with the stability of $\mathbf{x}^{(i)}$ for the discrete-time system given by the time-1 map φ_1. Now the eigenvalues of $D\varphi_1(\mathbf{x}^{(i)})$ are of the form e^λ, where λ is an eigenvalue of $D\mathbf{F}(\mathbf{x}^{(i)})$, and we observe that e^λ lies on the unit circle if and only if λ lies on the imaginary axis.

Thus the criteria are different from the discrete-time case, but the general idea remains the same; the stability of a fixed point for the non-linear system can be found in terms of the eigenvalues of its linearisation.

We note, though, that there are also various degenerate cases which may occur. For example, if both eigenvalues are purely imaginary, then the trajectories of the linear system are concentric circles around the fixed point, as shown in Figure 9.4. In this case, the non-linear effects may (or may not) qualitatively change the behaviour of

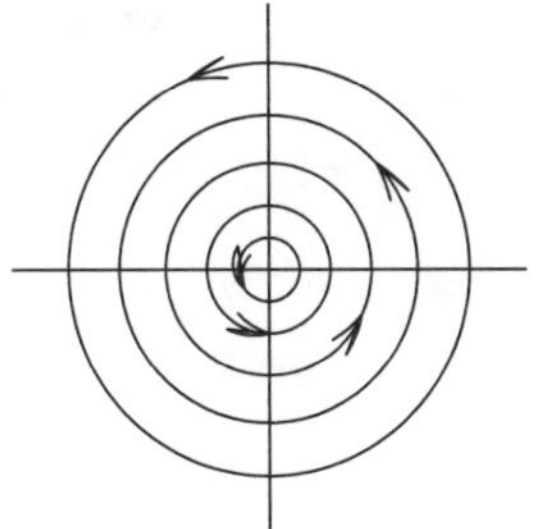

Figure 9.4. Trajectories around a centre.

the system in an arbitrarily small neighbourhood of the origin, and so we gain no information about the stability of the fixed point.

Exercise 9.2. Consider the following system of differential equations:

$$\dot{x} = ax(1 - x - by), \quad \dot{y} = ay(1 - y - bx).$$

For an arbitrary value of the parameters a and b, find the fixed points, and determine the type of their stability.

Lecture 34

a. The pendulum. One of the standard examples of a non-linear differential equation is the pendulum, a massless rod with one end fixed and a point mass at the other end. We let θ denote the angle by which the rod is displaced from the vertical, as shown in Figure 9.5(a).

Neglecting air resistance and other dissipative effects, the total energy of the pendulum is conserved. This quantity is given by $\frac{1}{2}mv^2 + mgh$, where m is the amount of mass at the end of the pendulum, v is the velocity of that mass, g is gravity, and h is the vertical displacement of the mass relative to the pivot point. Writing L for the length of the pendulum, we have $v = L\dot{\theta}$ and $h = -L\cos\theta$, so the total energy is

$$(9.7) \qquad E(\theta, \dot{\theta}) = \frac{1}{2}mL^2(\dot{\theta})^2 - mgL\cos\theta.$$

Because this quantity is constant in time, we may differentiate with respect to t and obtain

$$0 = mL^2\dot{\theta}\ddot{\theta} + mgL\dot{\theta}\sin\theta.$$

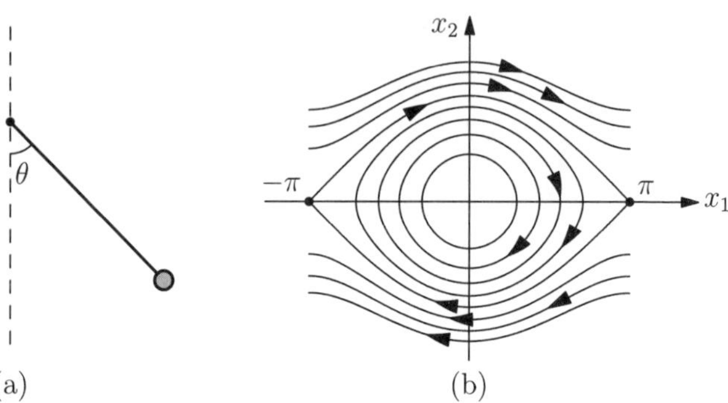

Figure 9.5. A pendulum.

Writing $a = g/L$, we see that the equation governing the motion of the pendulum is

$$\ddot{\theta} + a \sin \theta = 0. \tag{9.8}$$

We require *two* initial conditions, θ and $\dot{\theta}$, in order to specify a particular solution of (9.8). Thus the phase space of the pendulum is two-dimensional; if we write $x_1 = \theta$ and $x_2 = \dot{\theta}$, then (9.8) may be rewritten as

$$\dot{x}_1 = x_2, \tag{9.9}$$
$$\dot{x}_2 = -a \sin x_1,$$

or even more succinctly, in a single equation as $\dot{\mathbf{x}} = \mathbf{F}(\mathbf{x})$, where $\mathbf{F}(x_1, x_2) = (x_2, -a \sin x_1)$, which allows us to use the language of the previous lecture.

The fixed points of (9.9) occur at $(k\pi, 0)$ for $k \in \mathbb{Z}$; however, since adding 2π to x_1 does not change the physical state of the system, there are actually only two fixed points. One of these corresponds to a pendulum hanging motionless, pointing straight down (when k is even); the other corresponds to a pendulum balancing on its pivot, pointing straight up (when k is odd).

Intuitively, we feel that the first of these is "stable" while the second is "unstable"; we may attempt to confirm and clarify this intuition by linearising around the fixed points. Upon doing so, we discover that at $\mathbf{x} = (0,0)$, both eigenvalues of $D\mathbf{F}(\mathbf{x})$ are purely imaginary, and so the linearised system is a rotation, while at $\mathbf{x} = (\pi, 0)$, one eigenvalue is positive and one is negative, and so $(\pi, 0)$ is a saddle for the linearised system.

From our earlier discussion, it follows that $(\pi, 0)$ is a saddle for the non-linear system as well. This reflects our intuition that it should be "unstable", since a randomly chosen nearby trajectory will eventually be repelled; while there are trajectories which approach $(\pi, 0)$, they have zero Lebesgue measure.

At the origin, the eigenvalues have zero real part, and so we are in a degenerate case where the behaviour of the non-linear system is not determined by the linear part. However, we can use the fact that we know a conserved quantity for the system—the total energy. Rewriting (9.7) as

$$E(x_1, x_2) = \frac{1}{2}mL^2(x_2^2 - 2a\cos x_1)$$

shows that the trajectories of the pendulum move along level curves of $x_2^2 - 2a\cos x_1$. Near the origin, these are the closed trajectories shown in Figure 9.5(b), and so the origin is stable in the sense that nearby trajectories remain nearby, although they do not approach $\mathbf{0}$, just as with the trajectories in Figure 9.4.

What is the physical interpretation of the various trajectories in Figure 9.5(b)? Note that every trajectory intersects the vertical axis, which corresponds to the configuration where the pendulum is pointed straight down. The height at which the trajectory intersects the axis corresponds to the speed with which the pendulum is moving when it reaches this position.

If this speed is relatively small, then the pendulum will rise a litle ways, eventually reach a maximum height (equivalently, a maximum angle of displacement), and then fall again, passing back through the line $x_1 = 0$ with the same speed it initially had, but in the other direction, and so on; this corresponds to the closed nearly elliptical orbits around the origin in Figure 9.5(b). After some finite time T, the system is back where it started, and so for these trajectories we have $\mathbf{x}(t + T) = \mathbf{x}(t)$ for all $t \in \mathbb{R}$: these are *periodic orbits*.

If we start the pendulum off with a greater initial speed, then it will reach a greater height before reversing direction; it will also take longer to reach its maximum angle, and so the period T increases. At some critical initial speed, the maximum angle will be equal to π; that is, the pendulum has enough energy that gravity will not pull it back before it reaches the top.[2]

In fact, if the pendulum reaches the top in finite time, then it will have some momentum left over (even if only a very small amount), which will be enough to carry it over the top and into another complete rotation, so that it eventually reaches the bottom again, at which point it has exactly the same speed it began with. Thus there is some T such that $\mathbf{x}(t+T) = \mathbf{x}(t)+(2\pi,0)$ for all $t \in \mathbb{R}$; these orbits are not periodic from the point of view of the system in the plane, but since adding 2π to x_1 does not change the physical system, they still represent periodic behaviour of the pendulum itself.[3]

What happens, though, if the pendulum has *exactly* enough energy to reach the top; enough that gravity will not stop it short, but not enough that it will have any momentum left over? Then it will move more and more slowly as time goes on, but will never stop (in which case it would reverse direction) or reach the top (in which case it would have some momentum left and would keep going). In a manner of speaking, it reaches the top, but in infinite time.

There are two trajectories in Figure 9.5(b) which correspond to this situation; one runs from $(-\pi,0)$ to $(\pi,0)$, passing through a point on the positive x_2-axis, while the other runs from $(\pi,0)$ to $(-\pi,0)$, passing through a point on the negative x_2-axis. Each of these trajectories is the unstable curve for one fixed point, and the stable curve for the other. Although both of these have finite length as curves in $\mathbb{R}^2$, as trajectories it takes a solution of (9.9) an infinite amount of time to move the entire length of either one.

Trajectories such as these, which both originate and terminate in a fixed point, are known as *homoclinic* (if they begin and end in the same fixed point) or *heteroclinic* (if they begin and end in different fixed points). We have seen homoclinic orbits before, but the situation

---

[2]Many of us have tried to accomplish this on a swing set as a child.

[3]A more satisfactory model takes the phase space to be not $\mathbb{R}^2$, but the cylinder; that is, $\mathbb{R}^2$ wrapped up so that the x_1-axis becomes a circle. If we take $\mathbb{R}^2$ as the phase space, then the x_1-coordinate of high-energy trajectories tracks the number of revolutions the pendulum has undergone.

is different now, due to the fact that the orbit is a curve, rather than just a collection of points. In two dimensions, such orbits often act as separatrices between regions of different qualitative behaviour; in this case, the two heteroclinic[4] trajectories just described separate the orbits which "oscillate" (as we usually expect a pendulum to do) from the orbits which "spin".

b. Two-dimensional systems. We now turn our attention to general two-dimensional continuous-time systems. As with discrete-time systems, a useful first step in analysing any given system is to find its fixed points and periodic orbits, and then classify them by stability. We have already discussed the stability of fixed points, so we now consider periodic orbits.

In a discrete-time system, a periodic orbit is just a finite collection of points, and it could have any of the stabilities available to a fixed point; stable, unstable, saddle, etc. In contrast, a periodic orbit of a continuous-time system is a closed curve, and so has higher dimension than a periodic orbit for a map. This seemingly innocuous distinction is largely responsible for the absence of chaos in two-dimensional continuous-time systems, as we shall now see.

Proposition 9.2. *Let* $\gamma\colon \mathbb{R} \to \mathbb{R}^2$ *be a periodic solution of* (9.1), *and suppose that* γ *is* isolated; *that is, there exists some open neighbourhood* $U \subset \mathbb{R}^2$ *containing the curve* γ *such that* U *contains no other periodic orbits. Then either* γ *is stable and attracts every nearby trajectory, or it is unstable and repels every nearby trajectory.*

Proof. Fix a curve η which intersects the curve γ transversally (see Figure 9.6). Then we define a map $T_\eta\colon \eta \to \eta$ as follows: each point $\mathbf{x} \in \eta$ defines a unique solution of (9.1), which remains near γ, and so eventually intersects η again; T_η is defined to be the first point at which this intersection occurs. The curve η is called a *Poincaré section*, and T_η is the *Poincaré map* for η.

Now suppose that $T_\eta(\mathbf{x})$ is closer to γ than $\mathbf{x}$ is. We iterate the Poincaré map by continuing the solution curve through $T_\eta(\mathbf{x})$; this curve lies closer to γ than the solution curve through $\mathbf{x}$ does.

---

[4]If we take the cylinder as our phase space, then the fixed points $(-\pi, 0)$ and $(\pi, 0)$ coincide, and these are homoclinic trajectories.

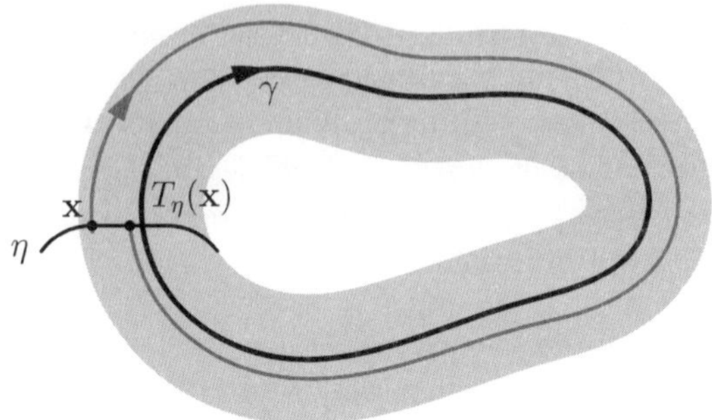

Figure 9.6. The Poincaré section for a transversal curve η.

In particular, $T_\eta^2(\mathbf{x})$ lies closer to γ than $T_\eta(\mathbf{x})$ does. This follows because solutions of (9.1) are unique, and so no two solution curves can cross each other; thus the second curve (starting at $T_\eta(\mathbf{x})$) cannot cross the first curve (starting at $\mathbf{x}$) and escape. This is the piece of the argument which makes explicit use of the fact that we are working in two dimensions.

Observe that fixed points of the Poincaré map correspond to periodic solutions of the ODE; thus because γ is an isolated periodic orbit, the map T_η has no fixed points except the intersection of γ and η. Since the trajectory of $\mathbf{x}$ under the Poincaré map moves monotonically along η, it must converge to a fixed point of T_η, which shows that the solution curve beginning at $\mathbf{x}$ approaches the periodic orbit γ. Similarly, any trajectory beginning close enough to γ approaches γ, and so the periodic orbit is stable.

A similar argument applies if $T_\eta(\mathbf{x})$ is further away from γ than $\mathbf{x}$ is, in which case γ is unstable. These are the only two options, which proves the result. $\qquad\square$

One consequence of Proposition 9.2 is that a periodic orbit for a two-dimensional continuous-time system cannot be a "saddle", and so the menagerie of possible local behaviours is tamer than it was for discrete-time systems.

There are a number of results of a more global character available for general two-dimensional continuous-time systems.

Proposition 9.3. *If γ is a periodic solution of (9.1), then the region enclosed by γ contains a fixed point of the system.*

Sketch of proof. One shows that this region is homeomorphic to a disc (this is the Schoenflies Theorem, a stronger version of the Jordan Curve Theorem), and then applies the Brouwer Fixed Point Theorem to the time-1 map. $\square$

Proposition 9.4. *Suppose that $\mathbf{F}\colon \Omega \to \mathbb{R}^2$ is such that*

$$(9.10) \qquad \frac{\partial F_1}{\partial x_1} + \frac{\partial F_2}{\partial x_2} \neq 0$$

for every $\mathbf{x} = (x_1, x_2) \in \Omega$. Then Ω contains no periodic solutions of the ODE (9.1).

Proof. Suppose γ is a periodic orbit in Ω, with $\gamma(t + T) = \gamma(t)$ for all t, and let $\Omega' \subset \Omega$ be the region enclosed by γ, so that $\gamma = \partial\Omega$. Suppose γ goes around Ω' counter-clockwise (the proof in the other case is similar). Then applying Stokes' Theorem, we have

$$\iint_{\Omega'} \frac{\partial F_1}{\partial x_1} + \frac{\partial F_2}{\partial x_2} \, dx_1 \, dx_2$$

$$= \int_\gamma F_1 \, dx_2 - F_2 \, dx_1$$

$$= \int_0^T F_1(\mathbf{x}(t))\dot{x}_2(t) - F_2(\mathbf{x}(t))\dot{x}_1(t) \, dt$$

$$= \int_0^T F_1(\mathbf{x}(t))F_2(\mathbf{x}(t)) - F_2(\mathbf{x}(t))F_1(\mathbf{x}(t)) \, dt$$

$$= 0.$$

However, since the original integrand is continuous and non-vanishing on Ω', it must be either positive everywhere or negative everywhere, and hence the integral cannot be zero. This contradiction implies that no periodic orbits exist in Ω. $\square$

Exercise 9.3. Consider the following system of differential equations:

$$\dot{x} = 2x + y + x^2, \quad \dot{y} = -6x - 2xy.$$

Find the fixed points, determine the type of their stability, and construct the phase portrait. Show that the system has no periodic solutions.

In order to state the general result regarding the absence of chaos in two dimensions, we adapt the notion of an ω-limit set to the continuous-time case.

Definition 9.5. If $\gamma\colon \mathbb{R} \to \mathbb{R}^2$ is a trajectory in $\mathbb{R}^2$ with $\gamma(0) = \mathbf{x}$, then the ω-*limit set* of $\mathbf{x}$ is

$$\omega(\mathbf{x}) = \{\mathbf{p} \in \mathbb{R}^2 \mid \mathbf{p} = \lim_{n\to\infty} \mathbf{x}(t_n) \text{ for some sequence } t_n \to \infty\}.$$

The α-limit set $\alpha(\mathbf{x})$ is defined similarly, but with the requirement that $t_n \to -\infty$ instead.

Exercise 9.4. Show that each ω-limit set $\omega(\mathbf{x})$ is closed and φ_t-invariant for all t, and that the same is true of the set $\Omega = \bigcup_{\mathbf{x}\in\mathbb{R}^2} \omega(\mathbf{x})$ of all points in $\mathbb{R}^2$ which lie in some ω-limit set.

Exercise 9.5. For the system of differential equations in Exercise 9.3, describe the ω-limit set of every point in $\mathbb{R}^2$.

A complete description of the possible ω-limit sets for flows in $\mathbb{R}^2$ is given by the following theorem, whose proof may be found in [**Liu03**].

Theorem 9.6 (Poincaré–Bendixson). *If γ is a bounded trajectory of a flow φ_t with initial condition $\gamma(0) = \mathbf{x}$, then one of the following occurs:*

(1) $\omega(\mathbf{x})$ is a union of fixed points and heteroclinic (or homoclinic) orbits.

(2) γ is a periodic orbit.

(3) $\omega(\mathbf{x})$ is a periodic orbit to which γ converges.

Remark. The latter two behaviours are quite easy to understand, and we have already seen examples of these. Regarding the first possibility, that $\omega(\mathbf{x})$ is a union of fixed points and hetero- or homoclinic orbits, we observe that in addition to the simple case where the trajectory of $\mathbf{x}$ approaches a fixed point $\mathbf{p}$, and hence $\omega(\mathbf{x}) = \{\mathbf{p}\}$, we may consider, for example, a continuous-time version of the figure-eight map in Figure 8.9: given a point $\mathbf{x}$ near the repelling fixed point $\mathbf{p}_2$, the ω-limit set of $\mathbf{x}$ is the fixed point $\mathbf{p}_0$ together with the homoclinic orbit surrounding $\mathbf{p}_2$. It is not difficult to construct similar examples for which $\omega(\mathbf{x})$ comprises an arbitrary number of fixed points and heteroclinic orbits.

The Poincaré–Bendixson theorem rules out the existence of chaos for two-dimensional flows by completely describing all the possible asymptotic behaviours, all of which are quite regular. Thus in order to observe chaos in a continuous-time system, we must look to higher dimensions.

Lecture 35

a. The Lorenz equations. For flows in $\mathbb{R}^3$, we do not have the proper topological context to make the theorems in the previous lecture work; in particular, a periodic orbit γ does not need to be the boundary of a φ_t-invariant region homeomorphic to a disc. Thus life can be much more interesting in higher dimensions. To illustrate this fact, we study a particular system of ODEs in $\mathbb{R}^3$, the *Lorenz equations*:

$$
\begin{aligned}
\dot{x} &= -\sigma x + \sigma y, \\
\dot{y} &= rx - y - xz, \\
\dot{z} &= xy - bz.
\end{aligned}
$$

(9.11)

Here r (called the *Reynolds number*) is the leading parameter, while σ and b will be fixed at $\sigma = 10$ and $b = 8/3$. In the next lecture, we will examine the behaviour of solutions of (9.11) as r ranges from 0 to some number $R > 25$.

But first, a little history. The equations (9.11) were first introduced in 1963 by Edward Lorenz, a meteorologist at M.I.T. who was studying the motion of the atmosphere. To a good approximation, the atmosphere is governed by the Navier–Stokes equations of fluid motion; however, these are enormously difficult to solve, and even an approximate numeric solution requires a powerful computer.

Four years earlier, Lorenz had begun using one of the more powerful computers available at that time[5] to simulate the motion of the atmosphere. More precisely, he considered a layer of fluid between two horizontal plates, where the lower plate is heated and the upper

---

[5]Sixty computations a second!

one is cooled. If the difference in temperature ΔT between the plates is small, then heat will flow by conduction from the bottom plate to the top plate while the fluid remains motionless, and this is a stable equilibrium configuration.

As ΔT increases, this equilibrium configuration becomes unstable, and a small disturbance is enough to cause convection cells to form in the fluid—these rotating vortices carry warm fluid from the bottom plate to the top plate and cooler fluid back down. As ΔT increases even further, these convection cells become unstable, and the fluid flow eventually becomes turbulent—that is, chaotic.

Initially, Lorenz considered a system of twelve equations in twelve variables which were obtained as Fourier coefficients of the functions in the Navier–Stokes equations; upon being given the initial conditions, the computer would calculate the (approximate) trajectory of the system. One day, Lorenz wanted to take a closer look at a part of the previous day's simulation, and so he entered as the initial condition the output from midway through the calculated trajectory. To his surprise, the results of this simulation, which should have matched the previous results perfectly, instead diverged quite quickly!

Lorenz soon realised what the problem was: the computer stored all initial data and intermediate calculations to an accuracy of six digits, but only printed out the first three digits. Thus when Lorenz entered the previous day's results, he unknowingly introduced a small error term, on the order of 10^{-4}. Rather than dying away or even remaining small as the system evolved in time, this error term grew exponentially, until it became large enough to make the two trajectories appear completely unrelated.

After further investigation, Lorenz was able to replace the system of twelve ODEs with the system (9.11) of three ODEs which now bears his name. Although this system does not capture all the details of the original system, it displays the same qualitative behaviour. The leading parameter r plays the role of ΔT; for small values of r, as we will see, trajectories of the system are quite simple. However, when r becomes sufficiently large, (9.11) displays one of the hallmarks of chaos, *sensitive dependence on initial conditions* at every point; that is, any two nearby trajectories eventually diverge exponentially, as described above.

b. Beyond the linear mindset. At the time Lorenz began his work, the prevailing opinion among physicists and mathematicians was that a chaotic signal, such as the one Lorenz observed in the solution of (9.11), could only be the result of a random noise from the external environment of the system.

Lorenz's examination of the system (9.11) demonstrated that a chaotic signal could be produced by a completely deterministic process, but because he published his results in the *Journal of Atmospheric Science*, it took some time for the mainstream of the mathematics and physics communities to become aware of them. However, the word eventually spread, and the new phenomenon of *deterministic chaos* has become an integral part of our understanding of the natural world.

c. Examining the Lorenz system. To begin our analysis of the system (9.11), we first find the fixed points at which $\dot{x} = \dot{y} = \dot{z} = 0$ and determine their stability via the Jacobian matrix $D\mathbf{F}$. It is helpful to observe that the system is invariant under the reflection in the z-axis given by

$$x \mapsto -x, \qquad y \mapsto -y, \qquad z \mapsto z,$$

and so solutions away from this axis come in symmetric pairs. In particular, any fixed point (x, y, z) with $x \neq 0$ or $y \neq 0$ has a twin $(-x, -y, -z)$ on the other side of the z-axis.

For $\sigma, b \neq 0$, the fixed point conditions $\dot{x} = 0$ and $\dot{z} = 0$ imply that $y = x$ and $z = x^2/b$. Then the condition $\dot{y} = 0$ may be written

$$rx - x - \frac{x^3}{b} = 0;$$

hence the system always has a fixed point at $\mathbf{0}$, and any other fixed point must satisfy

$$\frac{x^2}{b} = r - 1.$$

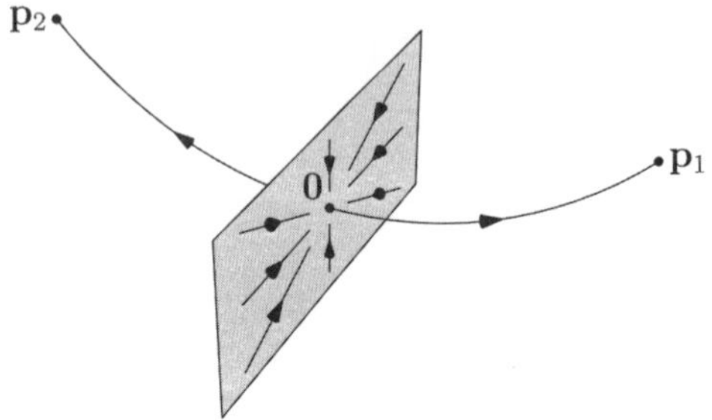

Figure 9.7. The fixed points for the Lorenz system when $r-1$ is small.

Therefore, for $0 < r < 1$, the only fixed point is the origin. At $r = 1$, a pitchfork bifurcation occurs, and two more fixed points appear:

$$\mathbf{p}_1 = \left(\sqrt{b(r-1)}, \sqrt{b(r-1)}, r-1 \right),$$

$$\mathbf{p}_2 = \left(-\sqrt{b(r-1)}, -\sqrt{b(r-1)}, r-1 \right).$$

To check the stability of these fixed points, we find the eigenvalues of the Jacobian matrix

$$DF(\mathbf{x}) = \begin{pmatrix} -\sigma & \sigma & 0 \\ r-z & -1 & -x \\ y & x & -b \end{pmatrix}.$$

At the origin, we have $x = y = z = 0$, and so

$$DF(\mathbf{0}) = \begin{pmatrix} -\sigma & \sigma & 0 \\ r & -1 & 0 \\ 0 & 0 & -b \end{pmatrix}$$

has $-b$ as one eigenvalue, and the other two eigenvalues are the roots of

$$\lambda^2 + (\sigma + 1)\lambda + \sigma(1 - r) = 0,$$

which is the characteristic polynomial of the 2×2 matrix in the upper-left corner of $DF(\mathbf{0})$. The constant negative eigenvalue $-b$ with eigenvector $(0, 0, 1)$ indicates that the vertical direction is always contracting at the origin.

For $0 < r < 1$, the other two eigenvalues are also negative, and the origin is an attracting fixed point; for $r > 1$, one of these eigenvalues

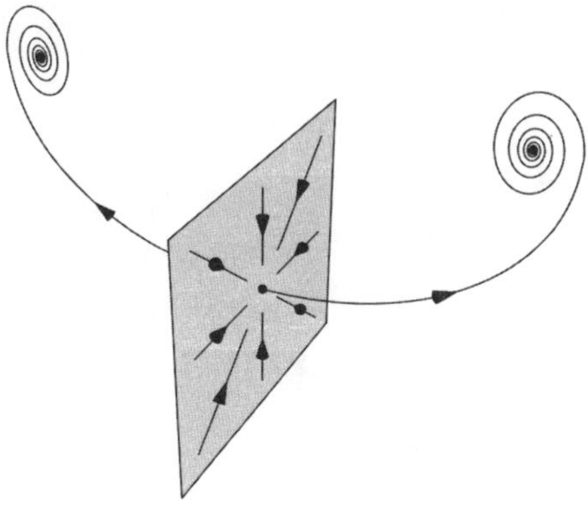

Figure 9.8. Changes in the behaviour of orbits as r increases.

is negative and the other is positive, and so the origin is a hyperbolic fixed point with two contracting directions and one expanding direction. Trajectories moving along the latter direction are attracted to one of the two fixed points $\mathbf{p}_1$ and $\mathbf{p}_2$, which exist precisely when $r > 1$; this is the situation shown in Figure 9.7.

In this parameter range, almost every trajectory is attracted to either $\mathbf{p}_1$ or $\mathbf{p}_2$; the basins of attraction of these two fixed points are separated by a surface through $\mathbf{0}$ which contains the z-axis, and which comprises all points whose trajectories approach $\mathbf{0}$. This is the stable separatrix through the origin; the long-term behaviour of a trajectory is determined by which side of the stable separatrix it lies on.

So far, all the computations have been relatively easy. Things are about to become rather more complicated, and as our purpose here is to give the flavour of the dynamics rather than to offer rigorous proofs, we will from now on merely quote the relevant results. A more complete exposition may be found, for example, in [**Spa82**].

For values of $r > 1$ close to 1, all the eigenvalues of $D\mathbf{F}(\mathbf{p}_1)$ and $D\mathbf{F}(\mathbf{p}_2)$ are real and negative; as r increases, two of the eigenvalues become complex. $\mathbf{p}_1$ and $\mathbf{p}_2$ remain stable, and in particular, still attract the trajectory along the unstable curve from the origin, but now trajectories approach these fixed points along the spirals shown in Figure 9.8.

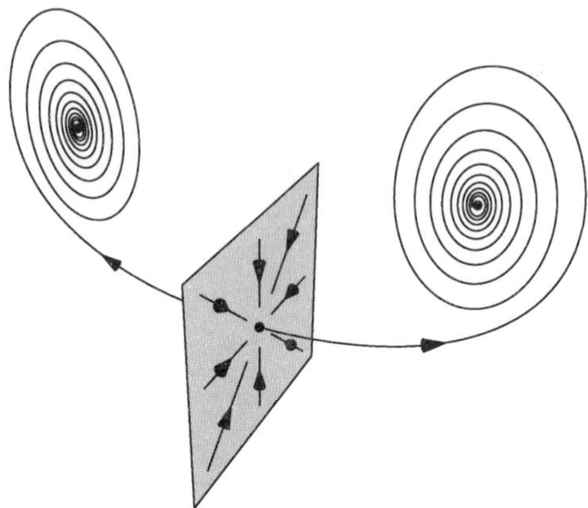

Figure 9.9. Weaker attraction and larger spirals as r increases still further.

As r increases still further, the spirals enlarge, and the trajectory along the unstable curve from the origin takes longer and longer to approach $\mathbf{p}_1$ or $\mathbf{p}_2$, as shown in Figure 9.9.

Through all of this, the system defined by (9.11) is Morse–Smale; all trajectories which begin on the stable separatrix (the "sheet" through the origin) approach $\mathbf{0}$, and all other trajectories approach either $\mathbf{p}_1$ or $\mathbf{p}_2$, depending on which side of the stable separatrix they begin on.

As r increases, the description of the system increasingly relies on numerical computations, and rigorous analytic results are no longer available. The use of computer simulations to carry out a detailed study of the Lorenz system began with the work of Oscar Lanford and shows that at some critical value $r = r_0 \approx 13.926$, things change. For $r < r_0$, the two halves of the unstable curve from the origin are *heteroclinic* orbits which approach $\mathbf{0}$ as $t \to -\infty$ and $\mathbf{p}_1$ or $\mathbf{p}_2$ as $t \to \infty$. As r approaches r_0, the spirals widen and these trajectories come closer and closer to the *stable* separatrix, shown as a vertical plane in Figure 9.9.

For values of r close to (but still less than) r_0, the trajectories along the unstable curve come close to the stable separatrix but are eventually attracted to one of the two fixed points $\mathbf{p}_1$ and $\mathbf{p}_2$. When

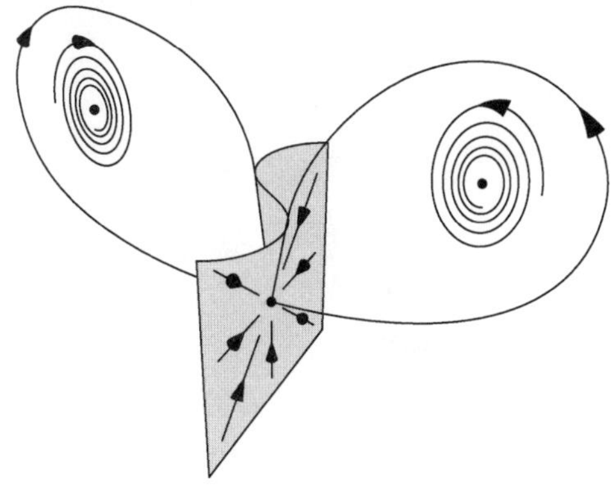

Figure 9.10. Appearance of homoclinic orbits at $r = r_0 \approx 13.926$.

r reaches r_0, this changes, and the trajectories become homoclinic; rather than spiraling in to $\mathbf{p}_1$ or $\mathbf{p}_2$, they are actually *contained in* the stable separatrix (which twists and turns accordingly) and approach the origin from a more or less vertical direction, as shown in Figure 9.10. Trajectories "inside" these homoclinic loops still spiral in to the fixed points $\mathbf{p}_1$ and $\mathbf{p}_2$.

For $r > r_0$, a completely new picture emerges. After circling one fixed point, the trajectories along the unstable curve from the origin return past the stable separatrix[6] and approach the other fixed point, as shown in Figure 9.11. The fixed points $\mathbf{p}_1$ and $\mathbf{p}_2$ are still stable, but now an unstable periodic orbit appears around each one—γ_1 and γ_2 in the figure. Trajectories "inside" γ_1 spiral in towards $\mathbf{p}_1$, while trajectories "outside" γ_1 pass the separatrix and approach $\mathbf{p}_2$; the unstable periodic orbit γ_2 plays a similar role vis-à-vis $\mathbf{p}_2$.

Although this simplified picture suggests that the system may be Morse–Smale at this point, things are actually rather more complicated than that. We will see in the next lecture that there is a horseshoe hidden in this picture and that, as a result, the system is

---

[6]Without intersecting it, which suggests that the geometry of the stable surface is quite complicated, as indeed it is, twisting and turning to avoid intersecting either itself or the unstable curves. The reader is encouraged to search the internet for animations of the stable manifold (in this case, it is a surface), which can convey the geometric complexity rather better than we are able to here.

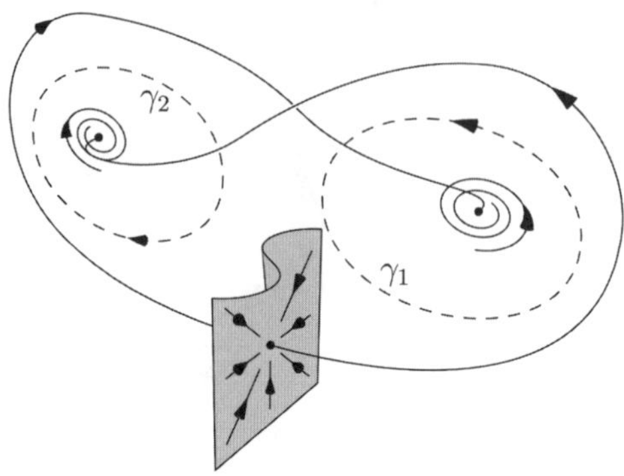

Figure 9.11. A change in behaviour when $r > r_0$.

no longer Morse–Smale (since it has infinitely many periodic points) and in fact exhibits intermittent chaos.

Lecture 36

a. Passing to a Poincaré map. The method described in the previous chapter for finding a horseshoe via a transversal homoclinic intersection does not work in the continuous-time case, where the stable and unstable manifolds cannot intersect transversally due to the uniqueness of solutions through a given point. While there is no universal mechanism for constructing a horseshoe in continuous-time systems, there are certain techniques which work in particular cases; thus we continue to examine the Lorenz system and demonstrate the existence of a horseshoe for some values of $r > r_0$.

More precisely, we will demonstrate the existence of a horseshoe for a certain Poincaré map of the Lorenz system; thus we begin with a few general words about Poincaré maps. As described in Lecture 34, the procedure for constructing such a map, which works for both continuous- and discrete-time systems, is as follows.

(1) Given a map $f\colon X \to X$ or a flow $\varphi_t\colon X \to X$, fix a subset $E \subset X$, called the *inducing domain* (in the discrete-time case) or the *Poincaré section* (in the continuous-time case). As a general rule, in the discrete-time case E is taken to be a set of positive measure

(with respect to some invariant measure), while in the continuous-time case E is taken to be a curve, surface, etc. through which a typical trajectory eventually passes.[7]

(2) For each $x \in E$, define a *first return time* by

$$\tau_E(x) = \begin{cases} \min\{n \in \mathbb{N} \mid f^n(x) \in E\} & \text{discrete-time,} \\ \inf\{t \in (0, \infty) \mid \varphi_t(x) \in E\} & \text{continuous-time.} \end{cases}$$

(3) If $\tau_E(x) < \infty$, define the *Poincaré map* at x by

$$T_E(x) = \begin{cases} f^{\tau_E(x)} & \text{discrete-time,} \\ \varphi_{\tau_E(x)} & \text{continuous-time.} \end{cases}$$

Note that the Poincaré map T_E depends on our choice of the section E. If we choose a smaller section $F \subset E$, then it may well happen that for some $x \in F$, the first return to E does not lie in F; we may have $T_E(x) \notin F$. In this case $\tau_F(x) > \tau_E(x)$, if indeed the orbit ever returns to F at all, and it is certainly the case that $T_F \neq T_E$. Thus if we wish to prove things by examining a Poincaré map, the section E ought to be chosen rather carefully.

We now carry out this selection for the Lorenz system, remarking before we do so that the existence of a section E for which the Poincaré map has the properties described below is a matter of detailed numerical analysis which lies beyond our ambitions at present.

Observe that the fixed points $\mathbf{p}_1$ and $\mathbf{p}_2$ both lie in the plane $z = r - 1$; we pass from a three-dimensional flow to a two-dimensional map by choosing a section of this plane as shown in Figure 9.12:

$$E \subset \{(x, y, z) \in \mathbb{R}^3 \mid z = r - 1\}.$$

Then we define the Poincaré map $T = T_E$ as described above.

Since the domain of T is two-dimensional, the map T is in some ways simpler than the flow φ_t; however, because not all trajectories which begin in E necessarily return to E, the return map T may not be defined on all of E. In particular, it may not be a continuous map on all of E; this is the price we pay for the simplification.

---

[7]More generally, the Poincaré section is usually an embedded submanifold of codimension one.

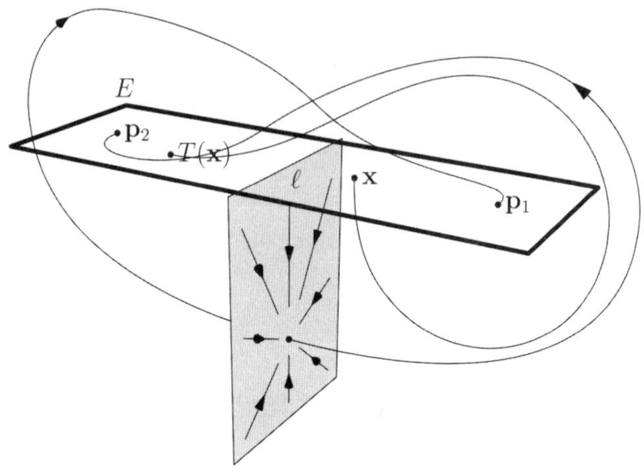

Figure 9.12. Defining the Poincaré section on E.

We immediately encounter this difficulty when we consider points in the intersection of E with the stable separatrix. These are both two-dimensional surfaces, and so their intersection is a curve ℓ, which is depicted schematically in Figure 9.12 as a line.[8] If $\mathbf{x}$ is any point on ℓ, then the trajectory $\varphi_t(\mathbf{x})$ approaches $\mathbf{0}$, and never crosses E again, so $T(\mathbf{x})$ is undefined.

Consider then a point $\mathbf{x} \in E$ which lies just to one side of ℓ; the trajectory $\varphi_t(\mathbf{x})$ will follow the stable separatrix towards the origin for some time before diverging and following the unstable curve out towards the edge of its range, eventually passing outside the edge of E, and then intersecting E again on the other side of ℓ, as shown in Figure 9.12.

As $\mathbf{x}$ approaches ℓ, the trajectory of $\mathbf{x}$ approaches the unstable curve through the origin, which is a heteroclinic trajectory, and it follows that $T(\mathbf{x})$ approaches $\mathbf{p}_2$. Thus any continuous extension of T to the line ℓ must have $T(\mathbf{x}) = \mathbf{p}_2$ for all $\mathbf{x} \in \ell$; however, an identical argument requires $T(\mathbf{x}) = \mathbf{p}_1$. Thus the Poincaré map has

---

[8]For simplicity of representation, Figure 9.12 depicts the stable separatrix as a plane, rather than the convoluted surface that it really is.

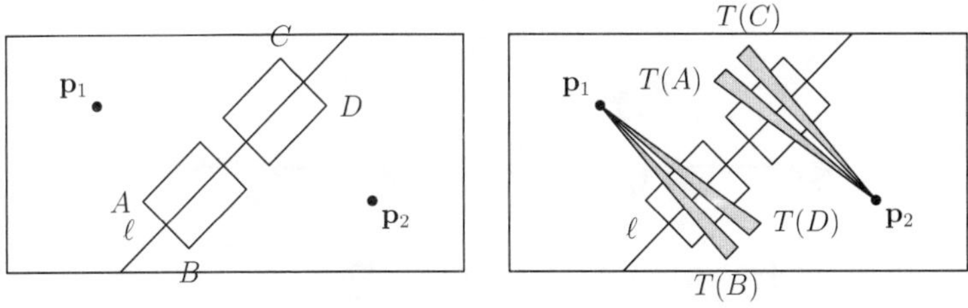

Figure 9.13. A horseshoe-like structure in the Poincaré section.

no continuous extension to the line ℓ, even though such an extension exists if we approach the line from only one side or the other.

b. Horseshoes in the Lorenz system. Despite the complications mentioned in the previous section, James Kaplan and James Yorke were able to demonstrate the existence of a horseshoe-like structure for the map T. They showed that for certain regions $A, B, C, D \subset E$ (carefully chosen with the help of a computer), the map acts as shown in Figure 9.13. To see this, we begin by considering the set of points whose trajectories remain in $R = A \cup B \cup C \cup D$; Figure 9.14 shows the topological structure of the part of the map which is significant for our purposes. The four darker trapezoids R_A, R_B, R_C, and R_D make up $f(R) \cap R$, the set of points in R with one preimage in R; these preimages are shown in Figure 9.15, where the union of the four rectangles is the set of points in R with one *forward* image in R.[9]

Taking the intersection of the four sets in Figure 9.15 with the four sets in Figure 9.14, we obtain eight trapezoids, whose union is the set of points in R with one forwards and one backwards image still in R. If we write, for example,

$$R_{A|C|D} = T(A) \cap C \cap T^{-1}(D)$$

for the set of all points lying in C whose first iterate is in D and which are the image of some point in A, then the set of points with

---

[9]This figure is schematic rather than quantitatively correct; it captures the topological behaviour which is observed in numerical simulations, even though the true regions do not have linear edges.

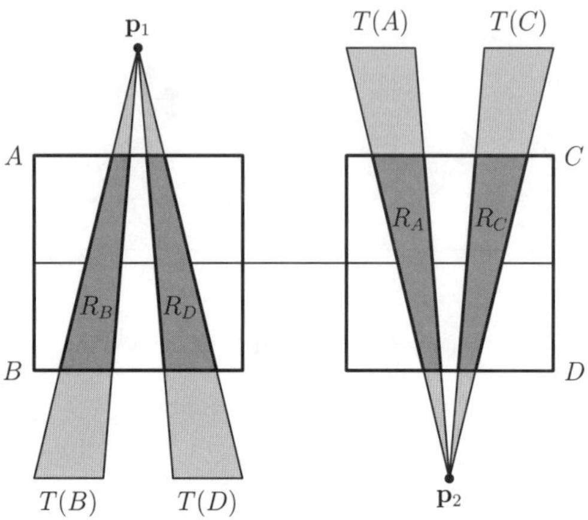

Figure 9.14. Points with one preimage.

one forwards and backwards image in R is

$$\bigcup_{(w_{-1}|w_0|w_1)} R_{w_{-1}|w_0|w_1},$$

where the w_j range over the alphabet $\{A, B, C, D\}$. Similarly, the set of points with two forwards and backwards iterates is

$$\bigcup_{(w_{-2}w_{-1}|w_0|w_1w_2)} R_{w_{-2}w_{-1}|w_0|w_1w_2},$$

and we may once again write the set of all points whose entire trajectories remain in R as

$$\Gamma = \bigcap_{n \geq 1} \bigcup_{(i_{-n}...i_{-1}.i_0i_1...i_n)} R_{i_{-n}...i_{-1}.i_0i_1...i_n}.$$

The invariant set Γ has many of the features of the Smale horseshoe. Topologically, both are totally disconnected maximal invariant sets for the relevant map; the careful reader will observe, however, that the set Γ we consider here is not closed due to the presence of the line of discontinuity ℓ. We may carry out the geometric construction described above to obtain a closed Cantor-like set, but this line, along with all its (countably many) preimages under T, must

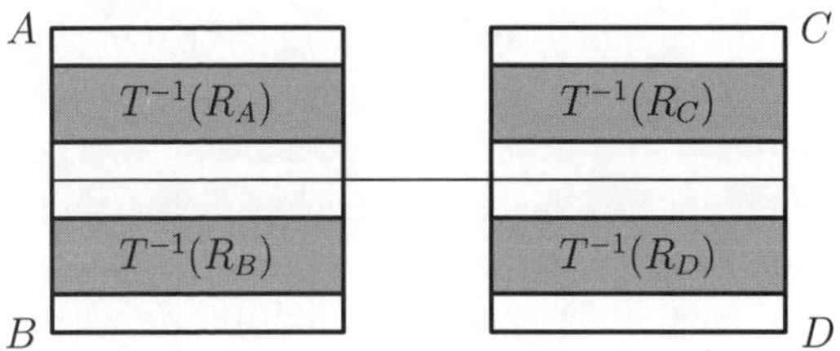

Figure 9.15. Points with one forward image.

be removed from that set to obtain Γ. Thus in what follows we only consider those points whose trajectory never falls on ℓ.

Dynamically, we have for Γ, just as we had for the Smale horseshoe, a stable and an unstable direction through each point $\mathbf{x} \in \Gamma$ (in fact, through each point in R); this may be seen as follows. Each of the regions A, B, C, and D is contracted horizontally and expanded vertically by the Poincaré map T, and so there exist two curves through $\mathbf{x}$, one stable and one unstable, with the following properties.

(1) If $\mathbf{y}$ and $\mathbf{z}$ lie on the stable curve of some point $\mathbf{x}$, then their orbits are asymptotic in positive time:

$$\lim_{n \to +\infty} d(T^n(\mathbf{y}), T^n(\mathbf{z})) = 0.$$

(2) If $\mathbf{y}$ and $\mathbf{z}$ lie on the unstable curve of some point $\mathbf{x}$, then their orbits are asymptotic in negative time:

$$\lim_{n \to -\infty} d(T^n(\mathbf{y}), T^n(\mathbf{z})) = 0.$$

(3) The tangent vectors to the stable and unstable curves at $\mathbf{x}$ lie close to the horizontal and vertical directions, respectively.

As always, the hyperbolic structure given by the existence of stable and unstable directions leads to chaos in one of its incarnations. This can be seen more explicitly by considering the symbolic dynamics associated to the map T; encoding a trajectory by its itinerary through the regions A, B, C, and D, we have a correspondence between points in Γ and sequences in the symbolic space

$$\Sigma_4 = \{A, B, C, D\}^{\mathbb{Z}}.$$

As usual, the dynamics of T are modeled by the dynamics of the shift $\sigma\colon \Sigma_4 \to \Sigma_4$. However, not all sequences in Σ_4 correspond to points in Γ; for example, we see from Figure 9.14 that $T(A) \subset C \cup D$, and so every time a point in Γ has an itinerary which includes the symbol A, it must be followed by either C or D. Thus what we have here is actually a *Markov* shift, a subshift of finite type with the following transition matrix:

$$\begin{pmatrix} 0 & 0 & 1 & 1 \\ 1 & 1 & 0 & 0 \\ 0 & 0 & 1 & 1 \\ 1 & 1 & 0 & 0 \end{pmatrix}.$$

The map $T\colon \Gamma \to \Gamma$ is topologically conjugate to this subshift of finite type,[10] and so all admissible itineraries for the subshift are encodings of trajectories of T.

This implies that T has a great many trajectories that appear random. However, because Γ is a horseshoe and has zero Lebesgue measure, the set of such trajectories is invisible from the point of view of the original system, since with probability 1, an arbitrarily chosen point will lie outside the horseshoe, and hence will eventually leave the region R in which chaotic behaviour is observed. Thus the chaos implied by the hyperbolic structure in this case is intermittent, just as with the Smale horseshoe; this sort of chaos *is* visible and occurs for a set of initial conditions with positive measure. With non-zero probability, a randomly chosen trajectory will appear chaotic for some period of time before leaving R and becoming more regular.

Finally, we may return to the Lorenz system itself by drawing the trajectories in $\mathbb{R}^3$ which connect each $\mathbf{x} \in \Gamma$ to its image $T(\mathbf{x})$; the resulting set is a *filled-in horseshoe*, which locally is homeomorphic to the direct product of $\mathbb{R}$ and Γ. This filled-in horseshoe plays the same role for the flow of the Lorenz system as Γ does for the Poincaré section; trajectories which venture near the filled-in horseshoe follow apparently chaotic trajectories for some finite period of time before wandering away and settling down, and so the Lorenz system displays intermittent chaos.

---

[10]Or rather, to this subshift with certain trajectories removed, corresponding to the line ℓ and its preimages.

Lecture 37

a. The Lorenz attractor. The scenario illustrated in Figures 9.13 and 9.14 occurs for a broad range of parameter values. The precise locations of the regions A, B, C, and D depend on the value of the leading parameter r, but the same topological outcome is observed, and the system has a horseshoe, leading to intermittent chaos.

At some value $r_1 \approx 24.05$, everything changes. As r increases beyond this value, the region $R = A \cup B \cup C \cup D$ becomes large enough to contain the fixed points $\mathbf{p}_1$ and $\mathbf{p}_2$, and indeed, to contain its own image $T(R)$; it becomes a trapping region. Taking the intersection of all the images of R, we obtain an attractor Λ for the map T. The dynamics of $T \colon \Lambda \to \Lambda$ are chaotic, and because Λ attracts nearby trajectories, this chaotic behaviour is observed for a set of initial conditions of positive measure, in contrast to the case in the previous lecture, and we now have persistent chaos.

In fact, we must be slightly more careful, since T is not defined on the line ℓ and so, strictly speaking, R is not a trapping region in the original sense. However, we do have

$$T(R \setminus \ell) \subset R,$$

and so we may define a nested sequence of sets R_n by $R_0 = R$, $R_{n+1} = T(R_n \setminus \ell)$; we have

$$R_0 \supset R_1 \supset R_2 \supset \cdots,$$

and the attractor is given by

$$(9.12) \qquad \Lambda = \bigcap_{n \geq 1} R_n.$$

Connecting each point $\mathbf{x} \in \Lambda$ to its image $T(\mathbf{x})$ by the corresponding trajectory in $\mathbb{R}^3$ gives an attractor for the Lorenz system itself; this was the object discovered by Lorenz, who originally studied the parameter value $r = 28$.

b. The geometric Lorenz attractor. Although the Poincaré map T is, conceptually speaking, a relatively simple object, any attempts to do actual calculations with it are quickly stymied by the fact that we do not have a convenient formula for T, but must rather integrate

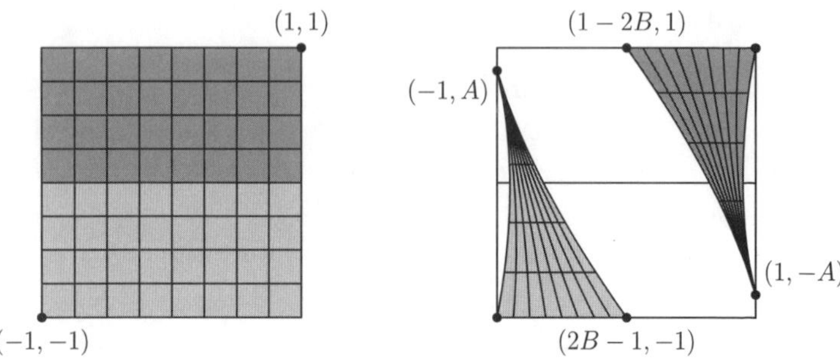

Figure 9.16. The geometric Lorenz map.

the original system for some variable period of time before obtaining $T(\mathbf{x})$, and so the map T is actually quite cumbersome to work with.

In order to bypass this difficulty, one approach is to study the *geometric Lorenz map*, which shares (or appears to share) many topological properties with the original Poincaré map, but which is given by an explicit set of formulae and thus is more amenable to concrete analysis. Although the exact structure of the Poincaré map T depends on the parameters r, σ, and b, there is a certain range of parameters within which the qualitative structure of the map can be shown to be well approximated by the geometric Lorenz map, which takes the square $R = [-1, 1] \times [-1, 1]$ into itself as follows:[11]

$$(9.13) \quad T(x, y) = \left((-B|y|^{\nu_0} + Bx\,\mathrm{sgn}\,y|y|^{\nu} + 1)\,\mathrm{sgn}\,y,\right.$$
$$\left.((1 + A)|y|^{\nu_0} - A)\,\mathrm{sgn}\,y\right),$$

where $\mathrm{sgn}\,y = y/|y|$ denotes the sign of y, and the parameters lie in the following ranges:

$$\frac{1}{2} < A \leq 1, \quad 0 < B \leq \frac{1}{2}, \quad \nu > 1, \quad \frac{1}{1 + A} < \nu_0 < 1.$$

Figure 9.16 shows the image of R under the action of T. Several features are immediately apparent. The two corners $(-1, -1)$ and $(1, 1)$ are fixed by T, and since the y-coordinate of $T(x, y)$ does not

---

[11]See [**Pes92**] for a more detailed discussion of the geometric Lorenz map and attractor.

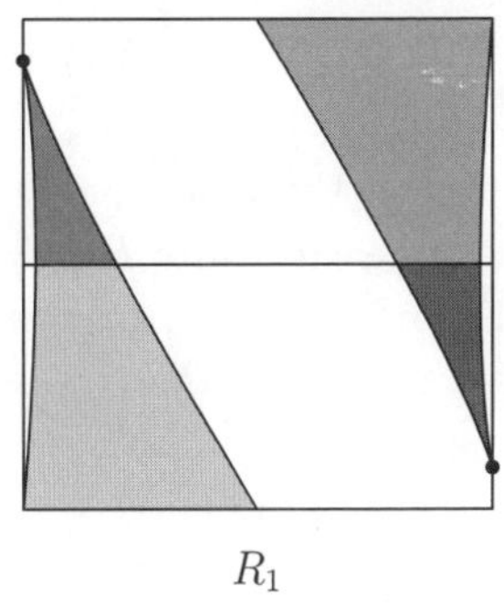
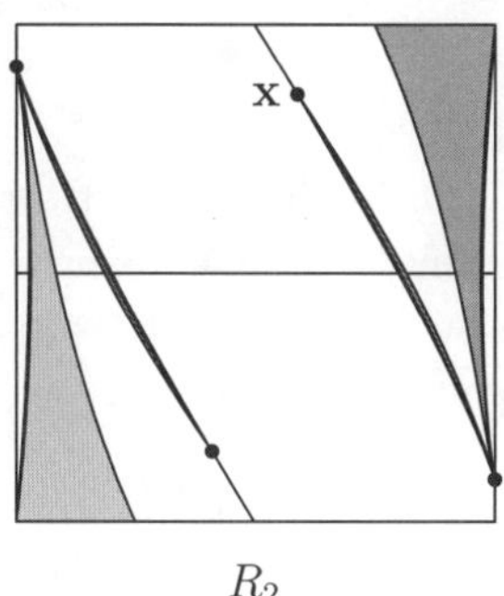

R_1 R_2

Figure 9.17. Constructing the geometric Lorenz attractor.

depend on x, T maps horizontal lines into horizontal lines. In particular, the lines $y = 1$ and $y = -1$ are mapped into themselves as follows:

$$T(x, 1) = (1 - B + Bx, 1),$$
$$T(x, -1) = (B - 1 + Bx, -1).$$

The map T is continuous everywhere except along the x-axis; the continuation of the map from the lower half of the square would take the entire x-axis to the point $(-1, A)$, while the continuation from the upper half would take it to $(1, -A)$. Furthermore, T is differentiable at every point (x, y) with $x \neq 0$ and $y \neq 0$, but the derivative of f goes to infinity as the point approaches the y-axis.

Since $0 < B \leq 1/2$, the map is contracting in the horizontal direction; since $A > 1/2$, it is expanding in the vertical direction, and so exhibits the same sort of hyperbolic structure at every point which we have already seen in the Smale–Williams solenoid, the Smale horseshoe, and so on. In both those cases, we found a maximal invariant set for the map on which the dynamics appear chaotic, and the same is true here. This set is referred to as the *geometric Lorenz attractor*, and because $T(R \setminus \ell) \subset R$, where ℓ is the x-axis, we may construct Λ explicitly as in (9.12).

Figure 9.17 shows the first two steps in the construction of Λ. $R_1 = T(R \setminus \ell)$ is the union of two triangles with curved sides; the image of $R_1 \setminus \ell$ comprises one triangle (lighter in the picture) and one biangle (skinnier and darker) inside each of these, for a total of four regions, whose union is R_2.

One might expect, then, that R_3 would be the union of eight regions, with one triangle and three biangles in each half of R_1, that R_4 will have one triangle and seven biangles in each half of R_1, and so on, always forming two "fans" with hinges at $(1, -A)$ and $(-1, A)$. Indeed, this is the general structure of the attractor, but things are not quite so cut-and-dried. Consider the biangle with one vertex at the point $\mathbf{x}$ in Figure 9.17. The bottom half of this biangle will be mapped to a biangle in the left half of R_1 with one vertex at $(-1, A)$, while the top half will be mapped to a biangle in the right half with one vertex at $(1, -A)$ and the other at $T(\mathbf{x})$, and so R_3 will have one triangle and seven biangles, as expected.

However, because the map is expanding in the vertical direction, the point $T(\mathbf{x})$ will lie somewhere below $\mathbf{x}$ and may actually lie below ℓ; if this is the case, then we have a biangle which does not split into two upon passing from R_3 to R_4, and so in general, R_n may not have $2^n - 2$ biangles, as naïve reasoning would suggest.

Another way of analysing the structure of Λ is to consider its cross-section on a horizontal line $\ell_a = \{(x, y) \mid y = a\}$. We see from Figures 9.16 and 9.17 that $R_0 \cap \ell_a$ is the entire interval $[-1, 1]$, while $R_1 \cap \ell_a$ is the union of two disjoint closed subintervals, and $R_2 \cap \ell_a$ is the union of one, three, or four disjoint closed subintervals, depending on the value of a. Thus for each fixed a, the cross-section $\Lambda \cap \ell_a$ is the result of a Cantor-like construction from which certain basic intervals have been deleted, corresponding to those steps in which one of the "fingers" of the fan does not cross ℓ_a (at the present step) or did not cross ℓ (at the previous step).

Ideally, we would like to have a rule that determines which basic intervals are deleted and which remain. For example, such a construction could conceivably be the result of a Markov rule with a particular transition matrix. However, it turns out that there is no simple rule in the present case. We may gain some perspective on this fact by considering a one-dimensional factor of the geometric Lorenz map; since T maps horizontal lines to horizontal lines, we can factor out

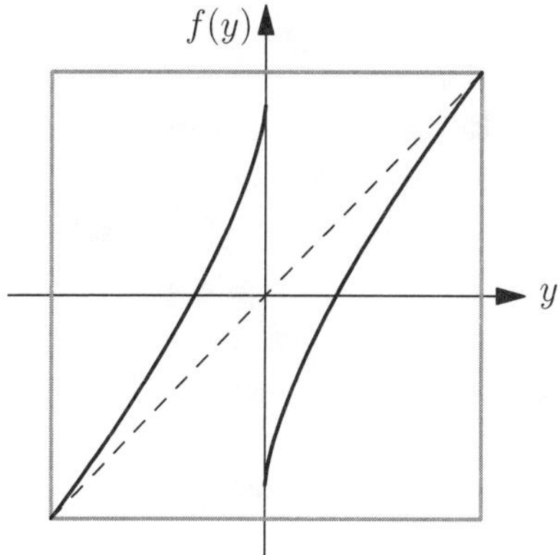

Figure 9.18. A one-dimensional factor of the geometric Lorenz map.

the x-coordinate and consider the one-dimensional map

$$f\colon [-1,1] \to [-1,1],$$

(9.14)

$$y \mapsto ((1+A)|y|^{\nu_0} - A)\operatorname{sgn} y,$$

whose graph is shown in Figure 9.18. This map is reminiscent of the piecewise continuous interval maps we have already discussed at length. However, this is *not* a Markov map, because the image of an interval of continuity is not a union of such intervals; $f([-1,0))$ contains part, but not all, of $(0,1]$. Thus while we can pass to symbolic dynamics via the partition $\{[-1,0),(0,1]\}$ and obtain a topological conjugacy between $f\colon [-1,1] \to [-1,1]$ and the shift σ on some invariant subset $Q \subset \Sigma_2^+$, the invariant subset Q will have a quite complicated topological structure which may not be given by any Markov rule.

c. Dimension of the geometric Lorenz attractor. The geometric Lorenz attractor Λ can be viewed as the limit set of a geometric construction in the square R. Although this construction is of a Cantor type, it is a far cry from the Cantor-like constructions we have seen in the previous chapters. In particular, computing the Hausdorff dimension of Λ is a daunting task.

Observe that for every $\mathbf{x} \in \Lambda$ there is a smooth curve $W^u(\mathbf{x})$ passing through $\mathbf{x}$ which lies in Λ. This is an unstable curve at $\mathbf{x}$, and it is the intersection of all basic sets of the construction which contain $\mathbf{x}$. Since Λ contains curves, we conclude that $\dim_H \Lambda \geq 1$.

One can also use the Non-uniform Mass Distribution Principle to obtain a better lower bound for the dimension. Indeed, if μ is a T-invariant measure supported on the attractor Λ, then $\dim_H \Lambda \geq \dim_H \mu$ and one can use (8.7) to compute $\dim_H \mu$.

Using this approach, it is possible to show that $\dim_H \Lambda$ is strictly bigger than 1. To do this, one can consider the one-dimensional map f associated to T given in (9.14), take an invariant measure ν for f, and then extend (or lift) it to obtain an invariant measure μ on Λ. However, finding the actual value of $\dim_H \Lambda$ is quite difficult.

d. Back to the Lorenz attractor, and beyond. As was already mentioned, connecting the points of the attractor for the Poincaré section of the Lorenz equations with the corresponding trajectories yields an attractor for the Lorenz system itself (although it was not until 1999 that the existence of this attractor was rigorously proved by Warwick Tucker). Figure 9.19 shows the attractor, viewed from several different angles; these images, first created by Oscar Lanford, have become iconic for chaos theory both because of their historic significance and because of the mathematical concepts they embody.

Despite the seminal role that it played in the study of chaos, interest in the Lorenz system began to wane in the 1980s, for a number of reasons. For example, given the origins of the model, the question

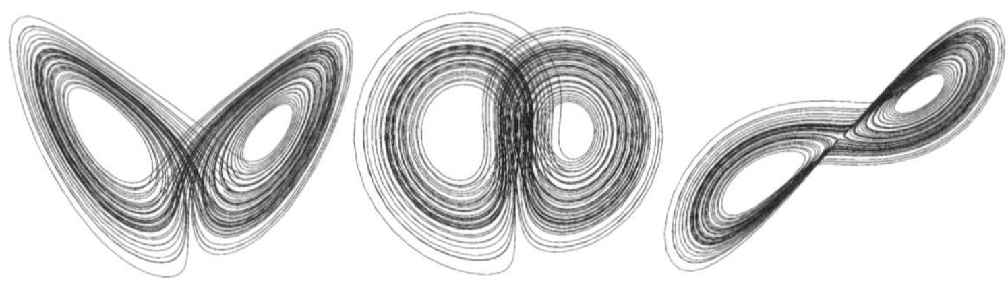

Figure 9.19. The Lorenz attractor in $\mathbb{R}^3$.

of its relevance naturally arises. How well do these equations approximate what actually occurs at the onset of turbulence? The Lorenz system (9.11) was obtained by restricting our attention to three particular terms in a more complicated system. If we add some of the other terms back into the model, do we still see the same sort of behaviour?

As it happens, the particular choice of x, y, and z in (9.11) was somewhat serendipitous. Studies of various other approximations to the Navier–Stokes equations failed to observe chaotic behaviour, which cast doubt on the relationship between (9.11) and the actual physical phenomenon. Nevertheless, the Lorenz system was the first example of deterministic chaos, and as such was tremendously important in its own right, whatever its relationship to a physical system.

Since Lorenz's original work, many other examples of deterministic chaos have been studied, and the mechanisms which produce it are now better understood. A typical chaotic system contains either a horseshoe (resulting from a homoclinic point) or an attractor (resulting from a trapping region); each of these is a fractal which is invariant under the action of the system, which is internally unstable (hence the chaotic behaviour), and which has zero volume (Lebesgue measure). Despite this last point, both horseshoes and attractors are observable via their effect on nearby trajectories; in the former case, this effect lasts for a finite period of time, producing intermittent chaos, while in the latter, the effect persists for all time, as nearby trajectories approach the attractor.

Depending on the context, one may see the Lorenz attractor and its many relatives referred to as "strange attractors", emphasising their fractal geometry, or as "chaotic attractors", emphasising the unpredictable nature of the observed dynamics, or as "hyperbolic attractors", emphasising the underlying dynamical instability of the attractor at each point. This final moniker has the advantage of giving pride of place to the force which drives both the fractal structure and the apparent randomness displayed by the attractor; namely, the presence at every point on the attractor of both stable and unstable directions, so that a saddle-like structure is ubiquitous. This is necessary because a trajectory with only stable directions will attract nearby trajectories, and hence cannot display chaotic behaviour. At the same time, a trajectory with only unstable directions will repel all nearby trajectories, and hence cannot be tightly intertwined with other such trajectories. This precludes the complicated geometric structure we observe in fractal sets.

Furthermore, the understanding of hyperbolicity as the driving impulse behind chaos makes it clear that in many cases, topological and geometrical considerations alone are enough to describe a chaotic system, without the need for an explicit formula; this greatly extends the theory's generality. For example, the geometric Lorenz map (9.13) can be generalised in a number of ways, such as choosing a line of discontinuity which is no longer horizontal; the resulting attractor is known as a *Belykh attractor*, after the Russian radiophysicist Vladimir Belykh, who first discovered it in his study of certain non-linear electrical circuits—a far cry from the study of the atmosphere which motivated the original system!

Indeed, the real significance of Lorenz's work lies far beyond its contribution to atmospheric science in the fact that it helped fling open the doors to whole new areas of mathematics, which have since found applications across the entire spectrum of scientific research. In the process, many strands of topology, geometry, and dynamics have been woven into a single fabric. The fractal sets first introduced by Cantor and studied by Besicovitch to answer questions in classical set and function theory were conscripted by Smale to serve a key role in a particular type of dynamical system. Eventually, through the work of Mandelbrot, Lorenz, and others, it became apparent that far from being pathologies with limited interest (and that only for mathematicians), these "fractals" in fact lie at the heart of many important phenomena in the natural world and still hold many deep mysteries which we have yet to understand.

Appendix

Lipschitz, Hölder, and other regularities

One of the fundamental ideas in analysis is the notion of a *continuous function*. Given two topological spaces X and Y, we denote the space of all continuous functions from X to Y by $\mathcal{C}^0(X, Y)$, and say that a member of this space is of class $\mathcal{C}^0$.

Within the class of continuous functions $\mathcal{C}^0(\mathbb{R}^p, \mathbb{R}^d)$, we have a further subclass of *continuously differentiable functions*—functions for which all the first partial derivatives exist and are continuous. We denote the space of such functions by $\mathcal{C}^1(\mathbb{R}^p, \mathbb{R}^d)$, and say that a member of this space is of class $\mathcal{C}^1$.

$\mathcal{C}^1$ functions have nicer properties than functions which are merely $\mathcal{C}^0$; however, there are also fewer of them, and so they are in some sense harder to come by. This trade-off is symptomatic of the whole notion of *regularity* of functions: there are many nested classes of functions, each more regular than the last, and hence having more properties for us to use, while at the same time comprising fewer functions, which are hence harder to come by.

For example, given $r \in \mathbb{N}$, we say that a function is of class $\mathcal{C}^r$ if all its partial derivatives up to order r exist and are continuous. Thus we have the following sequence of function spaces (for brevity of notation, we omit the domain and range):

$$\mathcal{C}^0 \supset \mathcal{C}^1 \supset \mathcal{C}^2 \supset \cdots \supset \mathcal{C}^r \supset \mathcal{C}^{r+1} \supset \cdots \supset \mathcal{C}^\infty.$$

Here $\mathcal{C}^\infty = \bigcap_r \mathcal{C}^r$ is the class of functions for which all partial derivatives of all orders exist and are continuous; such functions are often called *smooth*.[1]

Beyond $\mathcal{C}^\infty$ lie the *analytic functions*, those functions f for which the Taylor series of f at an arbitrary point $\mathbf{p}$ converges to f in a neighbourhood of $\mathbf{p}$. Such functions are also said to be of class $\mathcal{C}^\omega$.

There are also various regularity classes which lie in between the ones just mentioned. For example, given compact metric spaces (X, d) and (Y, ρ), a function $f \colon X \to Y$ is called a *Hölder function* (or *Hölder continuous*) if there exist $L > 0$ and $0 < \alpha \le 1$ such that for all $x, y \in X$, we have

$$\rho(f(x), f(y)) \le L d(x, y)^\alpha.$$

For $0 < \alpha < 1$, we denote the class of Hölder functions by $\mathcal{C}^\alpha(X, Y)$. The case $\alpha = 1$ is somewhat special; a Hölder function with $\alpha = 1$ is called a *Lipschitz function*; we denote the class of Lipschitz functions by $\mathrm{Lip}(X, Y)$.

The classes of Hölder and Lipschitz functions lie between $\mathcal{C}^0$ and $\mathcal{C}^1$, as the following exercises show.

Exercise A.1. Given $[a, b] \subset \mathbb{R}$, show that a Hölder continuous function on $[a, b]$ is uniformly continuous and that a $\mathcal{C}^1$ function on $[a, b]$ is Lipschitz.

Exercise A.2. Give an example of a Lipschitz function on $[a, b]$ which is *not* differentiable and of a Hölder function on $[a, b]$ which is *not* Lipschitz.

If a $\mathcal{C}^r$ function has partial derivatives of order r which are all Hölder continuous with exponent α, we say that it is of class $\mathcal{C}^{r+\alpha}$. Thus we have the following hierarchy of regularity:

$$\mathcal{C}^0 \supset \mathcal{C}^\alpha \supset \mathrm{Lip} \supset \mathcal{C}^1 \supset \mathcal{C}^{1+\alpha} \supset \mathcal{C}^2 \supset \cdots \supset \mathcal{C}^\infty \supset \mathcal{C}^\omega.$$

---

[1] However, one may also see the term *smooth function* used to refer to functions of lower regularity.

Exercise A.3. If f, g are Lipschitz functions, then so are $f + g$ and $f \cdot g$ (the product of f and g). Is this statement true if f and g are Hölder functions?

Exercise A.4. Suppose that $f \colon [a, b] \to \mathbb{R}$ satisfies a Hölder condition with exponent $\alpha > 1$. Show that f is a constant.

Vector spaces and linear maps

Given a real vector space V, a *norm* on V is a function $\|\cdot\| \colon V \to \mathbb{R}$ such that for every $v, w \in V$ and $\lambda \in \mathbb{R}$,

(1) $\|v\| \geq 0$, with equality if and only if $v = 0$;

(2) $\|\lambda v\| = |\lambda| \|v\|$;

(3) $\|v + w\| \leq \|v\| + \|w\|$.

The most familiar example is Euclidean space $\mathbb{R}^d$ with the norm

$$(\text{A.1}) \qquad \|\mathbf{v}\| = \sqrt{v_1^2 + \cdots + v_d^2}.$$

However, there are many other norms which we could put on $\mathbb{R}^d$. For example, we could let $(v_1, \ldots v_d)$ be the coordinates of $\mathbf{v}$ in some other basis, besides the usual orthonormal one, and then define a norm by (A.1). Or we could give different weights to different directions by fixing real numbers $c_1, \ldots, c_d > 0$ and taking

$$(\text{A.2}) \qquad \|\mathbf{v}\| = \sqrt{c_1 v_1^2 + \cdots + c_d v_d^2}.$$

The following exercise shows that whatever norm we choose on $\mathbb{R}^d$, it cannot be too different from the usual Euclidean one.

Exercise A.5. Let $\|\cdot\|_1$ and $\|\cdot\|_2$ be two norms on the finite-dimensional vector space $\mathbb{R}^d$. Show that there exist constants $m, M > 0$ such that for every $\mathbf{v} \in \mathbb{R}^d$, we have

$$m\|\mathbf{v}\|_1 \leq \|\mathbf{v}\|_2 \leq M\|\mathbf{v}\|_1.$$

Two such norms are said to be *equivalent*.

A vector space V is said to be the *direct sum* of two subspaces A and B if every vector $v \in V$ has a unique decomposition as $v = v_1 + v_2$, where $v_1 \in A$ and $v_2 \in B$, and we write $V = A \oplus B$. If $T \colon V \to V$ is a linear map which leaves A and B invariant, then the action of T on V can be decomposed into the action of T on each of the subspaces A and B.

Given a linear map $T\colon \mathbb{R}^d \to \mathbb{R}^d$, there exists a direct sum decomposition $\mathbb{R}^d = A_1 \oplus \cdots \oplus A_k$ called the *Jordan decomposition*, which has the property that each of the subspaces A_k (which are called *generalised eigenspaces*) is invariant under the action of T, and that denoting the restriction of T to A_k by T_k, the action of T_k on A_k has a particularly simple form, as follows. There exists a basis for A_k in which the matrix representing the map T_k is either of the form

$$\begin{pmatrix} \lambda & 1 & 0 & \cdots & 0 & 0 \\ 0 & \lambda & 1 & \cdots & 0 & 0 \\ \vdots & \vdots & \vdots & \ddots & \vdots & \vdots \\ 0 & 0 & 0 & \cdots & \lambda & 1 \\ 0 & 0 & 0 & \cdots & 0 & \lambda \end{pmatrix}$$

where $\lambda \in \mathbb{R}$ is the eigenvalue corresponding to A_k, or of the form

$$\begin{pmatrix} rR_\theta & \mathrm{Id} & 0 & \cdots & 0 & 0 \\ 0 & rR_\theta & \mathrm{Id} & \cdots & 0 & 0 \\ \vdots & \vdots & \vdots & \ddots & \vdots & \vdots \\ 0 & 0 & 0 & \cdots & rR_\theta & \mathrm{Id} \\ 0 & 0 & 0 & \cdots & 0 & rR_\theta \end{pmatrix}$$

where $r > 0$, Id is the 2×2 identity matrix, and R_θ is the rotation matrix

$$R_\theta = \begin{pmatrix} \cos\theta & -\sin\theta \\ \sin\theta & \cos\theta \end{pmatrix}.$$

The latter form corresponds to complex pairs of eigenvalues λ and $\bar{\lambda}$, with $\lambda = re^{i\theta}$.

Given a linear operator $T\colon \mathbb{R}^d \to \mathbb{R}^d$, the *norm* of T is

$$\|T\| = \sup\{\|T\mathbf{v}\| \mid \|\mathbf{v}\| = 1\};$$

this quantity of course depends on which norm we consider on $\mathbb{R}^d$. Using the Jordan decomposition, we see that $\|T\|$ is the maximum of the quantities $\|T_k\|$, and that each of these quantities is at least as large as $|\lambda_k|$, where λ_k is the eigenvalue corresponding to A_k. Thus $\|T\|$ is at least as large as the largest eigenvalue of T.

The reverse inequality does not always hold. Consider the map $T\colon \mathbb{R}^2 \to \mathbb{R}^2$ given by the matrix $\left(\begin{smallmatrix} \lambda & 1 \\ 0 & \lambda \end{smallmatrix}\right)$ for some $\lambda \in \mathbb{R}$. The vector $\mathbf{v} = (0, 1)$ has $\|\mathbf{v}\| = 1$ in the usual Euclidean norm, but

$$\|T\mathbf{v}\| = \|(1, \lambda)\| = \sqrt{1 + \lambda^2} > |\lambda|.$$

Thus $\|T\| \geq \|T\mathbf{v}\| > |\lambda|$, and we see that the norm of the linear operator may be strictly greater than the largest eigenvalue. However, all is not lost.

Exercise A.6. With T as above, consider the norm on $\mathbb{R}^2$ given by (A.2) with $c_1 = \delta$ and $c_2 = 1$, and show that

$$\|T\| = \sqrt{\delta + \lambda^2},$$

which can be made arbitrarily close to $|\lambda|$ by taking $\delta > 0$ arbitrarily small.

A similar procedure can be used on each Jordan block $T_k\colon A_k \to A_k$ to define a norm on A_k with respect to which $\|T_k\| \leq |\lambda_k| + \delta$; the direct sum of these norms gives a norm on $\mathbb{R}^d$ with respect to which $\|T\|$ is within δ of the largest eigenvalue. Thus we have the following result.

Proposition A.1. *Let $T\colon \mathbb{R}^d \to \mathbb{R}^d$ be a linear map, and let λ be the eigenvalue of T with maximum absolute value. Given any $\delta > 0$, there exists a norm on $\mathbb{R}^d$ with respect to which*

$$|\lambda| \leq \|T\| \leq |\lambda| + \delta.$$

Upon iterating the map T, Proposition A.1 shows that

$$\|T^j\mathbf{v}\| \leq (|\lambda| + \delta)^j \|\mathbf{v}\|$$

for every $j \geq 0$. Furthermore, by the result of Exercise (A.5), for *any* norm on $\mathbb{R}^d$, including the standard one, there exists a constant C such that

$$\|T^j\mathbf{v}\| \leq C(|\lambda| + \delta)^j \|\mathbf{v}\|$$

for every $j \geq 0$. This result is the key to the relationship between eigenvalues and contraction properties which plays such a prominent role in dynamics.

Hints to selected exercises

1.1. Find the fixed points of the map. Look for periodic points with short periods.

1.2. Try a linear coordinate change $h(x) = ax + b$, and solve the equation $g(h(x)) = h(f(x))$ to find appropriate values of the coefficients a and b. Bear in mind that the value of the parameter c will depend on the value of r.

1.5. Draw the graphs of f, f^2, etc.

1.6. Use the fact that the intersection of compact sets is compact, along with the symbolic representation of points in C.

1.11. Use the fact that E is totally disconnected and compact to show that given any $x \neq y \in E$, there exist *compact* disjoint open sets $U \ni x$ and $V \ni y$ such that $E = U \cup V$. Obtain a correspondence between such sets and cylinder sets in Σ_2^+, and pass to a correspondence between points in E and elements of Σ_2^+ (and hence with the middle-third Cantor set). Then show that this correspondence is a homeomorphism.

1.12. Find a cover of Z by open intervals whose total length is arbitrarily small.

1.14. Once the trajectory of x leaves the unit interval $[0, 1]$, it will inevitably go to either $-\infty$ or $+\infty$. Thus points in the middle third of this interval, which will leave the interval after one iteration, have a 50-50 chance of going either way. Argue inductively.

1.15. Show that every d_a-open ball contains a d'-open ball, and vice versa. Argue similarly for d''.

1.16. The diameter of the cylinders $C_{w_1 \dots w_n}$ goes to 0 as $n \to \infty$.

1.17. Whether the coding map is Hölder (or Lipschitz) will depend on the parameter a in the metric d_a. Whether the inverse coding map h^{-1} is Hölder will depend on the positioning of the basic intervals.

1.18. The abstract result follows from Exercise 1.11. To construct an explicit homeomorphism, it may be easier to first find a homeomorphism from Σ_k^+ to Σ_2^+.

2.1. Given an ε-cover $\mathcal{U}$ of Z, we may obtain a $\lambda\varepsilon$-cover of Z' by scaling each element of $\mathcal{U}$ by λ.

2.2. Since I_1 and I_2 are separated by a gap of length $1/3$, every ε-cover of C can be decomposed into the disjoint union of an ε-cover of I_1 and an ε-cover of I_2, for every $0 < \varepsilon < 1/3$.

2.7. Decompose $\mathbb{R}$ into countably many intervals such that f is Lipschitz on each.

2.12. Use the distance function $d(x, E) = \inf\{d(x, y) \mid y \in E\}$, where $E \subset X$ is arbitrary, and observe that if E is closed and $x \notin E$, then $d(x, E) > 0$. Use this to define the desired neighbourhoods of A and B as unions of balls around each point in A and B.

2.13. Given $x \neq y \in X$, consider the cover of X whose elements are $X \setminus \{x\}$ and $X \setminus \{y\}$, and apply the definition of topological dimension.

2.14. Consider the rational (or irrational) numbers. For the positive implication, given an arbitrary open cover, use compactness to pass to a finite subcover, and then use the totally disconnected property to obtain a refinement with multiplicity 1.

2.17. Construct a subset which is contained in Cantor sets with arbitrarily small ratio coefficients.

2.19. Obtain the set as the limit of a Moran construction with *nine* basic intervals at the first step, not just two.

2.23. Given $\varepsilon > 0$, divide A (or B) into two parts; one part where the points are sparse enough that each requires its own set of diameter ε to cover it, and another part where the points are dense enough that we may as well cover the entire interval containing that part.

2.25. Relate $L(Z, \varepsilon)$ to $N(Z, \varepsilon)$ and $N(Z, 2\varepsilon)$.

2.29. Moran's theorem does not apply directly since the scaling is different in different directions. Obtain the limiting Cantor set as the result of a different construction to which the theorem *does* apply.

3.1. Use Proposition 3.9 and the definition of Hausdorff dimension.

3.2. The hard part is showing that $\sum_k |I_k|^{\alpha_C} \geq 1$ whenever $\{I_k\}_k$ is a cover of C by open intervals. Figure 2.2 shows a cover with two sorts of sets, "good" (which cover a single basic interval) and "bad" (which cover bits and pieces of multiple basic intervals). First show that the inequality holds if every interval I_k is "good". Then deal with "bad" intervals as follows. If I is an interval in $[0, 1]$ and G is the largest gap from the Cantor construction which is contained in I, then we may write $I = L \cup G \cup R$, where L and R are the pieces of I which lie to the left and right of G. Prove that $|I|^\alpha \geq |L|^\alpha + |R|^\alpha$; this may be used to reduce to the case where all intervals are "good".

3.5. First show that $\mu(\sigma^{-1}(C_{w_1\ldots w_n})) = \mu(C_{w_1\ldots w_n})$ for an arbitrary cylinder $C_{w_1\ldots w_n}$.

3.7. A^n is the transition matrix for the map σ^n.

4.1. Decompose the Cantor tartan as the union of two direct products, so that the problem reduces to finding $\dim_H C \times [0, 1]$. Then the upper bounds can be obtained by exhibiting a "good" family of covers. For the lower bounds, construct an appropriate measure on $C \times [0, 1]$ and use the Uniform Mass Distribution Principle.

4.2. The Bowen balls in symbolic space are just cylinders, so covering Σ_k^+ with Bowen balls is the same as covering it with cylinders, which is reminiscent of the proof of Moran's theorem.

4.3. Use the variational principle and the result of Exercise 4.2 in conjunction with (4.14).

4.4. Given two symbols i and j, let A_n be the set of sequences $w \in \Sigma_k^+$ for which the symbol sequence i, j does not appear in the first n symbols. Show that $\mu(A_n) \to 0$ as $n \to \infty$.

4.6. Relate the balls $B(x, r)$ and the Bowen balls $B_f(x, n, \delta)$ to the basic sets $I_{w_1 \dots w_n}$.

5.6. The x- and y-components of the map act independently, so describe the action of each of the components of the map independently of the other.

5.7. Use the Intermediate Value Theorem.

5.8. Let X be the unit square, and map X to $\mathbb{R}^2$ so that the left-most third of X covers the right half of X, the right-most third covers the left half, and the middle is taken to somewhere outside of X.

6.1. Consider two sequences $x_n = f_c^n(0)$ and $x'_n = f_{c'}^n(0)$, and define *fundamental domains* by $I_n = [x_n, x_{n+1}]$ and $I'_n = [x'_n, x'_{n+1}]$ for $n \geq 0$. Define a homeomorphism $\phi \colon \mathbb{R} \to \mathbb{R}$ piecewise on each fundamental domain; ϕ can be taken to be linear on I_0, with $\phi(I_0) = I'_0$, and then the requirement that $\phi \circ f_c = f_{c'} \circ \phi$ determines ϕ on each of I_1, I_2, etc., via an inductive procedure. Extend ϕ to negative numbers by symmetry.

6.2. Use the definition of bifurcation value to show that if there is no bifurcation at c, then there is no bifurcation in a neighbourhood of c.

6.3. Draw the graphs, use the cobweb diagrams, find the fixed points and periodic points of period two, and determine their stability.

6.6. The given map can be rewritten as $(x^2 - 2)^2 - 2$.

6.7. Given $\theta \in \mathbb{R}$, let $g_\theta \colon z \mapsto e^{2\pi i \theta} z$ be rotation by θ. Use the fact that α is irrational to show that for suitable values of n, $f^n = g_\theta$ for arbitrarily small values of θ. Since the orbit of z_0 under f contains the orbit of z_0 under $f^n = g_\theta$, it suffices to show that given an arbitrary open arc on the circle $\{z \in \mathbb{C} \mid |z| = |z_0|\}$, a value of θ can be chosen such that the latter orbit enters the given open arc.

6.8. Recall that open sets are unions of cylinders, so it suffices to find a sequence w whose orbit enters every cylinder.

7.5. Estimate the pointwise dimension of the product measure in terms of the pointwise dimension of the Hausdorff measures on A and B, and then use Proposition 4.11 and the Non-uniform Mass Distribution Principle.

8.3. Use symbolic dynamics and the result of Exercise 6.8.

Suggested Reading

There are many books which cover topics in fractal geometry and/or dynamical systems (although few which consider their interaction in much detail). We mention a few titles which ought to be accessible to the reader of the present volume, as well as some more advanced works which are suitable for further in-depth study of the material. Finally, we mention some background references for basic material, and some popular, less technical, accounts of the subject. Complete references may be found in the bibliography.

Concurrent reading

An introduction to dimension theory, with many aspects of modern fractal geometry, may be found in

[Fal03] Kenneth Falconer. *Fractal Geometry: Mathematical Foundations and Applications*. John Wiley & Sons, Inc., Hoboken, NJ, 2003.

The book includes a more detailed discussion of the Hausdorff measure of various sets than we have given here, as well as some elements of the multifractal analysis which we do not consider here.

On the dynamical side of things, a very accessible treatment of the basic concepts in dynamics is given in

[Dev92] Robert Devaney. *A First Course in Chaotic Dynamical Systems: Theory and Experiment.* Addison-Wesley Publishing Company, Reading, MA, 1992.

Here the discussion focuses on some of the topological aspects of one-dimensional dynamics (with emphasis on bifurcation theory) as well as complex dynamics (that is, dynamical systems in which the variables are complex numbers, rather than real numbers), which we have not had space to consider here.

The reader with an interest in a more complete theoretical development of the topics in dynamical systems introduced here is encouraged to have a look at

[HK03] Boris Hasselblatt and Anatole Katok. *A First Course in Dynamics: With a Panorama of Recent Developments.* Cambridge University Press, New York, 2003.

Applications of dynamical systems to various areas of science are presented, along with the basic theory, in

[Str01] Steven Strogatz. *Nonlinear Dynamics and Chaos.* Westview Press, Boulder, CO, 2001.

A good account of the theory of chaos, with many examples of chaotic dynamical systems, can be found in

[ASY97] Kathleen T. Alligood, Tim D. Sauer, and James A. Yorke. *Chaos: An Introduction to Dynamical Systems.* Springer-Verlag, New York, 1997.

Further reading

A more advanced treatment of the rich and variegated connections between fractal geometry and dynamical systems, including the core results in multifractal analysis of dynamics, may be found in

[Pes98] Yakov Pesin. *Dimension Theory in Dynamical Systems: Contemporary Views and Applications.* University of Chicago Press, Chicago, IL, 1998.

A concise introduction to dynamical systems, written at the graduate level, is

[BS02] Michael Brin and Garrett Stuck. *Introduction to Dynamical Systems.* Cambridge University Press, Cambridge, 2002.

For the reader seeking a more encyclopedic treatment, the most comprehensive reference and text in dynamical systems available at the present time is

[KH95] Anatole Katok and Boris Hasselblatt. *Introduction to the Modern Theory of Dynamical Systems*. Cambridge University Press, Cambridge, 1995.

Our discussion of invariant measures and entropy is only the tip of the iceberg in the theory of measure-preserving transformations. A more complete discussion is given in

[Wal75] Peter Walters. *An Introduction to Ergodic Theory*. Springer-Verlag, New York–Berlin, 1975.

This includes, among other things, the traditional introduction of Kolmogorov–Sinai entropy, and a proof of the variational principle. For a discussion of entropy in its various guises, the reader is referred to

[Kat07] Anatole Katok. Fifty years of entropy in dynamics: 1958–2007. *J. Mod. Dyn.*, 1(4):545–596, 2007.

We have also only scratched the surface of bifurcation theory. A more complete account of that theory, along with many other topics, may be found in

[GH83] John Guckenheimer and Philip Holmes. *Nonlinear Oscillations, Dynamical Systems, and Bifurcations of Vector Fields*. Springer-Verlag, New York, 1983.

The FitzHugh–Nagumo model is just one of many important models in mathematical biology which may profitably be studied using techniques from dynamical systems. A good overview of the field is

[Mur93] J. D. Murray. *Mathematical Biology*. Springer-Verlag, Berlin, second, corrected edition, 1993.

The Lorenz equations discussed in Chapter 9 carry within them a much richer panoply of behaviours than we have had occasion to unveil. A more complete story is told in

[Spa82] Colin Sparrow. *The Lorenz Equations: Bifurcations, Chaos, and Strange Attractors*. Springer-Verlag, New York–Berlin, 1982.

Background reading

The basic concepts of point set topology, which we introduced briefly in Lecture 4, along with a more complete exposition of Lebesgue measure (and a host of other basic results and techniques) may be found in either of the following:

[Roy88]   Halsey L. Royden. *Real Analysis*. Prentice-Hall, Englewood Cliffs, NJ, third edition, 1988.

[Rud87]   Walter Rudin. *Real and Complex Analysis*. McGraw-Hill Book Co., New York, third edition, 1987.

Our discussions of measure theory (in Chapter 3), Jordan normal form and other concepts of linear algebra (in the Appendix), and the basic theory of ordinary differential equations (in Chapter 9) are all rather brief. Full details and proofs may be found in

[Hal78]   Paul R. Halmos. *Measure Theory*. Springer-Verlag, New York, 1978.

[HK71]   Kenneth Hoffman and Ray Kunze. *Linear Algebra*. Prentice-Hall, Inc., Englewood Cliffs, NJ, 1971.

[Liu03]   James Liu. *A First Course in the Qualitative Theory of Differential Equations*. Prentice-Hall, Inc., Upper Saddle River, NJ, 2003.

Popular references

There are also a number of books which touch on various subjects covered in this book at a less technical level, and which are targeted at a broader audience, either within the scientific community or beyond it. One such book, which helped to motivate the course in which the present work had its genesis, is

[Sch91]   Manfred Schroeder. *Fractals, Chaos, Power Laws: Minutes from an Infinite Paradise*. W. H. Freeman and Company, New York, 1991.

This requires some mathematics to follow, but covers an impressively broad range of topics, and is accessible to scientists from other fields.

For historical impact, it is hard to surpass

[Man82]   Benoit Mandelbrot. *The Fractal Geometry of Nature*. W. H. Freeman and Company, New York, 1982.

This requires a background similar to that required by Schroeder's book.

Finally, a very readable account of the historical development of chaos theory is given in

[Gle87]   James Gleick. *Chaos: Making a New Science.* Penguin Books, New York, 1987.

This book can be appreciated by specialists and laypersons alike.

Bibliography

[ASY97] Kathleen T. Alligood, Tim D. Sauer, and James A. Yorke. *Chaos: An Introduction to Dynamical Systems.* Springer-Verlag, New York, 1997.

[BM45] A. S. Besicovitch and P. A. P. Moran. The measure of product and cylinder sets. *J. London Math. Soc.*, 20:110–120, 1945.

[BS02] Michael Brin and Garrett Stuck. *Introduction to Dynamical Systems.* Cambridge University Press, Cambridge, 2002.

[Dev92] Robert Devaney. *A First Course in Chaotic Dynamical Systems: Theory and Experiment.* Addison-Wesley Publishing Company, Reading, MA, 1992.

[Fal03] Kenneth Falconer. *Fractal Geometry: Mathematical Foundations and Applications.* John Wiley & Sons, Inc., Hoboken, NJ, 2003.

[GH83] John Guckenheimer and Philip Holmes. *Nonlinear Oscillations, Dynamical Systems, and Bifurcations of Vector Fields.* Springer-Verlag, New York, 1983.

[Gle87] James Gleick. *Chaos: Making a New Science.* Penguin Books, New York, 1987.

[Hal78] Paul R. Halmos. *Measure Theory.* Springer-Verlag, New York, 1978.

[Hau18] Felix Hausdorff. Dimension und äußeres Maß. *Mathematische Annalen*, 79(1–2):157–179, 1918.

[HK71] Kenneth Hoffman and Ray Kunze. *Linear Algebra.* Prentice-Hall, Inc., Englewood Cliffs, NJ, 1971.

[HK03]   Boris Hasselblatt and Anatole Katok. *A First Course in Dynamics: With a Panorama of Recent Developments*. Cambridge University Press, New York, 2003.

[Kat07]   Anatole Katok. Fifty years of entropy in dynamics: 1958–2007. *J. Mod. Dyn.*, 1(4):545–596, 2007.

[KH95]   Anatole Katok and Boris Hasselblatt. *Introduction to the Modern Theory of Dynamical Systems*. Cambridge University Press, Cambridge, 1995.

[Liu03]   James Liu. *A First Course in the Qualitative Theory of Differential Equations*. Prentice-Hall, Inc., Upper Saddle River, NJ, 2003.

[Lya92]   A. M. Lyapunov. *The general problem of the stability of motion*. Taylor & Francis Ltd., London, 1992.

[Man82]   Benoit Mandelbrot. *The Fractal Geometry of Nature*. W. H. Freeman and Company, New York, 1982.

[Mur93]   J. D. Murray. *Mathematical Biology*. Springer-Verlag, Berlin, second, corrected edition, 1993.

[OP00]   D. R. Orendovici and Ya. B. Pesin. Chaos in traveling waves of lattice systems of unbounded media. In *Numerical methods for bifurcation problems and large-scale dynamical systems (Minneapolis, MN, 1997)*, volume 119 of *IMA Vol. Math. Appl.*, pages 327–358. Springer, New York, 2000.

[Pes92]   Ya. B. Pesin. Dynamical systems with generalized hyperbolic attractors: hyperbolic, ergodic and topological properties. *Ergodic Theory Dynam. Systems*, 12(1):123–151, 1992.

[Pes98]   Yakov Pesin. *Dimension Theory in Dynamical Systems: Contemporary Views and Applications*. University of Chicago Press, Chicago, IL, 1998.

[PS32]   L. Pontrjagin and L. Schnirelmann. Sur une propriété métrique de la dimension. *Ann. of Math.* (2), 33(1):156–162, 1932.

[PY04]   Ya. B. Pesin and A. A. Yurchenko. Some physical models described by the reaction-diffusion equation, and coupled map lattices. *Uspekhi Mat. Nauk*, 59(3(357)):81–114, 2004.

[Roy88]   Halsey L. Royden. *Real Analysis*. Prentice-Hall, Englewood Cliffs, NJ, third edition, 1988.

[Rud87]   Walter Rudin. *Real and Complex Analysis*. McGraw-Hill Book Co., New York, third edition, 1987.

[Sch91]   Manfred Schroeder. *Fractals, Chaos, Power Laws: Minutes from an Infinite Paradise*. W. H. Freeman and Company, New York, 1991.

[Sim97] Károly Simon. The Hausdorff dimension of the Smale–Williams solenoid with different contraction coefficients. *Proc. Amer. Math. Soc.*, 125(4):1221–1228, 1997.

[Spa82] Colin Sparrow. *The Lorenz Equations: Bifurcations, Chaos, and Strange Attractors*. Springer-Verlag, New York–Berlin, 1982.

[Str01] Steven Strogatz. *Nonlinear Dynamics and Chaos*. Westview Press, Boulder, CO, 2001.

[Wal75] Peter Walters. *An Introduction to Ergodic Theory*. Springer-Verlag, New York–Berlin, 1975.

Index

Titles in This Series